Comparative Biochemistry
of the
FLAVONOIDS

Comparative Biochemistry
of the
FLAVONOIDS

J. B. HARBORNE

Phytochemical Unit, The Hartley Botanical Laboratories
The University of Liverpool, England

1967

ACADEMIC PRESS · LONDON and NEW YORK

ACADEMIC PRESS INC. (LONDON) LTD.
Berkeley Square House
Berkeley Square
London, W.1

U.S. Edition published by
ACADEMIC PRESS INC.
111 Fifth Avenue
New York, New York 10003

PRINTED IN GREAT BRITAIN BY
SPOTTISWOODE, BALLANTYNE AND CO. LTD.
LONDON AND COLCHESTER

PREFACE

The plant flavonoids, mainly because of their intense colours but also for a variety of other reasons, continue to attract the attention of scientists from many different disciplines; and yet no comprehensive monograph on the biochemistry of this important group of pigments is available. The only comparable publication is "The Anthocyanin Pigments of Plants", written by the Cambridge biochemist Mrs. M. W. Onslow, which first appeared in 1916 and which ran to a second edition in 1925. While preparing a modern account of the anthocyanins, it immediately became apparent to the author that he, unlike Mrs. Onslow, would have to cover all the major flavonoid classes. The anthocyanins are too closely associated in occurrence and biosynthesis with the flavonols and flavones for them to be considered separately. In the 1920s, the main biological interest in the anthocyanins was in their inheritance since they provided a means of relating gene action to biochemical effect through their contribution to flower colour in higher plants. Today the main interest is in their systematic distribution and function and it is these aspects which receive most emphasis here.

The present volume on the flavonoids has three main aims: (1) to provide an outline of their structural variability and a key to their identification, (2) to describe their natural distribution in some detail and draw attention to their potential systematic importance, and (3) to summarise what is known of their inheritance, biosynthesis, function and economic importance. The material presented here should be a useful implement to organic chemists, plant biochemists and taxonomists and to botanists in general.

The first three chapters set out to deal with problems of flavonoid identification. Purely chemical matters are largely omitted, since two recent excellent publications on their chemistry, one edited by T. A. Geissman (1962) and the other written by F. M. Dean (1963), make their inclusion unnecessary. A special feature of these first chapters is the listing of the R_f values of as many flavonoids as are available at the present time. It is difficult to think of any other group of natural constituents in which the relationship between R_f and structure is so clearcut and where the R_f value plays such a large part in deducing the structure of a new compound. Much of the fascination of studying the flavonoid constituents stems from the biosynthetic versatility of plants in producing whole series of interrelated substances and the ease with which these structures can be tracked down quickly on a microscale by means of R_f values, spectral measurements and a few simple chemical operations.

Chapters 4 to 7 are devoted to detailed accounts of flavonoid occurrences in the many and various plant orders. Every effort has been made to include all reliable flavonoid records that have appeared in the literature; an addendum at the end of the book lists records from the most recent issues of journals.

Plants are arranged according to the latest revision (1964) of Engler's "Syllabus of the Plant Families". Modern taxonomic synonyms have been inserted where there is no ambiguity but the majority of plant names have been taken straight from the chemical literature and taxonomists are bound to find on occasion some difficulty in interpretation. The author in his survey work has usually had biological colleagues to call upon to identify and correctly name plants for him but most chemists and biochemists have not been so fortunate. By and large, the taxonomist should find that the data presented in Chapters 5 to 7 will indicate the amount of variation in flavonoid pattern in the plant groups of his interest. The results so far are regrettably fragmentary; few genera have been adequately, let alone completely, sampled for flavonoids. The present account is partly written in the hope that it will stimulate more carefully chosen intensive surveys among the more interesting families.

Chapters 8 and 9 cover the general biochemistry of flavonoids, their genetics, physiology, biosynthesis and function. There has been much steady progress in these fields in recent years and here, as elsewhere, it has been difficult to keep completely up-to-date with recent developments. The last chapter provides an outline of their chemotaxonomy. As E. C. Bate-Smith has said, the flavonoids are "privileged compounds" as taxonomic markers among the many different types of natural constituent and it is surely no accident that Holger Erdtman chose to look at flavonoids in his now classic survey of pine heartwoods. Future interest in the flavonoids certainly lies in relating their distribution to plant systematics and in determining more precisely their role in plant metabolism.

Anyone who has worked with the anthocyanins, as has the author, for a number of years must pay tribute to the pioneering contributions that have been made to the chemistry and biochemistry of these pigments by Sir Robert Robinson, O.M., the late Mrs. G. M. Robinson and their former students at Oxford. Although trained in the Robinson tradition, the author would make his own particular tribute to Professor T. A. Geissman, under whose tutelage as postdoctoral fellow he first strayed from the narrow path of Natural Products chemistry to the much wider pastures of Comparative Biochemistry. The author is also grateful to many former colleagues at the John Innes Institute, Hertford, and to Drs. E. C. Bate-Smith and T. Swain (both formerly of Cambridge) for their encouragement, interest or collaboration in a variety of flavonoid projects carried out over the years. He acknowledges the work of his research assistants (and especially Miss S. Margaret Davenport) who have directly or indirectly contributed greatly to this book. He must also express his gratitude to Professor V. H. Heywood for the hospitality of his department at the University of Liverpool while this book was written. Finally, and by no means least, he is grateful to the staff of Academic Press for seeing the present volume so expertly and speedily through the press.

JEFFREY B. HARBORNE

November, 1966

CONTENTS

Chapter 6

THE FLAVONOIDS OF THE SYMPETALAE

Chapter 7

FLAVONOIDS OF THE MONOCOTYLEDONEAE

Chapter 8

INHERITANCE AND BIOSYNTHESIS OF FLAVONOIDS IN PLANTS

Chapter 9

FUNCTION OF FLAVONOIDS

Chapter 10

CHEMICAL TAXONOMY OF FLAVONOIDS

CHAPTER 1

THE ANTHOCYANIN PIGMENTS

I. Introduction

The anthocyanins claim our attention because they are the most important and widespread group of colouring matters in plants. These intensely coloured, water-soluble pigments are known to be responsible for nearly all the pink, scarlet, red, mauve, violet and blue colours in the petals and leaves of higher plants. The anthocyanins are all based chemically on a single aromatic structure—that of the 3,5,7,3',4'-pentahydroxyflavylium cation, cyanidin (1). The colour of this substance is altered by the addition or removal of

(1) Cyanidin

hydroxyl groups or by methylation or glycosylation. Such modifications are known to be controlled in the flowers of many higher plants by single gene substitutions. The anthocyanins occurring in colour mutants of garden flowers have thus provided excellent material for relating biochemistry to gene action and most of our present knowledge of the biochemical genetics of higher plants is derived from such studies (see Chapter 8).

The primary function of the anthocyanins is undoubtedly to attract insects and birds to plants for purposes of pollination and seed dispersal. Flower and

1*

fruit colours are of considerable adaptive value in relation to such animal vectors, and flower colour, like flower shape, has clearly undergone modification under the influence of evolutionary forces. Besides providing much permanent pigmentation in plants, anthocyanins also appear transiently in young leaves and other organs in response to environmental change. Production in this instance seems to be directly related to the accumulation of excess carbohydrate in the plant.

From the chemical point of view, the anthocyanins have provided a considerable challenge, partly because of difficulties in their isolation and partly because of their relative lability. Although the anthocyanins which have sugars attached to them are sap soluble and reasonably stable, the aglycones (anthocyanidins) produced on acid hydrolysis are insoluble in water, unstable to light and are rapidly destroyed by alkali. The classical studies of Willstäter and Everest (1913), of Karrer and Widmer (1927, 1928, 1929) and of Robinson and his colleagues (1934, 1935) were at first concerned with establishing the chemical structures of the six anthocyanidins commonly found in garden flowers. The 3-glucosides, 3,5-diglucosides and 3-biosides of most of these anthocyanidins were then synthesised by Robinson's team and it was established that these glycosides were widespread in nature. Surveys (Lawrence et al., 1939) showed that most of the structural variation in anthocyanins resides not in the nature of the aglycone, but in the nature, number and position of attachment of sugar and other residues. In consequence of the lack of suitable methods of identifying the sugar components of anthocyanins on a small scale—the well-known Robinson distribution tests being of limited value here—the range of glycosidic variation could not be studied at this time.

The situation was dramatically altered by the introduction of paper chromatography for separating anthocyanins (Bate-Smith, 1948) and for identifying sugars (Partridge and Westall, 1948), and by the development of absorption spectroscopy for characterising these pigments (Harborne, 1957). By using these improved methods, anthocyanin characterisation can be carried out often with the small amount of pigment present in a single flower head (Harborne, 1958a). The resolution of the complex mixtures of pigments present in some plant species has also been possible for the first time. As a result, a large number of new glycosides of known anthocyanidins have been found. For example, of the 22 known pelargonidin glycosides, no less than 18 were discovered during the last 10 years. In the same period, eight new anthocyanidins have been discovered to be added to the eight previously known. Also, acylated anthocyanins have been investigated in more detail and earlier structures have been revised as a result.

Future interest in the plant anthocyanins lies mainly in determining the details of their biosynthesis (see Chapter 8) and in studies of their systematic distribution. While the anthocyanins in the most common garden flowers and fruits have been identified (see Chapters 5–7), there are still many groups of plants (e.g. tropical families such as the Bignoniaceae and Orchidaceae) where our knowledge of the pigments present is fragmentary. The main purpose of this chapter, therefore, is to consider the isolation and identification of anthocyanins and provide a comprehensive list of R_f values and sources. There are many earlier accounts of anthocyanins (e.g. Karrer, 1958; Wawzonek, 1951; Blank, 1958) but these are now seriously out of date, being based almost entirely on the work of the 1920–1940 period. Even the review of Hayashi (1962) fails to mention many of the more recently discovered glycosides. The present account is based on the author's own earlier review published in 1962.

A general point about nomenclature is necessary here. Anthocyanidins (and anthocyanins) are cations and are presumed to occur in living cells in association with anions of organic acids. At neutral pHs, anthocyanidins form unstable, colourless pseudobases and in alkali very labile anhydrobases so they are always isolated using mineral acid and are obtained as their chlorides. In the text and formulae, the presence of the anion is assumed and the pigments are referred to by the name of the cation. The term anthocyanin (from the Greek: anthos, flower and kyanos, blue) was first used for these pigments by Marquart in 1835.

II. The Anthocyanidins

A. SOURCES, ISOLATION AND PURIFICATION

Anthocyanidins can be obtained from two different plant sources, from the acid hydrolysis of naturally occurring coloured glycosides (the anthocyanins) or from acid treatment of colourless leucoanthocyanidins. Thus, delphinidin (2) is produced by the acid hydrolysis of delphin (3), the 3,5-diglucoside occurring in blue *Delphinium* flower spikes, or by acid treatment of leucodelphinidin (4) a flavan-3,4-diol widely distributed in plant leaves.

Treatment of plant tissue directly with boiling 2 N mineral acid may thus liberate anthocyanidins from both sources, and it is important to bear this in mind when carrying out preliminary surveys. In the case of flower petals, anthocyanidins are nearly always of glycosidic origin, since leucoanthocyanidins are restricted, with a few exceptions (e.g. some species of *Camellia*, *Primula* and *Armeria*), to leaf, bark or wood tissue. In leaf surveys, it is not so easy to determine the source of anthocyanidin, since visual anthocyanin colour is often masked by chlorophyll and mistakes may be made. For example, Hayashi and Abe (1955) wrongly concluded that cyanidin glycoside was present as a pigment in the fronds of the fern *Dryopteris erythrosora*, whereas in fact, the red colour is due to glycosides of apigeninidin and luteolinidin; a colourless leucocyanidin, giving cyanidin on acid treatment, is nevertheless abundantly present in the fern (Harborne, 1966c).

The types of anthocyanidins occurring as glycosides and the types "occurring" in leucoanthocyanidin form are different. Anthocyanins yield a range of aglycones (*see* Fig. 1·1): the six common anthocyanidins (pelargonidin, cyanidin, peonidin, delphinidin, petunidin and malvidin), aurantinidin, five rarer

methylated compounds and four 3-deoxyanthocyanidins. Leucoanthocyanidins normally yield only unmethylated pigments, cyanidin and delphinidin
commonly, with pelargonidin in relatively few cases. In addition, leucoanthocyanidins on acid treatment may yield one of four anthocyanidins which do
not occur naturally as glycosides. These are fisetinidin, robinetinidin, 3,7,4'-
trihydroxyflavylium and 3,7,8,3',4'-pentahydroxyflavylium.

(3) Delphin

(2) Delphinidin

(4) Leucodelphinidin

Anthocyanidins have been reported from time to time to occur naturally in
plant tissues *in the free state*, but to the author this seems unlikely, because of
their great insolubility and general instability. Such reports are probably due
to hydrolysis having occurred during extraction and isolation. Since acid-
containing solvents are nearly always used for these purposes, it is practically
impossible to avoid some hydrolysis occurring during these procedures. By
running chromatograms of extracts within minutes of extracting tissue with
ice-cold methanol containing 0·1% conc. HCl, hydrolysis can be minimised and
a true picture obtained. Re-examination in the way of flowers reputed to
contain free anthocyanidins [*Begonia* spp. (Bopp, 1957), *Camellia japonica*
(Hayashi and Abe, 1953) and *Lathyrus hirsutus* (Pecket, 1960)] showed that
only anthocyanins were present (Harborne, 1965c). It should be said that in
two cases the pigments present were pentosides, glycosidic types known to be
particularly acid-labile. Likewise, reports of fruits such as the grape containing
free aglycones have been found to be incorrect (Ribereau-Gayon, 1964).

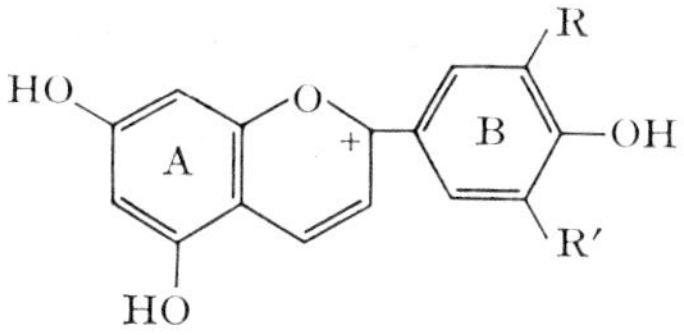

(5)

Apigeninidin (5), R=R′=H (Ap)
Luteolinidin (5), R=OH, R′=H (Lt)
Tricetinidin (5), R=R′=OH (Tr)
Columnidin (5), R=OH, R′=H, OH at
    6 or 8 position(Co).

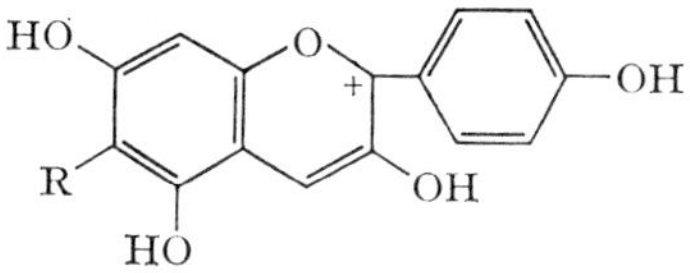

(6)

Pelargonidin (6), R=H (Pg)
Aurantinidin (6), R=OH (Au)

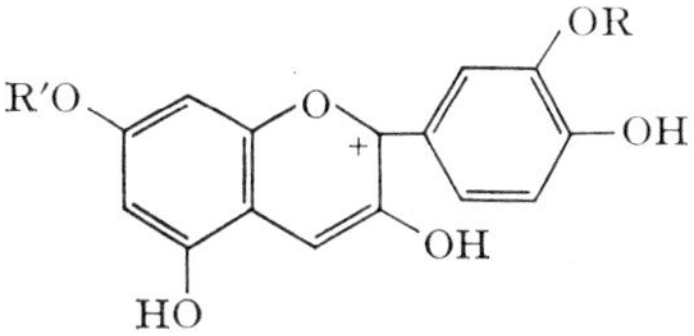

(7)

Cyanidin (7), R=R′=H (Cy)
Peonidin (7), R=Me, R′=H (Pn)
Rosinidin (7), R=R′=Me (Rs)

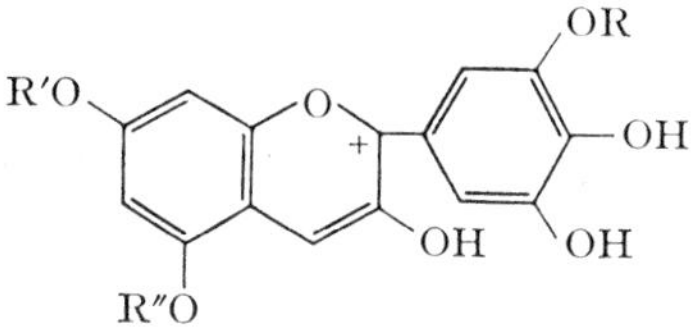

(8)

Delphinidin (8), R=R′=R″=H (Dp)
Petunidin (8), R=Me, R′=R″=H (Pt)
Pulchellidin (8), R″=Me, R=R′=H?
    (Pl)
Europinidin (8), R=R″=Me, R′=H?
    (Eu)

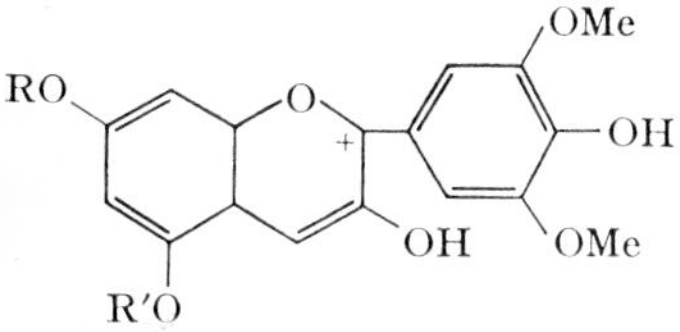

(9)

Malvidin (9), R=R′=H (Mv)
Hirsutidin (9), R=Me, R′=H (Hs)
Capensinidin (9), R′=Me, R=H? (Cp)

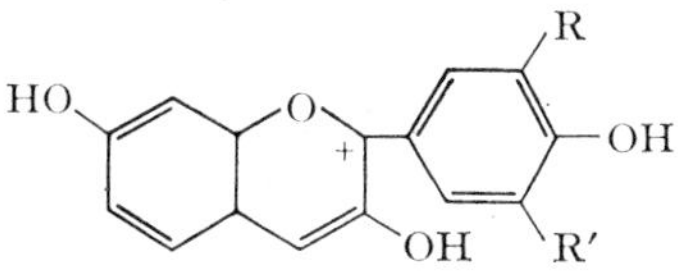

(10)

3,7,4′-Trihydroxyflavylium (10),
    R=R′=H
Fisetinidin (10), R=OH, R′=H
Robinetinidin (10), R=R′=OH

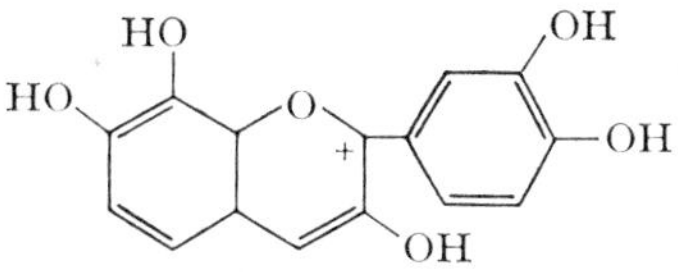

(11)

3,7,8,3′,4′-Pentahydroxyflavylium (11)

FIG. 1.1 Structures of the naturally occurring anthocyanidins.

To summarise, anthocyanidins are best obtained from pure anthocyanins (*see* Section III) by acid hydrolysis (2 N HCl at 100° for 40 min or 6 N HCl for shorter periods). The pigment will normally crystallise out from the cooled solution and may be collected in the usual way. If it fails to crystallise, it may be extracted into amyl alcohol, and this extract evaporated to dryness *in vacuo*. Anthocyanidins can be purified by re-crystallisation from hot conc. acid or better, by solution into hot ethanol, addition of 7% HCl (half a volume) and leaving to evaporate. The same procedure may be used on dried plant material, but the pigment so obtained is often contaminated with brown polymeric impurities and requires further (chromatographic) purification.

Chromatography also becomes necessary when dealing with the separation of anthocyanidin mixtures. On a small scale, such separations may be carried out on thick (Whatman No. 3) paper using Forestal or Formic solvent mixtures (*see* p. 7 for details) (Harborne, 1958a). Larger scale separations may be carried out on columns or thin layers of silica gel or silica gel–cellulose mixtures (Forsyth and Simmonds, 1957; Asen, 1965). Such separations are better carried out in a darkened laboratory, because the aglycones are subject to fading when exposed to light.

Anthocyanidins (and anthocyanins) may be synthesised either by condensing 2-*O*-benzoylphloroglucinaldehyde with an appropriately substituted acetophenone (*see* e.g. Robinson, 1934) or by reduction of the more easily accessible flavonols. The best reagents for this reduction are zinc and acetic anhydride (King and White, 1957) but some coloured dimeric products are always formed. This procedure, with modification, may also be used for preparing anthocyanins from flavonol glycosides (Krishnamurty *et al.*, 1963).

B. IDENTIFICATION

1. Physical Methods

The earlier colour and distribution tests, devised by Robinson and Robinson (1931) for distinguishing the six common anthocyanidins, have now been completely replaced by chromatographic and spectral techniques. The modern methods are more accurate and refined but still measure the same properties as did the earlier tests. Basically, anthocyanidins are identified by observing their colour in solution or on paper, their R_f value in the Forestal solvent and their spectral properties. Relevant data for all the naturally occurring pigments are collected in Table 1·1; data for related synthetic anthocyanidins and for leucoanthocyanidin-derived pigments are given in Table 1·2 for comparative purposes.

The main point to notice from Tables 1·1 and 1·2 is that all the known pigments are readily distinguished from each other by a combination of R_f values and λ_{max}. In general, the position a pigment occupies on a Forestal chromatogram and the position of its visible maximum are directly related in a regular way to the number of phenolic and methoxyl groups present in the molecule.

TABLE 1·1. R_f Values, Spectral Maxima and Colours of All the
Naturally Occurring Anthocyanidins

Pigment	Visible colour	R_f Values ($\times$ 100) in Forestal[1]	Formic	BAW	MeOH–HCl λ_{max} (mμ)	$\dfrac{E_{440}}{E_{max}}$ (as %)	AlCl$_3$ $\Delta\lambda$(mμ)
Apigeninidin	Yellow	75	44	74	277, 476	55	0
Luteolinidin	Orange	61	35	56	279, 493	45	52
Tricetinidin	Orange-red	46	28	38	281, 513	22	40
Columnidin	Orange-red	54	31	54	275, 511	23	39
Pelargonidin	Red	68	33	80	270, 520	39	0
Aurantinidin	Orange-red	53	24	52	286, 499	36	0
Cyanidin	Magenta	49	22	68	277, 535	19	18
Peonidin		63	30	71	277, 532	25	0
Rosinidin		76	39	77	— 524	—	0
Delphinidin	Purple	32	13	42	277, 546	16	23
Petunidin		46	20	52	276, 543	17	14
Pulchellidin		50	24	48	278, 543	9	—
Malvidin		60	27	58	275, 542	19	0
Europinidin		64	30	—	270, 542	16	38
Hirsutidin		78	36	66	— 536	23	0
Capensinidin		88	—	79	273, 538	12	0

[1]Solvents are: Forestal, acetic acid–conc.HCl–water (30 : 3 : 10); Formic, formic acid–conc–HCl–water (5 : 2 : 3); BAW, n-butanol–acetic acid–water (4 : 1 : 5, upper layer).

TABLE 1·2. R_f Values, Spectral Maxima and Colours of Anthocyanidins derived from Leucoanthocyanidins and of Synthetic Anthocyanidins

Pigment[1]	R_f Values ($\times$ 100) in Forestal	Formic	BAW	MeOH–HCl λ_{max} (mμ)	$\dfrac{E_{440}}{E_{max}}$ (as %)	AlCl$_3$ $\Delta\lambda$(mμ)
3,7,4′-Trihydroxy-flavylium	95	—	—	— 475	—	0
Fisetinidin	73	43	—	— 515	—	20
Robinetinidin	57	26	—	— 522	—	38
3,7,8,3′,4′-Penta hydroxyflavylium	55	30	—	— 530	—	50
Tricinidin	78	83	43	279, 509	32	0
Luteolinidin 3′-methyl ether	77	73	71	277, 493	56	0
7-O-Methylpelargonidin	81	57	—	— 510	34	0
7-O-Methylcyanidin	67	38	—	273, 533	24	+
7-O-Methyldelphinidin	44	20	56	274, 542	20	+
Morinidin	66	44	—	— 520	—	0
6-Hydroxycyanidin	30	12	39	283, 518	25	+

[1] Visible colours are similar to the most closely related naturally occurring anthocyanidin (see Table 1·1), i.e. the colour of tricinidin is like tricetinidin, 7-methylpelargonidin like pelargonidin, etc.

The R_f value in Forestal is a most important and reproducible property. Mobility decreases regularly with increasing hydroxylation (compare Pg 68, Cy 49 and Dp 32) but increases with methylation (compare Dp 32, Pt 46, Mv 60 and Hs 76). Similar relationships hold in the 3-deoxyanthocyanidin series. The formic acid solvent gives very similar R_f values to Forestal, but is used in addition because it is better for separating the aglycones clearly from their glycosides; if hydrolysis of an anthocyanin is for any reason incomplete, its presence on a Forestal chromatogram can sometimes produce misleading results. There is a shortage of other suitable solvents for separating anthocyanidins on paper. The m-cresol–acetic acid–water and n-butanol–HCl mixtures proposed by Bate-Smith (1954) are rather unsatisfactory for regular use. Butanol–acetic acid–water, if used with paper prewashed with 2N HCl (to prevent the pigment's fading), provides a reasonable third solvent. Phenol–water is useful for separating methylated from unmethylated pigments, but cannot be recommended for routine work.

Anthocyanidins can be separated successfully on thin layers of silica gel but the reproducibility of R_f values is not comparable with the results on paper chromatograms. A useful mixture for the T.L.C. of anthocyanidins is ethyl acetate–formic acid–2 N HCl (85:6:9). It is valuable for separating malvidin and peonidin, two pigments running close to each other on paper in Forestal. Successful separations of pigment mixtures depend on the presence of trace metals in silica gel G which complex with the catechol-derived pigments (cyanidin and delphinidin) and reduce their mobility; purer grades of silica gel give poor resolutions. In general, anthocyanidin colours fade rather rapidly on plates and results have to be analysed immediately after development. Thin-layer chromatography on cellulose (MN300) plates is superior to paper chromatography for two-dimensional separations; the solvents used by Nybom (1964) for such work are formic acid–conc. HCl–water (10:1:3) and amyl alcohol–acetic acid–water (2:1:1).

The colours of anthocyanidins on paper chromatograms in visible and u.v. light prove useful in identification. The three main types—pelargonidin, cyanidin and delphinidin—are clearly distinguished by their colours. 3-Deoxyanthocyanidins are immediately distinguished from other anthocyanidins by their different colours and greater stability. While the common anthocyanidins fade if run on unwashed paper in butanol–acetic acid–water, 3-deoxyanthocyanidins are unaffected. Anthocyanidins, like most other phenols, change colour on paper when ammonia is applied; a transient blue colour may be noted. Anthocyanidins fluoresce on paper only if the B-ring (for location of A and B rings see Fig. 1·1) has a single p-hydroxyl group and the 5-hydroxyl is methylated or glycosylated. The only anthocyanidin falling into this category is capensinidin, which has a 5-methoxyl group, but many anthocyanins fluoresce in this way (see Section III.A).

Anthocyanidins exhibit broad absorption maxima in the visible region and have less intense maxima in the u.v. at about 275 mμ (Table 1·2). Measurement of absorption spectra is one of the most valuable criteria for identification. A convenient solvent for spectral determinations is methanol containing 0·01% conc. HCl and all data in Table 1·1 are based on this. Changing to ethanol/HCl brings about a bathochromic shift of 10 mμ so that, when making comparisons, it is important to take this into account. Anthocyanidins which have catechol nuclei in their structures (e.g. cyanidin, delphinidin and petunidin) all exhibit a bathochromic shift of 25–35 mμ in the presence of aluminium ion at pH 2–4. The 3-deoxyanthocyanidin, luteolinidin, gives rather a larger shift ($\Delta\lambda = +52$ mμ) and so does 7,8,3′,4′-tetrahydroxyflavylium chloride, which has two catechol groups in its structure. All anthocyanidins give blue colours in alkaline solution but these normally fade very rapidly. Alkaline shifts may be measured in buffered solutions at pH 8–9. The only anthocyanidins to give *stable* alkaline spectra at higher pH are 3-deoxy-anthocyanidins.

The main facts relating anthocyanidin spectra to structure are as follows: (1) Introduction of a 3-hydroxyl group is always bathochromic; compare apigeninidin, 476 mμ, and pelargonidin, 520 mμ ($\Delta\lambda = +44$ mμ). (2) Introduction of successive hydroxyl groups into the B-ring is also always bathochromic; compare pelargonidin (520), cyanidin (535) and delphinidin (546 mμ). Note, however, that 2′-hydroxylation has no effect on spectrum, i.e. morinidin (3,5,7,2′,4′-pentahydroxyflavylium) has the same max as pelargonidin. (3) Introduction of a 6-hydroxyl group is usually hypsochromic; compare cyanidin (535) and its 6-hydroxy derivative (518) and pelargonidin (520) with its 6-hydroxy derivative (498 mμ) ($\Delta\lambda = -17$ and -22 mμ). In other systems, 6-hydroxylation may be either hypsochromic or bathochromic. Thus, 3,6,7,4′-tetrahydroxyflavylium has λ_{max} at 485 mμ (compare 3,7,4′-trihydroxyflavylium at 475 mμ) but 3,6,7,3′,4′,5′-hexahydroxyflavylium is at 512 mμ (compare 3,7,3′,4′,5′-pentahydroxyflavylium at 530 mμ). (4) Introduction of an 8-hydroxyl group may be either bathochromic (compare fisetinidin, 515, and its 8-hydroxy derivative, 530 mμ) or hypsochromic (compare cyanidin, 535 mμ, and 8-hydroxycyanidin, 518 mμ). (5) Methylation of the B-ring hydroxyl groups brings about small hypsochromic shifts (compare delphinidin, 546, with malvidin, 542 mμ). (6) Methylation of A-ring hydroxyl groups has a greater hypsochromic effect on spectra, but only if there is only one free B-ring hydroxyl. Thus, malvidin (542) and its 7-methyl ether, hirsutidin (536) differ by 6 mμ while cyanidin (535) and its 7-methyl ether (533) differ by only 2 mμ.

2. Chemical Methods

The anthocyanidins are intensely coloured, crystalline compounds which are not readily characterised by the classical methods of organic chemistry. Their melting points are not easy to determine and they often do not analyse satisfactorily since they form hydrates and lose hydrogen chloride on drying. They are relatively unstable to most chemical reagents and do not form easily characterised derivatives, although they can be methylated and acylated under appropriate conditions. In characterising anthocyanidins on a micro-scale, two procedures are useful for confirming identity: alkaline degradation and, for methylated pigments, controlled demethylation.

(a) *Alkaline degradation* is usually carried out by alkaline fusion or hydrolysis with aqueous or aqueous-alcoholic potash. Neither of these procedures is particularly suited to micro-scale work, because the yields are so low. Reductive cleavage with sodium amalgam in dilute alkali under nitrogen has given very good results on the 0·1–1 mg scale. The cleavage products obtained are best separated and identified by two-dimensional chromatography on layers of silica gel or cellulose (Harborne and Hurst, 1966). Degradation by oxidation with alkaline hydrogen peroxide has also been used (Karrer and de Meuron, 1932) but again is not a satisfactory procedure on a micro-scale. It gives a mixture of products and only works well with partly methylated pigments. The procedure however is very useful for isolating the sugar that is attached to the 3-hydroxyl group of an anthocyanin (*see* below).

(b) *Controlled demethylation* may be carried out by heating the methylated pigment in excess pyridinium chloride under nitrogen at 130–150° for periods of up to 5 hr. Heating for shorter periods removes all but a 7-methoxy group and samples taken at regular intervals will indicate the course of the demethylation and the number of intermediates produced.

C. The Common Anthocyanidins

PELARGONIDIN was first obtained by the hydrolysis of pelargonin, the major pigment of *Pelargonium zonale* (Willstäter and Bolton, 1914). It forms red-brown prisms and has λ_{max} in EtOH–HCl at 530 mμ (log ϵ 4·50) according to Sondheimer and Kertesz (1948). Convenient sources are scarlet radishes, strawberry fruits and petals of most cultivated geraniums (*Pelargonium*); a particularly rich source is the "dazzler" mutant of the Chinese primrose, *Primula sinensis*, which contains about 3% of the petal dry weight as pelargonidin glucoside (Harborne and Sherratt, 1961). At least 22 glycosides of pelargonidin are known. Leucopelargonidin is relatively rare but occurs in banana seed, corn endosperm, the testa of the French bean, the gum of *Eucalyptus calophylla* (Ganguly and Seshadri, 1959) and in some ferns (Harborne, 1966c).

CYANIDIN was first obtained by the hydrolysis of cyanin, the cornflower pigment of *Centaurea cyanus* (Willstäter and Everest, 1913). It forms red-brown needles, which melt with decomposition between 200 and 220°. Convenient sources are the leaf of the copper beech, petal of mauve *Antirrhinum majus* and blackberries or raspberries. Some 21 glycosides of cyanidin have been described. Leucocyanidin is by far the commonest of all leucoanthocyanidins and is widespread in higher plants.

PEONIDIN was first isolated as brown needles by the hydrolysis of peonin, the main pigment of *Paeonia* flowers (Willstäter and Nolan, 1915), and is still best obtained from this plant. There are only a few other plants which have peonidin as a major pigment. It is present, for example, with some cyanidin, in petals of *Rosa rugosa* and derived rose varieties such as "Roseraie de l'Hay" (Harborne, 1961). Some thirteen peonidin glycosides are known.

DELPHINIDIN was first obtained by hydrolysis of delphin, the principal colouring matter of the larkspur, *Delphinium consolida* (Willstäter and Meig, 1915). It forms dark violet plates or chocolate brown prisms, melts above 350° and has λ_{max} in EtOH–HCl at 555 mμ (log ϵ 4·49). Convenient sources include the juice of the pomegranate, the skin of the egg plant, *Solanum melongena*, and the petals of the bluebell or garden hyacinth (e.g. variety "Delft Blue"). Nine delphinidin glycosides have been described. Leucodelphinidin is a common leaf constituent in higher plants.

PETUNIDIN was first isolated as grey-brown plates from the petals of the garden *Petunia* by Willstäter and Burdick (1917). *Petunia* is not, however, a particularly good source of material, since most modern varieties with mauve, violet or blue petals mainly contain malvidin. Most "black" grapes and the red wines derived from them contain petunidin, mixed with varying amounts of delphinidin and malvidin. Sources of the pure pigment include petals of the garden *Anchusa*, leaf of the Chinese primrose and berries of various Solanaceous plants such as the deadly nightshade and the garden huckleberry, *Solanum guineese*. Nine petunidin glycosides are known.

MALVIDIN was first obtained by hydrolysis of the pigment of petals of the common mallow, *Malva sylvestris* (Willstäter and Zollinger, 1917). It forms dark brown needles, melts above 300° and has λ_{max} in MeOH–HCl at 274 (log ϵ 4·13) and 547 (log ϵ 4·50) mμ (Koeppen and Basson, 1966). Besides the mallow, it may be isolated in a fairly pure state (after acid hydrolysis) from the petals of the purple loosestrife *Lythrum salicaria*, many *Epilobium*, *Rhododendron ponticum*, *Clarkia elegans* and *Tibouchina semidecandra*. Ten glycosides of malvidin are described.

D. RARER ANTHOCYANIDINS

HIRSUTIDIN (7-*O*-methylmalvidin), dark red prisms, was first obtained from flowers of *Primula hirsuta* by Karrer and Widmer (1927). It occurs in the flowers of about 10 other *Primula* species (*see* Chapter 6, p. 198), but has also been reported in petals of *Catharanthus roseus* (Forsyth and Simmonds, 1957). The only known glycoside is the 3,5-diglucoside (hirsutin).

ROSINIDIN (7-*O*-methylpeonidin), dark red plates, was isolated from the flowers of *Primula rosea* by the author in 1958 (Harborne, 1958b, 1960c). The only other known source is the garden polyanthus; certain blue forms contain mixtures of rosinidin, hirsutidin, malvidin and peonidin. Like hirsutidin, rosinidin occurs as the 3,5-diglucoside.

CAPENSINIDIN (5-*O*-methylmalvidin) was isolated from the sky-blue corollas of *Plumbago capensis* as the 3-rhamnoside by Harborne (1962a). Its structure as 5,3',5'-trimethoxy-3,7,4'-trihydroxyflavylium is based on its general properties and the fact that on controlled demethylation, it yields malvidin, petunidin and delphinidin. The presence of a 5-O-methyl group in capensinidin is indicated by its vivid fluorescence in u.v. light, its low 440/max ratio and the position of its visible maximum (538 mμ) when compared with malvidin (546 mμ). Its occurrence with azaleatin (5-O-methylquercetin) supports the proposed structure but the presence of large quantities of this flavonol in *P. capensis* petals has prevented the isolation of sufficient capensinidin for elementary analysis.

PULCHELLIDIN was found in petals of *P. pulchella* by Harborne (1966a). Its spectral and chromatographic properties agree with it being the 5-O-methyl ether of delphinidin and thus an isomer of petunidin. On demethylation, it yields delphinidin. Like capensinidin, it is accompanied in the flowers by the flavonol, azaleatin. Isolation of pulchellidin in quantity is impossible, because the petals of *P. pulchella* are so small that proof of its structure must await synthesis.

EUROPINIDIN is the third new anthocyanidin to be isolated from the genus *Plumbago*. It occurs in petals of *P. europca* and in those of the closely related *Ceratostigma plumbaginoides* (Harborne, 1966a). Its colour, spectral and chromatographic properties suggest it is a dimethyl ether of delphinidin. It appears, from the 440/max ratio, to have a 5-O-methyl substituent and furthermore it cannot be the 3′,5′-dimethyl ether, which is the known malvidin. That it may be the 5,3′-dimethyl ether is supported by its co-occurrence with 5-O-methylmyricetin, but further work is required to confirm this suggestion.[1]

AURANTINIDIN (6- or 8-hydroxypelargonidin) was isolated from the yellow-orange petals of *Impatiens aurantiaca* by Clevenger (1964) and its properties have been studied in the author's laboratory. In its colour properties, it is most similar to pelargonidin; one of its glycosides has exactly the same vivid yellow fluorescence in u.v. light as pelargonidin 3,5-diglucoside. Molecular weight determination by mass spectra indicates that it has one more hydroxyl group than pelargonidin; its chromatographic behaviour and its failure to be affected by demethylating reagents show that methoxyl groups are absent. Like other 3-hydroxyanthocyanidins, it is rapidly destroyed by aqueous ferric chloride and it is in fact more unstable chemically than pelargonidin, indicating the presence of a catechol nucleus. That the extra hydroxyl group is in the A-ring is apparent from its colour properties and its difference from cyanidin and morinidin. This was confirmed by reductive cleavage, which gave *p*-hydroxyphenylpropionic acid and pyrogallol. Aurantinidin must, therefore, be either 6- or 8-hydroxypelargonidin. Both isomers were synthesised by standard procedures and were found to have identical R_f values and spectral properties so it has not yet proved possible to determine which isomer aurantinidin is. Similarly, the preparation of 6- and 8-hydroxycyanidin, by reductive acetylation of quercetagetin and gossypetin respectively, gave pigments which could not be distinguished from each other.

[1] A low 440/max ratio would also be shown by a 5-deoxyanthocyanidin and the possibility that europinidin is based on, say, 3,7,8,3′,4′,5′-hexahydroxyflavylium has not yet been ruled out.

APIGENINIDIN (3-deoxypelargonidin) was the first 3-deoxyanthocyanidin to be isolated; it was found by Robinson and Robinson (1932) in the bright scarlet corollas of *Rechsteineria cardinalis* (formerly *Gesneria cardinalis*). It forms red prisms, which melt above 300°. It occurs fairly widely in the Gesneriaceae as a petal pigment (*see* Chapter 6, p. 189) and is present in the fronds of several ferns (*see* p. 116). It has also been reported in the leaf of *Sorghum vulgare* (Stafford, 1965) and in the petals of *Chiranthodendron pentadactylon* (Pallares and Garza, 1949); it seems unlikely that it is really present in the latter plant as a "trigalloyl triglucose" derivative, as has been reported. The only fully characterised apigeninidin glycoside is the 5-glucoside, gesnerin, but two other derivatives are known (for R_f values *see* Table 1·4).

LUTEOLINIDIN (3-deoxycyanidin) was first isolated from flowers of *Rechsteineria cardinalis*, where it occurs with apigeninidin as the 5-glucoside, and from the sepals of *Kohleria eriantha* (Harborne, 1960c). It does not readily crystallise and is obtained as a crimson powder. In most of its known sources, it is accompanied by apigeninidin, i.e. in plants of the Gesneriaceae, in *Sorghum* and in ferns. It is present on its own in the moss *Bryum cryophyllum,*, where it occurs as the 5-mono- and 5-diglucosides (Bendz *et al.*, 1962).

TRICETINIDIN (3-deoxydelphinidin) was isolated from processed tea leaf by Roberts and Williams (1958). Its structure was confirmed in this laboratory by comparing it with synthetic material, obtained by demethylation of tricinidin (3-deoxymalvidin). Neither tricinidin nor any other methylated 3-deoxyanthocyanidin has yet been found to occur in nature as a pigment.

COLUMNIDIN was found in the leaf and petal of several *Columnea* spp. and other members of the Gesneriaceae (Harborne, 1966a, c). Its general properties (Table 1·1), its molecular weight (287), and the fact that it gives 3,4-dihydroxyphenylpropionic acid and pyrogallol on reductive cleavage, proves that it must be either 5,7,8,3',4'- or 5,6,7,3',4'-pentahydroxyflavylium. Its spectral properties, when compared with luteolinidin, favour the former structure but confirmation by means of synthesis is still required.

III. The Anthocyanins

A. GENERAL ACCOUNT

1. Isolation and Purification

Serious attempts to isolate anthocyanins from plants were made as long ago as 1849 by Morot, but success was not achieved until 1913 when Willstäter and Everest extracted dry cornflower petals with alcoholic hydrochloric acid and reprecipitated the pigment with large volumes of ether; further precipitations and crystallisations gave pure cyanidin 3,5-diglucoside. This method (and variants employed, for example, lead salts or picrates) was rapidly applied by Willstäter and his co-workers to the isolation of anthocyanins from many other plant sources. The procedure works well when relatively large quantities of a single anthocyanin are present in a flower and was used in all laboratories up to 1940.

This method of isolation has a number of obvious drawbacks. Besides being laborious and wasteful in terms of plant material, it fails to separate mixtures of anthocyanins. Moreover, labile acyl and sugar residues in the anthocyanins may be lost during recrystallisation and pigments obtained by precipitation sometimes tend to retain flavones and free sugars as persistent impurities. The great advantage of paper chromatographic methods, which were first used in 1948 by Bate-Smith, is that they overcome all these difficulties. Paper chromatography is particularly successful in separating anthocyanin mixtures and has been used for isolating as many as 15 anthocyanins from a single plant source. As a method, it has the particular advantage that hydrolysis products, which are inevitably produced during isolation, can be eliminated at each stage of purification.

The general procedure for the separation of anthocyanins on paper is briefly as follows. Plant tissue (preferably fresh) is extracted with methanolic HCl and the extract concentrated *in vacuo* at 30–40° and the concentrate applied directly to several sheets of thick filter paper and chromatography is then carried out in BAW, BuHCl or 1% HCl (for key to solvent abbreviations, *see* p. 30). The anthocyanins appear as clear discrete coloured bands which are then cut out from the dried papers and the pigments eluted with methanolic acetic acid. The eluates are collected and concentrated and the process is repeated at least twice, preferably with different solvent systems. By judicious choice of chromatographic solvents, individual pigments are cleanly separated from co-occurring anthocyanins, from their hydrolysis products and from other plant constituents. Sufficient material can be handled by paper chromatography to yield pigment in crystalline form if so desired.

The only precaution that must be taken is the avoidance of solvents containing mineral acid during the final stages of purification. The acid reacts with substances in the filter paper to produce arabinose, which may then cause erroneous results when the sugars of the anthocyanins are analysed. Solvent mixtures containing acetic acid can safely be used and it is as well to wash papers before use with dilute acetic acid.

Many other chromatographic techniques have been applied to the separation of anthocyanins but, in the author's view, none offers any advantage over paper; most, in fact, are not as good. Chromatography on columns of cellulose has been used by several workers (e.g. Chandler and Harper, 1958) but the results are variable (cf. Endo, 1957) and such columns are of limited capacity and produce dilute eluates which, with anthocyanins, are difficult to concentrate without loss. Column chromatography on silica gel, magnesol or polyamide is useful for the preliminary purification of crude plant extracts but cannot be recommended for the resolution of anthocyanin mixtures. Paper electrophoresis will separate the two anthocyanins in cherry fruit extracts

(Markakis, 1960) but the method has no obvious advantage to offer over the simple descending technique.

Thin-layer chromatography of anthocyanins has recently been developed (Birkofer *et al.*, 1962; Hess and Meyer, 1962) and probably offers a satisfactory alternative to paper. Asen (1965) has successfully obtained pure anthocyanins in quantity after separation on 1 mm thick layers of a silica gel–cellulose (2:1) mixture. The solvents he used were the same as for paper chromatography (BuHCl and 1% HCl); in addition, a mixture of acetone and $N/2$ HCl (1:3) proved useful.

2. Identification

(a) *The glycosidic variation.* A considerable range of anthocyanins are now known and 21 classes of glycoside, excluding those with acylated substituents, have been described (Table 1·3). In seven of these classes (e.g. 3-glucoside, 3-rutinoside), glycosides based on all of the six common anthocyanidins have

TABLE 1·3. The Known Classes of Anthocyanidin Glycosides

General Class	Individual types	Known aglycones[1]
3-Monosides	3-Glucoside[2]	All six
	3-Galactoside	All six
	3-Rhamnoside	All six
	3-Arabinoside	Cy, Pn
3-Biosides	3-Rutinoside	All six
	3-Sambubioside	Pg, Cy, Pn, Dp
	3-Lathyroside	Pg, Cy, Pn
	3-Gentiobioside	Pg, Cy, Pn, Pt, Mv
	3-Sophoroside	Pg, Cy, Pt
3-Triosides	3-Gentiotrioside	Pg, Pn, Pt, Mv
	3-(2^G-Glucosylrutinoside)	Pg, Cy
	3-(2^G-Xylosylrutinoside)	Cy
3,5-Diglycosides	3,5-Diglucoside	All six
	3-Rhamnoside-5-glucoside	All six
	3-Galactoside-5-glucoside	Pg, Cy
	3-Rutinoside-5-glucoside[2]	All six
	3-Sambubioside-5-glucoside[2]	Pg, Cy
	3-Sophoroside-5-glucoside[2]	Pg, Cy
	3-Arabinoside-5-glucoside	Cy
3,7-Diglycosides	3,7-Diglucoside	Cy
	3-Sophoroside-7-glucoside	Pg

[1] Only the six common anthocyanidins are referred to here.

[2] This indicates that the glycoside is also known with an acyl attachment. For aglycone abbreviations, *see* Fig. 1·1.

been characterised and, in the remaining classes, the as yet unknown pigments could, no doubt, be found if the right plant were analysed. In spite of the considerable variability, there are some structural restrictions in the pigments known at present. Thus, sugars are always found in the 3-position, except in the case of the 3-deoxyanthocyanins when the sugar is usually in the 5-position. If a second position in the molecule is glycosylated, it is nearly always the 5- rather than the 7-position (only two 3,7-glycosides are known as compared to 24 3,5-glycosides) and substitution on a B-ring hydroxyl is not recorded. Furthermore, the 5- (or 7-) sugar is always glucose, never rhamnose, even when the 3-sugar is rhamnose. Finally, only a limited number of sugars are involved: four monosaccharides, five disaccharides and three trisaccharides. All the di- and trisaccharides have at least one glucose (or in one instance galactose) unit and the linkages are $\beta1\rightarrow2$, $\beta1\rightarrow6$ or $\alpha1\rightarrow6$.

(b) *General approach.* Anthocyanins, when isolated as their chlorides, are intensely coloured compounds which do not usually melt sharply. Some anthocyanins, e.g. 3-glucosides and 3,5-diglucosides, readily crystallise from aqueous HCl, others, especially those containing three sugar residues, are more water-soluble and only crystallise with difficulty. Even when obtained pure, anthocyanins are usually hydrated and do not give meaningful results on elementary analysis; attempts to remove the waters of crystallisation usually lead to loss of some hydrogen chloride. A few anthocyanins form crystalline picrates, but no suitable derivatives for their general characterisation are known. The procedure of methylation and hydrolysis, commonly used with flavones to determine the positions of sugar substituents, is of little value here. Anthocyanins are, thus, not readily characterised by the usual techniques of organic chemistry.

To overcome the above difficulties, Robinson and Robinson (1931, 1932) devised a series of colour and distribution tests. Their identification tests depend on the facts that the distribution coefficient of an anthocyanin between amyl alcohol and aqueous acid is related to the nature and number of its sugars and that anthocyanins alter their colour with changing pH. Valuable as these tests are, more recent work has shown that they are liable to error, probably because of their sensitivity to impurities and the subjective element involved in determining by eye the colour and distribution of a pigment. Use is now made of the more precise and refined techniques of paper chromatography and spectrophotometry to measure these characteristic properties.

The anthocyanins, of which over a 100 are known, are now identified by their R_f values and spectral properties, and also on the basis of their behaviour towards enzymes and of quantitative and qualitative studies of their hydrolysis products. If a known glycoside is present, then direct comparison with authentic material is usually possible. If the glycoside is a new one, then the fact that its R_f values differ from those of the known glycosides is usually sufficient to confirm its novelty. Synthesis, though desirable, is not usually possible because of the scarcity of starting materials. However, some glycosides can be obtained from the corresponding flavonol glycosides by reductive acetylation (*see* p. 6).

(*c*) *Spectral properties*. Simple anthocyanins in acid solution have two main absorption maxima, one in the visible region between 465 and 550 mμ and a smaller one in the u.v. at about 275 mμ (*see* Fig. 1·2 and Table 1·4). The position of the visible peak varies slightly according to the solvent used; thus cyanidin 3-rutinoside in ethanol/HCl has λ_{max} at 533, in methanol/HCl at 523 and in aqueous HCl at 507 mμ. Measurements have been standardised by using methanol/HCl.

TABLE 1·4. Spectral Properties of Anthocyanins

Glycoside[1]	λ_{max} in MeOH–HCl (mμ)	$\dfrac{E_{u.v.max}}{E_{vis.max}}$ (as %)	$\dfrac{E_{acyl peak}}{E_{vis.max}}$ (as %)	$\dfrac{E_{440}}{E_{vis.max}}$ (as %)
Ap 5-glucoside	273, 477	37	under 30	45
Lt 5-glucoside	277, 495	28	under 30	23
Pg 5-glucoside	— 513	—		15
Pg 7-glucoside	270, 508	51		42
Pg 3-glucoside	270, 506	64	under 20	38
Pg 3,5-diglucoside	269, 504	45		21
Pg, 3,7-diglucoside	279, 498	69		42
Cy and Pn 3-glucoside	274, 523	60		24
Cy and Pn 3,5-diglucoside	273, 524	44	under 20	13
Rs 3,5-diglucoside	278, 519	—		12
Dp, Pt and Mv 3-glucoside	276, 534	54		18
Dp, Pt and Mv 3,5-diglucoside	273, 533	42	under 20	11
Hs 3,5-diglucoside	273, 532	49		9
Cp 3-rhamnoside	278, 533	56		12
Monardein (*p*-coumaric)	289, 313, 507	—	60	21
Salvianin (caffeic)	285, 329, 507	—	48	21
Matthiolanin (*p*-coumaric and ferulic)	289, 328, 509	—	92	20
Raphanusin C (*p*-coumaric)	278, 310, 523	—	60	17
Rubrobrassicin (diferulic)	282, 333, 530	—	88	14
Hyacinthin (*p*-coumaric)	284, 310, 527	—	64	21
Petanin (*p*-coumaric)	282, 310, 538	—	71	11
Tibouchinin (*p*-coumaric)	280, 305, 536	—	64	10

[1] Abbreviations of the anthocyanidins (Pg=pelargonidin, etc) are given in Fig. 1·1. Only glucosides are given, but other glycosides have identical values, since the nature of the sugar has no effect on spectral properties. The structure of the acylated glycosides are given in Table 1·5. Typical log ϵ values for anthocyanins at the visible max. are as follows: Pg 3-glucoside, 4·50; Pg 3,5-diglucoside, 4·51; Cy 3-glucoside, 4·47; Cy 3-galactoside, 4·48; and Dp 3,5-diglucoside, 4·37.

The data in Table 1·4 show that the position of the visible maximum is a clear indication of the hydroxylation pattern of the anthocyanidin from which the anthocyanin is derived. Introduction of a sugar residue into the 3-position of anthocyanidins has a regular hypsochromic effect on spectra; thus all pelargonidin 3-glycosides have λ_{max} at about 505 mμ, all cyanidin and peonidin 3-glycosides at 520–526 mμ and all delphinidin derivatives at 532–537 mμ (Fig. 1·2). The nature of the sugar substitution has no effect on spectra. As

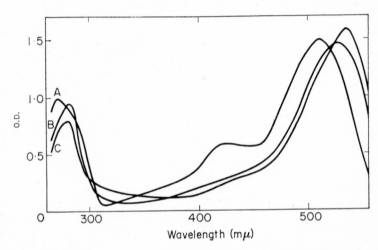

FIG. 1·2. Absorption spectra of anthocyanins: curve A, pelargonidin 3-rhamnoside; curve B, cyanidin 3-rhamnoside; curve C, delphinidin 3-rhamnoside. Solvent: MeOH/HCl.

with the anthocyanidins, anthocyanins which have free O-dihydroxylic groupings undergo bathochromic spectral shifts of 25–35 mμ in the presence of aluminium ion at pH 2 to 4.

The two most common classes of anthocyanin, the 3- and 3,5-diglycosides, have similar spectral maxima but show differences in intensity in the 400–460 mμ region and in the u.v. Optical density comparisons show (Fig. 1·3 and Table 1·4) that 3,5- and 5-glycosides have only 50% of the absorption at 440 mμ (when compared with the colour maximum) as do the 3-glycosides and the free anthocyanidins. Similarly, the intensity of the short-wave maximum is lower in the case of the 3,5-diglycosides than with the 3-glycosides. Such measurements are very useful for distinguishing the two classes of glycoside. Correlated with differences in spectral characteristics are differences in fluorescence; 5- and 3,5-diglycosides of pelargonidin, peonidin, rosinidin, malvidin and hirsutidin fluoresce in solution and appear as intense fluorescent spots on chromatograms when examined in u.v. light. 3,5-Glycosides of the other anthocyanidins give only very weak fluorescence, while all 3-glycosides give dull colours under the same conditions.

Pigments containing sugars in the 5-, 7- and 3,7-positions have different absorption maxima from each other and from the 3- and 3,5-glycosides so that unusually substituted anthocyanins can be readily detected by spectral means. The pelargonidin 3-sophoroside-7-glucoside in *Papaver*, for example, was immediately recognised (Fig. 1·4) as a novel pigment from its difference from all the other known pelargonidin derivatives.

The spectra of acylated anthocyanins (Fig. 1·3 and Table 1·4) show two peaks in the u.v. region, due to the superimposition of the absorption of the acyl group, normally a cinnamic acid, upon the pigment absorption. The

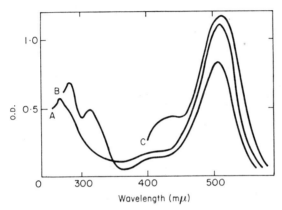

Fig. 1·3. Absorption spectra of anthocyanins: Curve A, pelargonidin 3,5-diglucoside; curve B, pelargonidin 3-*p*-coumaroylglucoside-5-glucoside (monardein); curve C, pelargonidin 3-glucoside. Solvent : MeOH/HCl.

position of the new peak between 310 and 335 mμ indicates the nature of the cinnamic acid present and the intensity of this peak when compared with that of the colour maximum is a measure of the number of cinnamic acid units that are present in the pigment complex.

Other spectroscopic techniques have not yet been widely applied to anthocyanins. Infra-red spectra of a few anthocyanins have been measured (Koeppen and Basson, 1966) but the method is only useful for distinguishing acylated from unacylated pigments. Nuclear magnetic resonance spectroscopy has yet to be used with anthocyanins.

(*d*) *Chromatographic properties*. The R_f value is the most important single characteristic for the identification of an anthocyanin. For this reason, values for all the available known pigments are collected together in Table 1·6 at the end of this chapter and should be consulted for details of the relationships between R_f value and structure. The situation may be summarised by saying that the various glycosides of any one anthocyanidin can be differentiated by means of their R_f values in the four common solvent systems and that the mobility of a pigment indicates very clearly the nature and number of sugar and other substituents it contains.

In the case of simple glycosides, as the number of sugar units present increases the pigments have decreasing R_f values in butanol–acetic acid and increasing values in aqueous acid. Compare, for example, the 3-glucoside, 3-gentiobioside and 3-gentiotrioside of pelargonidin, which have R_f values ($\times$ 100) in BAW of 44, 30 and 25 and in 1% HCl of 14, 21 and 35. Addition of an acyl group has the opposite effect of a sugar substitution and acylated

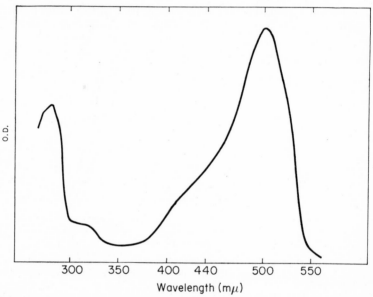

Fig. 1·4. Absorption spectrum of pelargonidin 3-sophoroside-7-glucoside, the pigment of *Papaver orientale* petals. Solvent: MeOH/HCl.

pigments have lower R_f values in aqueous solvents but higher ones in butanol mixtures (cf. pelargonidin 3,5-diglucoside (31 in BAW, 23 in 1% HCl) and its *p*-coumaroyl derivative monardein (40 in BAW, 19 in 1% HCl)). Finally, the nature of the anthocyanidin modifies R_f value slightly but in a very regular fashion (compare the R_f values of the 3-glucosides of the six common anthocyanidins in BAW, which are 44 (Pg), 41 (Pn), 38 (Cy), 38 (Mv), 35 (Pt), and 26 (Dp)) and the behaviour of an unknown delphinidin glycoside, for example, can be accurately predicted if the corresponding cyanidin and pelargonidin derivatives are available.

(*e*) *Hydrolysis.* After determining R_f values and spectral properties of an anthocyanin, its structure is confirmed by studying its hydrolysis products. This may be carried out with acid or enzyme and the rate of hydrolysis usually gives some information regarding the nature of the substituents. Anthocyanins, because they carry a net positive charge, are more resistant to acid hydrolysis than other flavonoid glycosides and simple 3-glucosides and 3-galactosides require $\frac{1}{2}$ hr at 100° in N HCl; triglycosides and acylated glycosides require up to 1 hr for complete hydrolysis (Harborne, 1965a). This resistance to hydrolysis may be turned to advantage, since the intermediates are easily isolated by stopping the reaction at different time intervals. The identification of those intermediates is useful for determining whether the sugars present are attached to one or more phenolic hydroxyl groups. Thus, 3-diglycosides only yield one, whereas 3,5-diglycosides give two intermediates and 3- and 3,5-triglycosides produce two and four intermediates respectively.

Acid hydrolysis removes acyl groups from complex pigments, but the yield of acyl groups is low due to the instability of cinnamic acids in hot acid. Alkaline hydrolysis at room temperature (under nitrogen, if anthocyanins with catechol groups are being studied) specifically removes an acyl group quantitatively leaving the sugars of the anthocyanin intact. Enzymic hydrolysis, using a specific anthocyanase present in *Aspergillus niger* (Huang, 1955), is useful for distinguishing simpler glycosides, which are rapidly hydrolysed (1–2 hr at pH 4 and 37°), from 3-rhamnosides or 3-rhamnoside-5-glucosides, which are slowly hydrolysed (4 hr), and from acylated pigments, which are unattacked (Harborne, 1965a). In the enzymic hydrolysis of anthocyanin to sugar and anthocyanidin, the aglycone is spontaneously transformed to a colourless pseudobase which is rapidly destroyed by molecular oxygen.

Sugars may also be obtained from anthocyanins, with destruction of the aglycone, by oxidation with alkaline hydrogen peroxide (Karrer and Widmer, 1927). The method, which works well on a micro-scale (Chandler and Harper, 1961), yields only the sugar attached in the 3-position. The identification of acylated sugars from alkaline peroxide oxidation of acylated anthocyanins effectively proves (Harborne, 1964a) that the acyl group in these pigments is attached to the 3-sugar and not to a phenolic hydroxyl group, as was at one time suggested.

Anthocyanidins obtained by the acid hydrolysis of anthocyanins are identified by the methods described earlier in this chapter (Section II.B). In determining the aglycone:sugar ratios of anthocyanins, anthocyanidin produced after quantitative hydrolysis can be determined spectrophotometrically. Sugar concentration is better measured after chromatographic separation followed by reaction with aniline hydrogen phthalate.

The sugars of anthocyanins can be identified by the usual procedure of paper chromatography. The presence of glucose and galactose can be confirmed by use of glucose and galactose oxidase respectively and these and other monosaccharides by measuring the spectral maxima of the coloured products formed after reaction with resorcinol–H_2SO_4 or aniline hydrogen phthalate (Harborne, 1960b). Disaccharides can be satisfactorily separated after long development on paper but high-voltage electrophoresis is better for distinguishing, if a glucosylglucose is involved, which of the eight possible isomers is present (Harborne, 1963b). An acyl residue (usually *p*-coumaric, caffeic or ferulic acid) is identified after its isolation by spectroscopy, chromatography and mixed melting point.

B. MONOSIDES

Four classes of anthocyanidin 3-monosides are known: 3-glucoside, 3-galactoside, 3-rhamnoside and 3-arabinoside. A xyloside (of petunidin) has been reported in the flowers of *Lavandula pedunculata* (Labiatae) but the position of the sugar is not known for certainty (Maroto, 1950). All the naturally occurring monosides have the sugar attached to the 3-position and this is apparently related to the need for pigment stability. The 5-, 7- and 4'-glucosides of pelargonidin have been synthesised by Leon *et al.* (1931) but none is as stable as the 3-glucoside. The recent report that delphinidin occurs as the 7-galactoside in the leaves of *Bladhia sieboldii* (Yeh and Huang, 1961) is suspect, since the evidence presented for this unusual substitution is un-convincing. If 5- or 7-glycosides do occur naturally, their presence in a plant should immediately be apparent from their instability and different colour reactions, R_f values and spectral properties.

3-Glucosides are by far the most common of all anthocyanins and have been synthesised by Robinson and his co-workers (cf. Robinson, 1934). Cyanidin 3-glucoside (12) is the commonest anthocyanin occurring as it does in many flowers, leaves and fruits and particularly in autumnal leaves. The 3-glucosides of pelargonidin, cyanidin, peonidin, delphinidin, petunidin and malvidin can all be isolated from the flowers of one or other of the colour varieties of the Chinese primrose, *Primula sinensis* (Harborne and Sherratt, 1961), or from the seedcoat of colour forms of the French bean, *Phaseolus vulgaris* (Feenstra, 1960). Convenient sources of individual 3-glucosides include the strawberry (pelargonidin), blackberry (cyanidin), *Verbena* flower (delphinidin) and grape (malvidin). It should be noted that peonidin 3-glucoside cannot be obtained

(12) Cyanidin 3-glucoside   (13) Cyanidin 3-galactoside

from the American cranberry, as was earlier reported (Grove and Robinson, 1931); the pigments here are 3-galactosides and 3-arabinosides (Zapsalis and Francis, 1965).

3-Galactosides are much less common than 3-glucosides. The best known, cyanidin 3-β-galactoside (13), was isolated from the skin of "Grimes Golden Delicious" apple by Sando in 1937 and also occurs in the leaves of the purple copper beech (Robinson and Smith, 1955), where it is accompanied by traces of pelargonidin 3-galactoside. By far the best sources of 3-galactosides are the berries of various *Vaccinium* species (cranberries, bilberries, blueberries, etc.) which contain all but the pelargonidin derivative; cyanidin 3-galactoside also occurs in other plants of the Ericaceae. 3-Galactosides are not easy to distinguish from 3-glucosides and some mistakes have occurred in identification. Thus, the pigment of the wild strawberry, *Fragaria vesca*, thought to be pelargonidin 3-galactoside (Robinson, 1934), is in fact the 3-glucoside (Sondheimer and Karash, 1956), and likewise the so-called galactoside of malvidin in *Primula sinensis* flowers, called "primulin" by Scott-Moncrieff (1936) and incorrectly identified by Bell and Robinson (1934), is the 3-glucoside (Harborne and Sherratt, 1957, 1961).

3-Rhamnosides are rare; they probably have an α-linkage, since they are not attacked by anthocyanase, and are likely to be analogous in their configuration to flavonol 3-rhamnosides, which are α-linked. 3-Rhamnosides of the six common anthocyanidins may all be found in flowers of different colour varieties of the sweet pea (*Lathyrus odoratus*) in conjunction with other glycosides (Harborne, 1963b). The 3-rhamnosides of pelargonidin, cyanidin and delphinidin also occur in the pink petals of *Plumbago rosea* (Plumbaginaceae) and that of malvidin in the blue flowers of *Ceanothus*.

Pelargonidin 3-arabinoside is not known, but the 3-arabinosides of the other five anthocyanidins probably occur in *Vaccinium* berries although they have not all been isolated as yet. They are accompanied in the berries by much larger amounts of the corresponding 3-galactosides. Cyanidin 3-arabinoside is known to be the L-sugar and to have the α-configuration, since this pigment isolated from cocoa cotyledons (*Theobroma cacao*) has been directly compared with synthetic material (Forsyth and Quesnel, 1957). There is a close stereochemical relationship between L-arabinose and D-galactose which may account for the remarkable co-occurrence of arabinosides and galactosides in *Theobroma*, *Vaccinium* and *Rhododendron* and also in many grasses (*see* Chapter 7).

C. BIOSIDES

Five 3-biosides are known (3-rhamnosylglucoside, 3-xylosylglucoside, 3-xylosylgalactoside, 3-sophoroside and 3-gentiobioside) and the linkages in these disaccharides are now fairly well established (*see* Fig. 1·5). The 3-rhamnoglucosides, the most common class of 3-bioside, contain rutinose, 6-α-L-rhamnosido-D-glucose. Cyanidin 3-rutinoside (antirrhinin) (14) is the best known bioside and was isolated from mauve flowers of *Antirrhinum majus* by Scott-Moncrieff (1930) who obtained it in four different hydrated forms ($1\frac{1}{2}$, $2\frac{1}{2}$, $3\frac{1}{2}$ and $4H_2O$). Pelargonidin, cyanidin and delphinidin 3-rutinoside may conveniently be obtained from flowers of *Antirrhinum* and tulip and from the fruit of the aubergine, *Solanum melongena*. Petunidin 3-rutinoside seems to be only accessible in a rather rare colour form of the cultivated potato (Harborne, 1960), but peonidin 3-rutinoside occurs (with cyanidin derivatives) in the fruit of the sloe, *Prunus spinosa*, and in the domestic plum. Rhamnosylglucosides with other sugar–sugar links may exist. A rhamnosylglucoside of pelargonidin with different R_f values from the 3-rutinoside was reported in

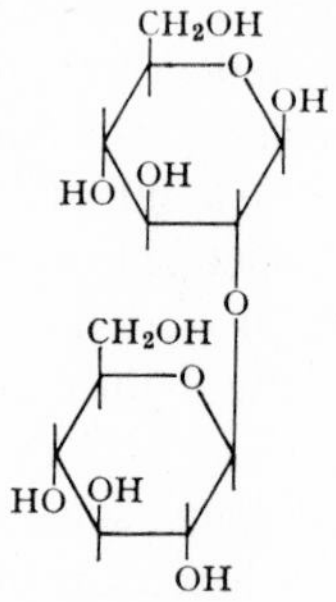

Sophorose

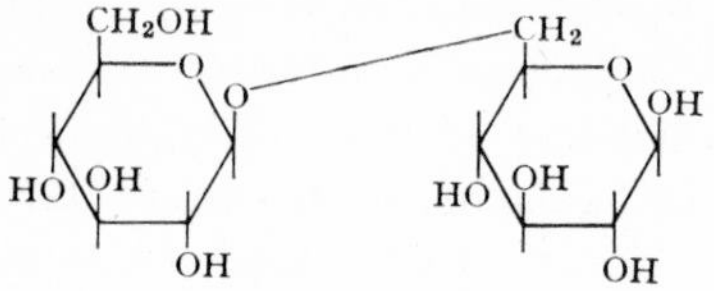

Gentiobiose

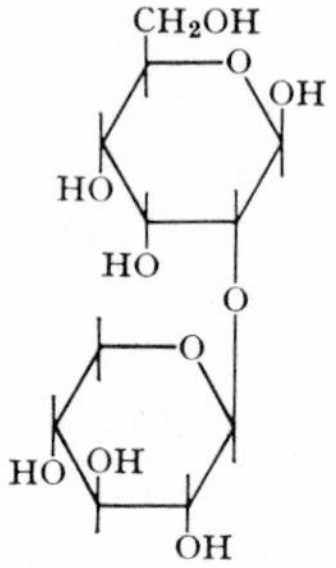

Lathyrose

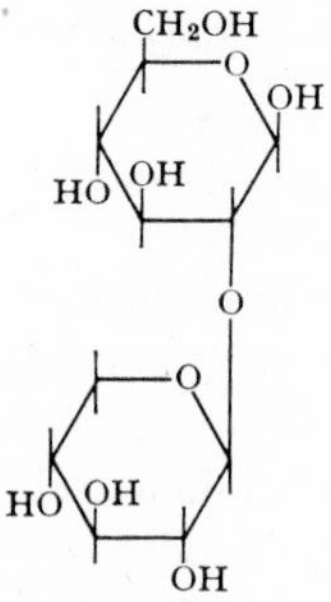

Sambubiose

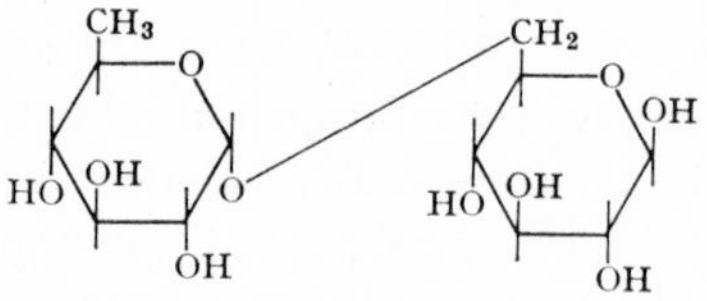

Rutinose

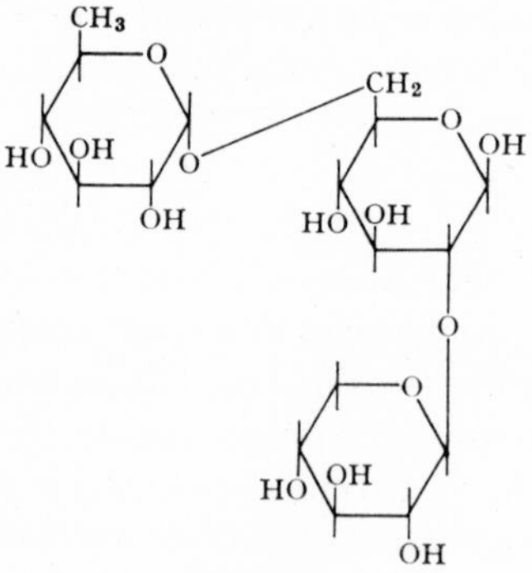

2^G-Xylosylrutinose

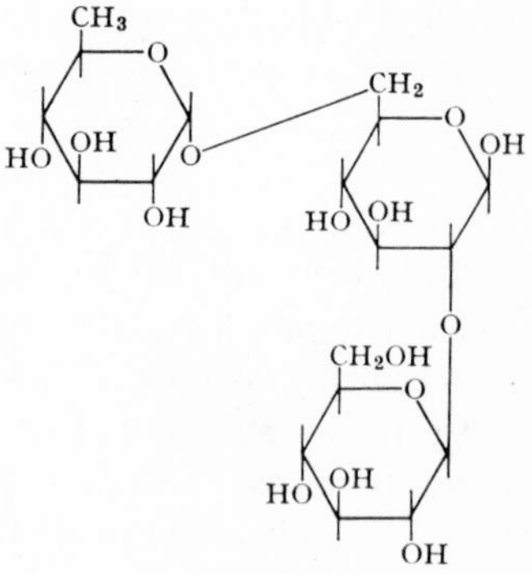

2^G-Xylosylrutinose

2^G-Glucosylrutinose

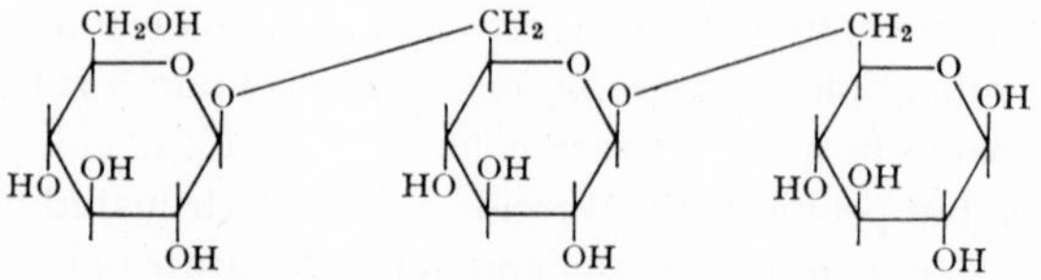

Gentiotriose

FIG. 1·5. Di- and trisaccharides present in anthocyanins.

Tulipa by Halevy (1962), but it is not clear yet in what way it differs from the 3-rutinoside.

The disaccharide present in the cyanidin 3-xylosylglucoside in elderberries, *Sambucus nigra*, was first thought to be primeverose (6-β-D-xylosido-D-glucose (Reichel *et al.*, 1957)) but has since been shown to be 2-β-D-xylosido-D-glucose (sambubiose) (Reichel and Reichwald, 1960). Cyanidin 3-sambu-

(14) Antirrhinin   (15) Mecocyanin

bioside has also been found in the leaves of *Begonia* and in the petals of *Streptocarpus*, in which plant it occurs with the pelargonidin derivative. A delphinidin 3-xylosylglucoside has been reported in the pericarp of *Daphniphyllum macropodum* (Shibata and Ishikura, 1964) but no xylosylglucoside of a methylated anthocyanidin has yet been described. The closely related disaccharide, lathyrose (2-β-xylosido-D-galactose), is found, linked to pelargonidin, in petals of sweet pea, *Lathyrus odoratus*. A xylosylrhamnoside of cyanidin has also been reported (Shibata, 1958) but re-investigation suggests that it was an impure form of the 3-rutinoside (*see* Chapter 7, p. 237).

Two series of anthocyanidin 3-diglucosides are known. Those with gentiobiose (β,1→6 linkage) are rare, while those with sophorose (β,1→2 linkage) are fairly common. The 3-gentiobiosides of peonidin, petunidin and malvidin occur only in flowers and leaves of *Primula sinensis*, while pelargonidin 3-gentiobioside also occurs in flowers of *Tritonia* cv. "Prince of Orange" and cyanidin 3-gentiobioside also in petals of *Petunia* (Birkofer *et al.*, 1963). The mecocyanin isolated from poppy flowers is not the 3-gentiobioside (Grove and Robinson, 1934) but the 3-sophoroside of cyanidin (15). This pigment and the pelargonidin 3-sophoroside occur widely in *Papaver* (*see* Chapter 5, p. 150), as well as in many other genera. Convenient sources of cyanidin 3-sophoroside are most varieties of raspberry (e.g. "October"), redcurrant (e.g. "Red Lake") and sour cherry (e.g. "Wye Morello"). 3-Sophorosides based on delphinidin no doubt occur in nature but they do not appear to have been described.

Five classes of 3,5-dimonosides are now known: 3,5- and 3,7-diglucosides, 3-galactoside-5-glucoside, 3-rhamnoside-5-glucoside and 3-arabinoside-5-glucoside. The 3-glucoside-5-arabinoside reported by Nordström (1956) in *Dahlia variabilis* is, in fact, cyanidin 3,5-diglucoside(Harborne and Sherratt 1957). The 3,5-diglucosides of the six common anthocyanidins have all been synthesised by Robinson and his co-workers. The 3,5-diglucosides of pelargonidin, cyanidin, delphinidin and malvidin are all widely occurring but are still obtained quite conveniently from their original sources, i.e. *Pelargonium*

2

zonale, Centaurea cyanus, Delphinium consolida and *Malva sylvestris* respectively. Peonidin 3,5-diglucoside is less common, but occurs in *Rosa* and *Pelargonium* as well as in *Paeonia*. Petunidin 3,5-diglucoside occurs in only trace amounts in the garden *Petunia* and is better obtained from flowers of *Anchusa*, where it is the only pigment. The only naturally occurring 3,7-diglucoside is that of cyanidin (16) which occurs in red varieties of *Petunia* (Birkofer *et al.*, 1963).

(16) Cyanidin 3,7-diglucoside   (17) Petunidin 3-rhamnoside-5-glucoside

All six 3-rhamnoside-5-glucosides occur in quantity in flowers of different colour forms of the sweet pea and the delphinidin, petunidin and malvidin derivatives also occur in most wild *Lathyrus* species and also in *Pisum, Vicia* and *Cicer*. Petunidin 3-rhamnoside-5-glucoside (17) has been found in the flowers of the unrelated *Limonium* (Plumbaginaceae). Accompanying 3-rhamnoside-5-glucosides in the sweet pea are pelargonidin and peonidin 3-galactoside-5-glucosides; the two pigments are present in such small quantities that their identification is still tentative. The sole representative of the 3-arabinoside-5-glucoside class is the cyanidin derivative, isolated recently by Asen and Budin (1966) from petals of the *Rhododendron* cultivar "Red Wing".

D. TRIOSIDES

Three 3-triosides are known, one type with a linear trisaccharide, the gentiotrioside, and two with branched trisaccharides in their structure. No less than four classes of anthocyanidin have two sugars attached to the 3-position and a third in the 5- or 7-position: the 3-rutinoside-5-glucosides, the 3-sambubioside-5-glucosides, the 3-sophoroside-5-glucosides and a single 3-sophoroside-7-glucoside. The 3-gentiotriosides of pelargonidin, peonidin, petunidin and malvidin occur with the 3-gentiobiosides in *Primula sinensis* but have not been found anywhere else in nature. By contrast, anthocyanins with the branched trisaccharides, 2^G-glucosylrutinose and 2^G-xylosylrutinose, have been found in no less than four genera, *Rubus, Ribes, Begonia* and *Clivia*. Such anthocyanins are closely related to the 3-rutinosides, 3-sophorosides and 3-sambubiosides and not surprisingly each occurs in close association with the appropriate pair of 3-biosides. All three of the known pigments (two cyanidin derivatives and pelargonidin-(2^G-glucosylrutinoside) (18)) are easily obtained from raspberries (cvs. "September", "Tweed" or "Preussen"), but the cyanidin derivatives are also conveniently isolated from red currants (cvs. "Earliest of the Fourlands", "Fay's Prolific" or "Red Lake") or leaves and petals of common *Begonia* cultivars (Harborne and Hall, 1964b).

3-Rutinoside-5-glucosides of all six common anthocyanidins have been described and they occur fairly widely, being present in an acylated form in many plants of the Solanaceae (potato, tomato, petunia) and in *Viola*. All six may be isolated unacylated from garden hybrids of *Streptocarpus*. Cyanidin 3-sambubioside-5-glucoside accompanies the 3-sambubioside in elderberries as a minor component but may be isolated in quantity (as an acyl derivative) from purple flowers of *Matthiola incana* (Cruciferae); the pelargonidin derivative is present in scarlet flowers of the latter plant. The best sources of 3-sophoroside-5-glucosides are, again, plants of the Cruciferae, the pelargonidin and cyanidin derivatives occurring in the red and purple radish respectively and the cyanidin derivative in red cabbage. These pigments occur in the above plants in an acylated form but they are found occasionally elsewhere (e.g. *Gladiolus*, *Pisum*) unacylated. Finally, there is the 3-sophoroside-7-glucoside of pelargonidin (19), the only representative of its class, which occurs in *Papaver orientale, P. nudicaule* and also in three *Watsonia* species. It is quite different from the other pelargonidin glycosides in its colour, spectral max and R_f so it is readily distinguished from the corresponding 3-sophoroside, with which it usually occurs.

(18) Pelargonidin 3-(2^G-glucosylrutinoside)   (19) Pelargonidin 3-sophoroside-7-glucoside

E. ACYLATED GLYCOSIDES

Considerable difficulty has been experienced in the past in determining the structure of acylated anthocyanins, mainly because of their lability and close association with each other (they often occur as mixtures) and with unrelated plant constituents. Following the use of chromatographic methods of separation and purification, a better understanding of their constitution has been obtained and most structures advanced during the 1930's have had to be revised or changed. The structure of most acylated pigments that have been fully characterised at the present time are indicated in Table 1·5. A number of other acylated glycosides have been described in less detail and there is no doubt that many others are awaiting discovery in plants.

All acylated glycosides that have been examined in detail have their acyl group attached to the sugar in the 3-position and not to one of the free phenolic hydroxyl groups present in the pigment. The evidence for this rests partly on spectral considerations but mainly on the results of hydrolytic and oxidative

TABLE 1·5. The Structure of Acylated Anthocyanins

Trivial name	Nature and number of acyl groups	Sugar unit[1]	Source[2]
PELARGONIDIN DERIVATIVES			
Monardein	*p*-Coumaric (1)	3G5G	*Monarda didyma* petals
Salvianin	Caffeic (1)	3G5G	*Salvia splendens* petals
Pelanin	*p*-Coumaric (1)	3RG5G	*Solanum tuberosum* tubers
Matthiolanin	*p*-Coumaric (1) and ferulic (1)	3XG5G	*Matthiola incana* petals
Raphanusin A	*p*-Coumaric (1)	3GG5G	} *Raphanus sativus* roots
Raphanusin B	Ferulic (1)	3GG5G	
CYANIDIN DERIVATIVES			
Perillanin	⎤	3G5G	*Perilla ocimoides* leaf
Cyananin	⎬ *p*-Coumaric (1)	3RG5G	*Solanum tuberosum* petals
Raphanusin C	⎦	3GG5G	} *Raphanus sativus* roots
Raphanusin D	Ferulic (1)	3GG5G	
Rubrobrassicin C	Ferulic (2)	3GG5G	*Brassica oleracea* var. *rubra* leaves
Hyacinthin	*p*-Coumaric (1)	3G	*Hyacinthus orientalis* bulbs
PEONIDIN DERIVATIVES			
Peonanin	*p*-Coumaric (1)	3RG5G	*Solanum tuberosum* tubers
DELPHINIDIN DERIVATIVES			
Awobanin	*p*-Coumaric (1)	3G5G	*Commelina communis* petals
Delphanin	*p*-Coumaric (1)	3RG5G	*Solanum tuberosum* petals
PETUNIDIN DERIVATIVES			
Petanin	*p*-Coumaric (1)	3RG5G	*Solanum tuberosum* tubers
Guineesin	*p*-Coumaric (2)	3RG5G	*Solanum guineese* berries
MALVIDIN DERIVATIVES			
Coumaroyloenin	⎤	3G	*Vitis vinifera* berries
Tibouchinin	⎬ *p*-Coumaric (1)	3G5G	*Tibouchina semidecandra* petals
Negretein	⎦	3RG5G	*Solanum tuberosum* tubers

[1] Abbreviations: 3G, 3-glucoside; 3G5G, 3,5-diglucoside; 3RG5G, 3-rutinoside-5-glucoside; 3XG5G, 3-sambubioside-5-glucoside; 3GG5G, 3-sophoroside-5-glucoside.

[2] In most cases the first known source is given; for other sources, *see* Tables 5·1, 6·1 and 7·1 in later chapters.

cleavage. Negretein (20), for example, the pigment of the purple "Congo" salad potato, yields a *p*-coumaroylrutinose on oxidation with alkaline H_2O_2 and on acid hydrolysis gives *p*-coumaroylrhamnose but no *p*-coumaroyl-glucose. The position of attachment of the acyl group to the sugar hydroxyl is difficult to determine because of the lability of this ester linkage and because of the propensity of acyl groups on sugars for changing their positions during chemical treatment. Birkofer *et al.* (1965) have been able to show however, that *p*-coumaric acid in negretein, petanin and peonanin is attached to the

(20) Negretein

4-hydroxyl group of the rhamnose and that the same acid in monardein and salvianin is attached to the 6-hydroxyl of the glucose in the 3-position. Furthermore, they have obtained evidence that the *p*-coumaric (or ferulic) acid in raphanusin A (or B) is attached to the 6-hydroxyl group of the glucose moiety nearest to the anthocyanidin (*see* 21).

(21) Raphanusin B

The acyl group of these pigments is nearly always *p*-coumaric acid (Table 1·4) but caffeic and ferulic acid both appear on occasion. Other aromatic acids (*p*-hydroxybenzoic, protocatechuic, gallic, sinapic) have been reported as being present in certain pigments but closer examination has shown that they have been mistakenly identified. For example, delphinin, the pigment of *Delphinium consolida*, was supposed to contain *p*-hydroxybenzoic acid (Willstäter and Meig, 1915) but later work (Harborne, 1964a) has shown that it is not even acylated. Again, rubrobrassicin-C, the main red cabbage pigment, was reported to have sinapic acid in its structure (Chmielewska *et al.*, 1955). The pigment certainly retains a sinapic acid impurity right up to the final stages of purification but the pure pigment gives only diferuloylglucose on oxidation (Harborne, 1964). The possibility that aliphatic organic acids are attached to anthocyanins must be borne in mind (cf. *Sorghum*, p. 245) but no positive evidence of this is yet available. Malonic acid is certainly not present in monardein or salvianin, as was once thought (Karrer and Widmer, 1927).

The possibilities of structural variation in the acylated anthocyanin series are clearly considerable but fortunately those known at present have only one (and rarely two) acyl groups. Furthermore, the great majority of known pigments are 3,5-diglycosides and those that are 3-glycosides are all 3-glucosides.

IV. Table of R_f Values and Sources

The following Table (1·6) lists the R_f values and sources of practically all the known anthocyanins. Most of the data have been published before, in a review (Harborne, 1958a) and in subsequent research papers; they are collected together here as an aid to anthocyanin identification. A few words on nomenclature are required. Trivial names were given to most simple glycosides isolated before 1940, e.g. antirrhinin for cyanidin 3-rutinoside, callistephin for pelargonidin 3-glucoside, etc., but fortunately the habit has not been extended to the present day. The problem of finding suitable names for the 100 or so pigments now known is considerable and, if available, they would provide a burden to the newcomer to the field who has enough trouble memorising the names of the anthocyanidins. Furthermore, while some names of glycosides are obviously convenient and are related to the structure of the aglycone, e.g. pelargonin for pelargonidin 3,5-diglucoside, cyanin for cyanidin 3,5-diglucoside, etc., others based on generic or species names can lead to confusion. The use of fragarin for example, for pelargonidin 3-galactoside is unfortunate since the pigment of *Fragaria* has subsequently been shown to be the 3-glucoside.

In the following tables, trivial names are omitted for the glycosides, but are used for the di- and trisaccharides (e.g. gentiobiose for glucosyl-β,1→6-glucose) since they are well established and well known. No attempt is made to indicate the configuration of the sugar or the linkage, as these have not always been established. However, the monosaccharides are probably all in the pyranose form and are as follows: D-glucose, D-galactose, and D-xylose (all β-linked), but L-rhamnose and L-arabinose (both α-linked). Trivial names are retained for the acylated pigments, for convenience and also because many of their structures have not yet been completely elucidated.

R_f values are given only for paper chromatography in spite of the fact that thin-layer chromatography has been applied to anthocyanins and is used in some laboratories instead of paper. This choice is deliberate, because paper is still the most convenient medium and gives more reproducible results. Also there is no evidence that significantly better resolutions have been achieved on silica gel, as compared to paper. Anthocyanin mobilities on silica gel differ somewhat from those on paper but R_f values on thin layers of cellulose are very similar to those obtained on paper and it is a simple matter to relate the data collected here to such conditions.

The R_f values quoted are based on those obtained in the author's laboratory, but it is clear that many other laboratories have been able to obtain almost identical data with little difficulty. As always, markers should be run on all chromatograms, so that adjustments can be made for differences in temperature or technique. The results here are for descending chromatography on No. 1 Whatman paper at about 20°. The four solvents employed are: BAW, *n*-butanol–acetic acid–water (4:1:5, top layer) (used within 1–2 hr of mixing); BuHCl, *n*-butanol–2 N HCl (1:1, top layer) (paper equilibrated for 24 hr in

TABLE 1·6. R_f Values and Sources of All Known Anthocyanins

| Glycoside | BAW | R_f values (× 100) in | | | Source |
		BuHCl	1% HCl	HOAc–HCl	
A. PELARGONIDIN					
3-Rhamnoside	71	64	22	53	*Plumbago rosea* petals
5-Glucoside	51	49	18	57	Synthetic
7-Glucoside	46	51	15	—	Synthetic
3-Rhamnoside-5-glucoside	46	24	39	70	*Lathyrus odoratus* petals
3-Glucoside	44	38	14	35	*Callistephus chinensis* petals
3-Galactoside	39	37	13	33	*Fagus sylvatica* leaves
3-Rutinoside	37	30	22	44	*Antirrhinum majus* petals
3-Sambubioside	37	34	31	60	*Streptocarpus hybrida* petals
3-Sophoroside	36	30	38	65	*Papaver rhoeas* petals
3-(2^G-Glucosylrutinoside)	33	15	63	73	*Rubus idaeus* berries
3-Lathyroside	35	31	34	60	*Lathyrus odoratus* petals
3,5-Diglucoside	31	14	23	45	*Pelargonium zonale* petals
3-Galactoside-5-glucoside	31	16	23	46	*Lathyrus odoratus* petals
3-Gentiobioside	30	26	21	47	*Primula sinensis* petals
3,7-Diglucoside	30	10	38	70	Synthetic
3-Rutinoside-5-glucoside	29	13	40	58	*Streptocarpus hybrida* petals
3-Sophoroside-5-glucoside	23	10	60	68	*Gladiolus gandavensis* petals
3-Gentiotrioside	25	10	35	52	*Primula sinensis* petals
3-Sambubioside-5-glucoside	24	18	43	70	*Matthiola incana* (Ac) petals
3-Sophoroside-7-glucoside	18	04	73	84	*Papaver orientale* petals
Monardein	40	46	19	53	*Monarda didyma* petals
Pelanin	37	43	27	67	*Solanum tuberosum* tubers
Salvianin	37	37	17	48	*Salvia splendens* petals
Raphanusin A	34	34	49	73	*Raphanus sativus* roots
Raphanusin B	34	26	49	73	*Raphanus sativus* roots
Matthiolanin	34	29	37	61	*Matthiola incana* petals

Colours in u.v. light: 3-glycosides—dull orange-red; 3,5-diglycosides—fluorescent yellow; 3,7-diglycosides—dull orange-yellow; acylated glycosides—bright yellow.

TABLE 1·6—*continued*

Glycoside	R_f values ($\times 100$) in				Source
	BAW	BuHCl	1% HCl	HOAc–HCl	
B. CYANIDIN					
3-Rhamnoside	63	65	14	50	*Plumbago rosea* petals
5-Glucoside	44	39	07	42	Synthetic
3-Arabinoside	42	32	05	27	*Theobroma cacao* leaves
3-Glucoside	38	25	17	26	*Chrysanthemum indicum* petals
3-Galactoside	37	24	07	26	*Fagus sylvatica* leaves
3-Rutinoside	37	25	19	43	*Antirrhinum majus* petals
3-Sambubioside	36	24	24	51	*Sambucus nigra* berries
3-Rhamnoside-5-glucoside	34	21	36	64	*Lathyrus odoratus* petals
3-Sophoroside	33	22	34	61	*Papaver rhoeas* petals
3-Lathyroside	31	15	29	55	*Lathyrus odoratus* petals
3-Arabinoside-5-glucoside	29	08	18	42	*Rhododendron* "Red Wing" petals
3,5-Diglucoside	28	06	16	40	*Centaurea cyanus* petals
3-(2^G-Xylosylrutinoside)	28	15	47	68	*Begonia metallica* leaves
3-(2^G-Glucosylrutinoside)	26	11	61	73	*Begonia coccinea* petals
3-Rutinoside-5-glucoside	25	08	36	59	*Streptocarpus hybrida* petals
3-Gentiobioside	20	10	14	46	*Primula sinensis* leaves
3,7-Diglucoside	20	05	17	52	*Petunia hybrida* petals
3-Sambubioside-5-glucoside	19	10	41	54	*Sambucus nigra* berries
3-Sophoroside-5-glucoside	17	09	54	62	*Pisum sativum* pods
Hyacinthin	33	63	04	24	*Hyacinthus orientale* bulbs
Perillanin	35	34	11	43	*Perilla ocimoides* leaves
Cyananin	32	26	22	62	*Solanum tuberosum* petals
Raphanusin C	34	21	40	68	*Raphanus sativus* roots
Raphanusin D	33	15	39	63	*Raphanus sativus* roots
Rubrobrassicin C	21	13	39	—	*Brassica oleracea* var. *rubra* leaves

Colours in u.v. light: 3-glycosides—dull magenta; 3,5-diglycosides—bright red.

C. PEONIDIN

3-Rhamnoside	67	69	18	47	*Lathyrus odoratus* petals
3-Arabinoside	48	42	09	36	*Vaccinium macrocarpon* berries
5-Glucoside	45	31	08	30	Synthetic
3-Rhamnoside-5-glucoside	44	23	39	65	*Lathyrus odoratus* petals
3-Glucoside	41	30	09	33	*Primula sinensis* petals
3-Galactoside	39	28	10	32	*Vaccinium macrocarpon* berries
3-Sambubioside	34	25	38	—	*Lathyrus odoratus* petals
3-Lathyroside	34	25	38	63	*Lathyrus odoratus* petals
3-Rutinoside	34	14	16	41	*Verbascum phoenicium* petals
3,5-Diglucoside	31	10	17	44	*Paeonia officinalis* petals
3-Galactoside-5-glucoside	30	10	17	44	*Lathyrus odoratus* petals
3-Rutinoside-5-glucoside	29	12	37	60	*Magnolia lennei* petals
3-Gentiobioside	25	10	19	—	*Primula sinensis* leaves
3-Gentiotrioside	10	09	36	—	*Primula sinensis* leaves
Peonanin	34	31	22	62	*Solanum tuberosum* tubers

Colours in u.v. light: 3-glycosides—dull magenta; 3,5-diglycosides—fluorescent pink.

D. DELPHINIDIN

3-Rhamnoside	37	51	13	34	*Plumbago rosea* petals
3-Rutinoside	30	15	11	37	*Solanum tuberosum* tubers
3-Sambubioside	—	15	—	44	*Daphniphyllum macropodum,* peri-carps
3-Glucoside	26	11	03	18	*Verbena hybrida* petals
3-Galactoside	23	11	03	18	*Empetrum nigrum* berries
3-Rhamnoside-5-glucoside	21	10	26	51	*Lathyrus odoratus* petals
3-Rutinoside-5-glucoside	20	06	37	61	*Viola × wittrockiana* petals
3,5-Diglucoside	15	03	08	32	*Verbena hybrida* petals
Awobanin	30	22	05	32	*Commelina communis* petals
Delphanin	31	24	31	59	*Solanum tuberosum* petals

Colours in u.v. light: 3-glycosides—dull purple; 3,5-diglycosides—bright purple.

2*

TABLE 1·6—continued

Glycoside	BAW	R_f values ($\times 100$) in BuHCl	1% HCl	HOAc–HCl	Source
E. PETUNIDIN					
3-Rhamnoside	40	42	10	36	Lathyrus odoratus petals
3-Rutinoside	35	16	13	42	Solanum tuberosum tubers
5-Glucoside	35	27	03	30	Synthetic
3-Glucoside	35	14	04	22	Primula sinensis petals
3-Galactoside	33	13	04	20	Vaccinium angustifolium berries
3-Rhamnoside-5-glucoside	28	09	31	55	Lathyrus odoratus petals
3-Sophoroside	—	17	36	66	Petunia hybrida petals
3-Gentiobioside	26	07	12	—	Primula sinensis petals
3,5-Diglucoside	24	04	08	32	Anchusa petals
3-Rutinoside-5-glucoside	23	06	37	61	Atropa belladonna petals (Ac)
3-Gentiotrioside	21	05	29	—	Primula sinensis petals
Petanin	32	26	19	59	Solanum tuberosum tubers
Guineesin	40	67	30	—	Solanum guineese berries

Colours in u.v. light: as for delphinidin glycosides.

Glycoside	BAW	BuHCl	1% HCl	HOAc–HCl	Source
F. MALVIDIN					
5-Glucoside	43	24	04	22	Synthetic
3-Rhamnoside	39	40	11	39	Lathyrus odoratus petals
3-Glucoside	38	15	06	29	Primula polyanthus petals
3-Galactoside	36	15	06	29	Vaccinium uliginosum berries
3-Rutinoside	35	16	15	45	Sinningia speciosa petals
3,5-Diglucoside	31	03	13	42	Malva sylvestris petals
3-Rhamnoside-5-glucoside	31	10	34	61	Lathyrus odoratus petals
3-Rutinoside-5-glucoside	30	05	40	63	Streptocarpus hybrida petals
3-Gentiobioside	22	10	15	—	Primula sinensis petals
3-Gentiotrioside	16	06	31	—	Primula sinensis petals
Tibouchinin	40	42	10	—	Tibouchina semidecandra petals
Negretin	36	28	20	64	Solanum tuberosum tubers

Colours in u.v. light: 3-glycosides—dull purple; 3,5-diglycosides—fluorescent cerise.

G. RARE ANTHOCYANIDINS

Capensinidin 3-rhamnoside	41	40	30	72	*Plumbago capensis* petals
Pulchellidin 3-glucoside	40	39	12	32	*Plumbago pulchella* petals
Europinidin 3-glucoside	27	19	06	31	*Plumbago europea* petals
Hirsutidin 3,5-diglucoside	25	02	32	56	*Primula hirsuta* petals
Rosinidin 3,5-diglucoside	21	10	36	72	*Primula rosea* petals
Aurantinidin 3-sophoroside	08	23	47	65	*Impatiens aurantiaca* petals
Aurantinidin 3,5-diglucoside	05	09	17	35	*Impatiens aurantiaca* petals

Colours in u.v. light: capensinidin, hirsutinidin, aurantinidin and rosinidin 3,5-diglucosides fluoresce; the remainder have dull colours.

H. 3-DEOXYANTHOCYANIDINS

Apigeninidin glycosides:

5-Glucoside	41	38	22	55	*Rechsteineria cardinalis* petals
Adiantum 1	47	52	31	64	*Adiantum vietchianum* fronds
Dryopteris 1	41	41	53	77	*Dryopteris erythrosora* fronds

Luteolinidin glycosides:

5-Glucoside	31	27	13	40	*Rechsteineria cardinalis,* petals
5-Diglucoside	27	52	44	68	*Bryum cryophyllum*
Adiantum 2	36	36	15	46	*Adiantum vietchianum* fronds
Dryopteris 2	31	32	44	70	*Dryopteris erythrosora* fronds
Pteris 1	42	66	30	60	*Pteris quadriaurita* fronds
Pteris 2	48	71	38	70	*Pteris quadriaurita* fronds

Tricetinidin glycoside:

Dryopteris 3	26	24	30	63	*Dryopteris erythrosora* fronds
Columnidin 5 (?) -glucoside	27	25	02	12	*Columnea × banksii* petals

Colours in u.v. light: apigeninidin glycosides—intense yellow fluorescence; luteolinidin glycosides—bright orange; other glycosides—dull orange-red.

tank containing lower layer); 1% HCl, water–conc. HCl (97:3); and HoAc–HCl, acetic acid–conc. HCl–water (15:3:82).

The problem of obtaining authentic anthocyanins for purposes of comparison with an unknown is a real one. Some sources have been mentioned in earlier sections of this chapter and one source for each pigment is reported in the table. In the case of common glycosides, it is not possible to list all sources here but tables in later chapters of this book (*see* pp. 127 and 186) may be consulted for alternative sources of the pigments listed.

The author, in his work with anthocyanins, has had the inestimable advantage of a supply of some of the synthetic anthocyanins prepared in Sir Robert Robinson's laboratory at Oxford and also of some of the natural pigments isolated by earlier biochemists at the John Innes Institute. These samples are now mainly exhausted so that anyone entering the field for the first time is forced to rely on isolating pigments anew. This, however, presents few problems, since chromatographic techniques are so readily available. A representative collection of anthocyanins can be obtained from plant material available to any householder, who has a garden of a reasonable size or is in reach of a florist and a supermarket. Flowers of a few key plants (e.g. sweet pea, dahlia, pelargonium) and fruits of a few others will provide many of the best-known pigments. Finally it should not be forgotten that the complex anthocyanins will provide good yields of simpler ones on partial acid hydrolysis. 3-Glucosides may thus be obtained from 3,5-diglucoside and no less than five simpler glycosides are produced by controlled acid hydrolysis of complex anthocyanins such as negretein.

CHAPTER 2

FLAVONE AND FLAVONOL PIGMENTS

I. Introduction

Of the various classes of naturally occurring compounds based on the flavonoid skeleton, the flavones and flavonols are collectively the most abundant group, providing somewhere between 200 and 300 known substances. They are sometimes known as anthoxanthins (i.e. yellow flower pigments) but this term is not one that is much used today. They differ from the anthocyanins in being more highly oxidised, the structures of the two parent compounds being shown below:

(1) Flavone   (2) Flavonol

From the point of view of colour, anthocyanins are much more important but nevertheless flavones and flavonols do make a significant contribution either as yellow pigments in their own right or else as co-pigments to anthocyanins, when they have a bluing effect on flower colour.

The distinction between flavones (1) and flavonols (2) is an arbitrary one, since flavonols are simply a class of flavone in which the 3-position is substituted by a hydroxyl group. It is a convenient division because such a large

number of structures have been isolated. Furthermore, the two groups differ
in their spectral and colour properties and are usually distinguishable by
chromatographic means. Finally, phytochemical surveys indicate that the
simple difference in structure between flavones and flavonols is one that is of
considerable phylogenetic significance.

Flavonol glycosides often correspond closely in structure with anthocyanins
(compare two very common pigments in each class, rutin (3) and antirrhinin
(4)) and the two groups of pigment are frequently found together, especially
in flower petals, and are intimately connected biosynthetically. The most

(3) Rutin   (4) Antirrhinin

common structural types in each group are the same and the flavonol analogues
of pelargonidin, cyanidin and delphinidin are widespread in nature. However,
flavonols do differ from anthocyanins in being more variable structurally;
many more flavonols are known, both as aglycones and as glycosides.

Flavone glycosides (e.g. 5) are related to the 3-deoxyanthocyanins (e.g. 6)
as flavonol glycosides are to anthocyanins, but flavones are more widely
distributed than 3-deoxyanthocyanins. Structural variation in the flavone
series is again considerable, and many partly and fully methylated derivatives
have been described. Flavones differ from most other common flavonoid types
in occurring in plants as C-glycosyl derivatives (see Section III.B). and as
dimers, the biflavonyls (see Chapter 3).

(5) Luteolin 5-glucoside   (6) Luteolinidin 5-glucoside

Flavones and flavonols are easier to identify than anthocyanins since they
are more stable. Most workers in the classical period (i.e. from Perkin and von
Kostanecki in the 1890's to Zemplen and Bognar in the 1940's) studied pigments which crystallised out from concentrated plant extracts. More recent
studies have been devoted to water-soluble flavones and ones which occur as
complex mixtures and which are not usually obtainable in quantity. Micro-
methods of identification have thus been developed (Nordström and Swain,

1953) and widely exploited. The methods are similar to those used with anthocyanins (see Chapter 1) and are based on the combined use of paper chromatography and u.v. spectroscopy. Other physical methods (e.g. infra-red spectroscopy) have been used occasionally and are important for determining the structures of the more complex flavonoids. A special contribution of nuclear magnetic resonance (n.m.r.) spectroscopy has been the solution of structural problems associated with the C-glycosylflavones (Hillis and Horn, 1965).

Many excellent accounts of the naturally occurring flavones and flavonols have already appeared (see especially Geissman, 1962, and Dean, 1963). In the restricted space available here, it is only possible to cover phytochemical aspects adequately. This involves listing all the flavones and flavonols known at the present time (March, 1966). The need for this is indicated by the fact that there are at least twice as many flavonol glycosides known today as were recorded by Hattori (in Geissman, 1962). Methods of identification will be mentioned and R_f and spectral data are included, where possible. The distribution of flavones and flavonols is described in detail in Chapters 4–7.

II. The Flavones

A. Apigenin, Luteolin and Tricin

There are only two common flavones, apigenin (7) and luteolin (8). They can be recognised on Forestal chromatograms as dark brown spots which change to bright yellow-green or yellow on fuming with ammonia and which are more mobile than the three common flavonols.

(7) Apigenin   (8) Luteolin

APIGENIN was first isolated as the aglycone of apiin, a glycoside present in parsley seed (von Gerichten, 1901). It is a pale yellow powder, m.p. 348–350° (acetate, m.p. 185–187°). At least six O-glycosides as well as a larger number of C-glycosyl derivatives have been described (see Section III).

LUTEOLIN was first isolated as the colouring matter of weld, Reseda luteola (Moldenhauer, 1856). It melts at 330–331° (acetate m.p. 226–227°). Both compounds are conveniently obtained from parsley seed, following acid hydrolysis, but they have to be separated chromatographically. Fairly pure apigenin can be obtained more directly from flowers of chrysanthemums, daisies or zinnias and, likewise, luteolin from carrot or foxglove leaves.

Two monomethyl ethers of apigenin are known: the 7-methyl ether, genkwanin (m.p. 285–287°), which can be isolated from petals of Daphne genkwa or bark of *Prunus serrata*, and the 4'-methyl ether, açacetin (m.p. 261–263°), which is conveniently obtained from flowers of the common toadflax, *Linaria vulgaris*. Acacetin is easily distinguished from apigenin by R_f (Table 2·1) and it is different in its colour when treated with ammonia. The 7,4'-dimethyl ether of apigenin (m.p. 174°) is reported to be present in birch buds (Bauer and Dietrich, 1933).

Two methyl ethers of luteolin have been described, the 3'-methyl ether, chrysoeriol, and the 4'-methyl ether, diosmetin. They have similar R_f values

TABLE 2·1. R_f Values, Spectral Properties and Sources of the Common Flavones and their Derivatives

Flavone	R_f (× 100) in			Source
	BAW[2]	Forestal	PhOH	
APIGENIN	89	83	88	*Apium graveolens* seed
7-Me ether: genkwanin	—	—	—	*Daphne genkwa* petal
4'-Me ether: acacetin	91	91	88	*Robinia pseudacacia* leaf
6-OH deriv.: scutellarein	—	—	—	*Scutellaria altissima* leaf
LUTEOLIN	78	66	66	*Reseda luteola* leaf
3'-Me ether: chrysoeriol	82	77	90	*Eriodictyon glutinosum* leaf
4'-Me ether: diosmetin	85	80	86	*Diosma crenulata* leaf
6-OH deriv.:	54	53	42	*Catalpa bignonioides* leaf
TRICETIN	56	37	28	Synthetic
3',5'-DiMe ether: tricin	73	72	87	*Triticum dicoccum* leaf

	Spectral properties[3]					
	λ_{max} in EtOH		$\varDelta\lambda+$ NaOAc	$\varDelta\lambda+$ AlCl$_3$	$\varDelta\lambda+$ NaOEt	$\varDelta\lambda+$ H$_3$BO$_3$
	Band I	Band II	Band I	Band II	Band II	Band II
APIGENIN	269	336	9	45	61	0
Genkwanin	269	335	−1	50	60	0
Acacetin	276	332	2	13	41	0
Scutellarein	286	339	—	—	48	0
LUTEOLIN	255, 268[1]	350	14	40	54	25
Chrysoeriol	252, 269	350	20	35	58	0
Diosmetin	253, 268	345	10	40	39	0
6-Hydroxy deriv.	285	349	+	26	decomp.	21
TRICETIN	250, 269	356	20	42	decomp.	26
Tricin	248,[1] 269	355	20	24	65	0

[1] Indicates shoulder or inflection.

[2] BAW = butanol–acetic acid–water (4:1:5); Forestal = conc. HCl–acetic acid–water (3:30:10); PhOH = water-saturated phenol.

[3] Only the most important data are given. Two main bands in neutral solution are generally of approximately the same intensity. Spectral shifts are obtained by adding appropriate reagent to the ethanolic solution of the flavone (cf. Jurd, 1962). All values are given in millimicrons (mμ). Log ϵ values for flavones are in the range 4·1 to 4·4; values for apigenin are 4·31 (269 mμ) and 4·32 (336 mμ).

but fortunately differ in their colour reactions and spectral properties (Table 2·1). While diosmetin appears on a paper chromatogram as a dark spot changing only slightly in colour with ammonia, chrysoeriol is a very characteristic bright yellow-green in u.v. light in the presence of ammonia. Convenient sources of chrysoeriol are leaves of lucerne, where it occurs as the 7-glucuronide along with tricin, and petals of most ivory varieties of the snapdragon, *Antirrhinum majus*. Diosmetin is not as widespread as was once

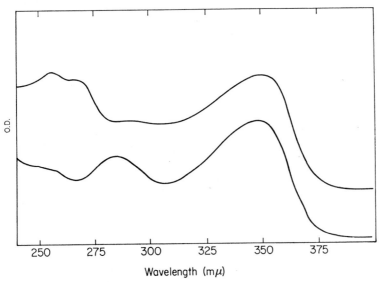

FIG. 2·1. Absorption spectra of luteolin (top curve) and 6-hydroxyluteolin (bottom curve). Solvent: 95% EtOH.

thought, and has been incorrectly reported to occur in *Dahlia variabilis* petals, in *Apium graveolens* seed (cf. Nordström and Swain, 1953) and in various labiates (Hörhammer and Wagner, 1962). It can however be obtained from leaves of *Mentha* and flowers of *Columnea* species (Harborne, 1966c).

Introduction of a 6-hydroxyl group into apigenin and luteolin produces compounds which have very distinctive spectral properties (e.g. λ_{max}^{EtOH} at 285 mμ instead of at 269 mμ; Fig. 2·1 and Table 2·1) and which appear on chromatograms as dark absorbing spots unaffected by ammonia. Neither 6-hydroxyapigenin (scutellarein) nor 6-hydroxyluteolin has been reported often in plants but both may have been overlooked in leaf surveys because of their dull colours. Scutellarein may be isolated from leaves of several species of *Scutellaria* (e.g. the common skullcap, *S. altissima*), of the greater knapweed, *Centaurea scabiosa*, and of the plantain *Plantago asiatica*. 6-Hydroxyluteolin is present in the leaves of the tree *Catalpa bignonioides*.

Curiously, the flavone tricetin (9), which corresponds in structure to the widely distributed anthocyanidin delphinidin and flavonol myricetin, has not yet been isolated from a plant. Its properties are given in Table 2·1 in case it is discovered in future surveys. Its 3′,5′-dimethyl ether, TRICIN (10), is known, although it is a rare constituent. It has a very distinct colour in u.v. light on paper when treated with ammonia—bright almost fluorescent yellow. It differs from most of the common flavones in its spectral properties in that

(9) Tricetin   (10) Tricin

the short u.v. peak is less intense than the long wave-length band. The best source of tricin is seed of *Orobanche ramosa* a parasite of tobacco, where it occurs free as the main phenolic constituent. In leaves of wheat, *Triticum*, it is present as the 5-glucoside but is accompanied by large amounts of glyco-flavones and in lucerne *Medicago sativa*, where it occurs as the 7-glucuronide, it is accompanied by chrysoeriol 7-glucuronide.

B. Rarer Structures

There are some 34 rarer flavones, which can be conveniently divided into five classes (Table 2·2). Most of them have characteristic u.v. absorption spectra and their maxima are included in Table 2·2 as far as they are available. R_f values have not generally been recorded for the rarer flavones and in the systems used for common flavones and their derivatives (e.g. BAW, PhOH and Forestal) they have high R_f values and do not separate well from each other. Flavone itself has an R_f of 99 in BAW and Forestal and appears as a dull purple absorbing spot. Again, the C-methylflavone sideroxylin has R_f 92 in both BAW and PhOH. In detecting and separating these more highly substituted flavones, thin-layer chromatography on silica gel is the most effective procedure. The C-methylflavones, sideroxylin and eucalyptin, separate well on silica gel in chloroform–acetic acid (9:1) or toluene–ethyl formate–formic acid (5:4:1); R_f values ($\times 100$) are 45, 68, 35 and 64 respectively (Hillis and Isoi, 1965). Again, 5,7,2′,4′-tetrahydroxyflavone in *Artocarpus* heartwood can be separated on silica gel using acetone–benzene (1:3) from the related substance with a C_5-isoprenoid residue attached at C-6 (Radhakrishnan *et al.*, 1965).

Spectral data are important for distinguishing the more complex flavones. Measurements of spectral shifts produced when inorganic salts are added to the cell solution are also of considerable diagnostic value (*see* below). The simple record of maxima, as given in Table 2·2, is not completely satisfactory since the shape and intensity of the bands vary considerably; the direct comparison of spectral curves is more informative. A few general points about the maxima however, may be noted. In 2'-hydroxyflavones, the long wavelength band (designated band II) is usually of weaker intensity than the short wave-length band (band I) and in a substance such as wightin, band II is a shoulder, not a peak. 6-Hydroxylation characteristically causes band I to shift to longer wave-lengths but methylation of this or adjacent hydroxyl groups reduces the size of the shift (compare the values for scutellarein (286), hispidulin (277), and mikanin (270 mμ). Finally, C-methylation of the flavone nucleus likewise has a bathochromic effect on band I (compare sideroxylin (max at 280) with genkwanin (max at 270)).

Furanoflavones have not been listed in Table 2·2 since they are outside the scope of the present work. Nevertheless, these substances, which are only found in seed and rootbark of *Pongamia* species (*see* under Leguminosae in Chapter 6), are reported to have quite characteristic spectral properties (Khanna and Seshadri, 1963).

Infra-red spectral determinations have been mainly used with flavones for "fingerprint" identification, rather than for structural determination, because interpretation is generally difficult and shifts are unpredictable. However, infra-red data may be useful in at least two ways. Flavones lacking a 5-hydroxyl group are readily distinguished from those with a 5-hydroxyl because the frequency of the carbonyl band absorption is at 1620–1639 cm^{-1} instead of at 1653 cm^{-1} or above. Also, flavones unsubstituted in the B-ring differ considerably in the 680–870 cm^{-1} region (CH deformation vibrations) from 4'-substituted flavones; e.g. wogonin has two bands at 690 and 765 whereas apigenin has a single band at 831 cm^{-1} (Wagner, 1965).

Nuclear magnetic resonance spectral measurements have been used for elucidating the structures of the more highly substituted flavones. One example of the type of data obtainable is shown below (taken from Hillis and Isoi, 1965):

Proton shifts (T.M.S., $\delta = 0$) for
sideroxylon dissolved in CDCl$_3$ and pyridine

It will be noted that protons of the 6- and 8-C-methyl groups can be distinguished from each other and from those of the 7-O-methyl group.

The importance of the newer physical methods for flavone identification can be gauged by the fact that most of the compounds listed in Table 2·2 have been identified within the last 10 years. Chrysin (5,7-hydroxyflavone) is

TABLE 2·2. List of the Rarer Naturally Occurring Flavones

Trivial name	Structure	Spectral maxima	Source
GROUP I: Simple			
Flavone	Parent compound	250, 297	*Primula farinosa*
—	5-OH	272, 337	*Primula imperialis*
Primetin	5,8-diOH	259, 289, 340	*Primula modesta*
Chrysin	5,7-diOH	270, 314	*Pinus* heartwood
Tectochrysin	5-OH, 7-OMe	269, 310	*Pinus* heartwood
—	7,4′-diOH	225, 253^1, 329	*Trifolium repens* leaf
—	7,3′,4′-triOH	237, 313^1, 342	*Trifolium repens* leaf
—	5,6-diOMe	—	*Casimiroa edulis*
GROUP II: 6- or 8-Hydroxyflavones			
Baicalein	5,6,7-tri-OH	276, 324	*Scutellaria baicalensis* root
Oroxylin-A	5,7-diOH, 6-OMe	—	*Oroxylum indicum* bark
Wogonin	5,7-diOH, 8-OMe	—	*Scutellaria baicalensis* root
Hispidulin	5,7,4′-triOH, 6-OMe	277, 338	*Ambrosia hispida* plant
Pectolinaringenin	5,7-diOH, 6,4′-diOME	275, 335	*Cirsium oleraceum* leaf
Mikanin	5-OH, 6,7,4′-triOMe	256, 270, 361	*Mikania cordata* leaf
Pedalitin	5,7,3′,4′-tetraOH, 6-OMe	—	*Sesamum indicum* leaf
Dinatin	5,6,7,3′-tetraOH, 4′-OMe	—	*Digitalis lanata* leaf
Sinensetin	5,6,7,3′,4′-pentaOMe	240, 265, 328	*Citrus sinensis* peel

GROUP III: 2'-Hydroxyflavones

—	5,7-diOH, 2'-OMe	272, 318	*Scutellaria epilobifolia*
Echioidinin	5,2'-diOH, 7-OMe	268, 340	*Andrographis echioides*
—	5,6,2'-triOMe	234, 268, 328	*Casimiroa edulis*
Wightin	5,3'-diOH, 7,8,2'-triOMe	272, 330–340[1]	*Andrographis wightiana*
Norartocarpetin	5,7,2',4'-tetraOH	251, 271, 288,[1] 355	*Artocarpus heterophyllus*

GROUP IV: C-Methylflavones

Strobochrysin	6-Me, 5,7-diOH	276, 327	*Pinus* heartwood
—	6-Me, 5-OH, 7,4'-diOMe	—	*Eucalyptus torelliana*
Eucalyptin	6,8-diMe, 5-OH, 7,4'-diOMe	281, 287, 323	
Sideroxylin	6,8-diMe, 5,4'-diOH, 7-OMe	280, 293[1], 332	*Eucalyptus sideroxylon*

GROUP V: 5,6,7,8-Hydroxylated flavones

—	5,6,7,8-tetraOMe	271, 305	*Lindera lucida*
Xanthomicrol	5,4'-diOH, 6,7,8-triOMe	282, 296, 336	*Satureia douglasii*
Demethoxysudachitin	5,7,4'-triOH, 6,8-diOMe	283, 337	*Citrus sudachi*
Tangeretin	5,6,7,8,4'-pentaOMe	271, 322	*Citrus deliciosa*
Sudachitin	5,7,4'-triOH, 6,8,3'-triOMe	283, 349	*Citrus sudachi*
Nobiletin	5,6,7,8,3',4'-hexaOMe	248, 272, 332	*Citrus aurantium*
Lucidin	5,7-diOH, 6,8-diOMe, 3',4'-O_2CH_2	285, 343	
—	5,6,7,8-tetraOMe, 3',4'-O_2CH_2	251, 271,[1] 335	*Lindera lucida*

GROUP VI: Alkaloidal flavones

Ficine	5,7-diOH, 8-(*N*-methylpyrrolidyl)	275, 329	*Ficus pantoniana*

[1] Indicates shoulder or inflection.

the simplest compound that can be formed from the condensation of acetate–malonate and cinnamic acid (*see* Chapter 8) and it is interesting that practically all the known flavones can be derived directly or indirectly from chrysin by further substitutions. The only exceptions are 5,6-dimethoxy- and 5,8-dihydroxyflavone, two substances which may well be formed by other routes. Chrysin can clearly undergo hydroxylation in either the A- or B-ring and the many hydroxy derivatives produced frequently undergo *O*- or *C*-methylation. Judging from the Group II flavones (Table 2·2), the 6-, 8- or 4'-hydroxyls are preferred positions for *O*-methylation but clearly in *Citrus* (Group V flavones) there is no limit to the number of hydroxyl groups that may be methylated. Methylation seems to be linked to hydroxylation in that the more highly substituted flavones are rare (e.g. 6-hydroxyluteolin) or are not known to occur naturally (e.g. 6,8-dihydroxyluteolin) whereas occurrences of their methyl ethers are well documented. Only a few of the various possible methyl ethers of flavones such as 6-hydroxyluteolin have been found so far, but many more are no doubt present in plants awaiting isolation.

III. Flavone Glycosides

A. *O*-Glycosides

Of the 12 or so classes of flavone *O*-glycosides that have been described most have the sugar attached to the 7-hydroxyl, the most acidic grouping; 7-*O*-glucosides and 7-*O*-rutinosides are common types. 7-*O*-Glucuronides are more restricted in their occurrence, being found mainly in plants of the genera *Antirrhinum, Digitalis, Erigeron, Medicago* and *Scutellaria*. The disaccharide, apiosylglucose, is found mainly in association with flavones but has also been found attached to an isoflavone (*see* Chapter 3). The 7-apiosylglucosides of apigenin and luteolin occur in seeds of parsley and celery (Umbelliferae).

TABLE 2·3. Properties of Apigenin and Luteolin Glucosides

Pigment	M.p.	λ_{max} in EtOH (mμ)	$\Delta\lambda$ alk. (mμ)	R_f values ($\times$ 100) in BAW	15% HOAc
APIGENIN					
5-Glucoside	295°	—	—	—	—
7-Glucoside	178°	268, 335	63	65	25
4'-Glucoside	—	270, 334	57	65	25
7,4'-Diglucoside	—	273, 320	53	14	62
LUTEOLIN					
5-Glucoside	280°	255, 264, 352	51	82	07
7-Glucoside	252–254°	255, — 353	54	44	15
3'-Glucoside	243–245°	256, 268, 350	52	40	21
4'-Glucoside	177–178°	— 270, 341	39	68	34

Sugars are attached to other hydroxyls than the 7- and, in the case of apigenin, 5-, 4'- and 7,4'-glycosides have been described (Table 2·3). Apigenin 7-glucoside is fairly common, especially in the Compositae (see Table, 6·9 Chapter 6) but the 5-glucoside is restricted to the leaves of *Amorpha fruticosa*. The 7- and 4'-glucosides occur as an inseparable mixture in petals of *Dahlia variabilis* (Nordström and Swain, 1953). Apigenin 7,4'-diglucuronide occurs in *Antirrhinum* petals and the related 7,4'-diglucoside (11) in yew pollen.

(11)   (12)

In the case of luteolin, the four isomeric monoglucosides can readily be differentiated by R_f, u.v. spectra and melting point (Table 2·3). Whether mixtures of two of them, e.g. the 7- and the 4'-glucoside (cf. apigenin series), would separate if they occurred together in the same plant is not known. Of the four glucosides, the most recent discovery is the 3'-glucoside (12), which occurs in the leaves of *Dracocephalum thymiflorum* (Litvinenko and Sergienko, 1965). The widely distributed 7-glucoside has been known for a long time; a convenient source is the leaf of the carrot, *Daucus carota*. It does not, however, occur in *Digitalis purpurea* leaves as reported by Hukuti (1936); the 7-glucuronide and 7-glucosylglucuronide are present instead (Harborne, 1964a). The 5-glucoside was first reported in the seed of *Galega officinalis* (Barger and White, 1923) but has since been noted in *Dahlia* petals (Nordström and Swain, 1953). Luteolin 4'-glucoside occurs in flowers of the Spanish broom, *Spartium junceum*, and *Gnaphalium affine* and in leaves of *Acer cissifolium* (Spada and Cameroni, 1958; Aritomi, 1965).

Other glycosides of luteolin reported include a 7-diglucoside in *Dahlia* petals and yew pollen, a 7-xylosylglucoside in *Salix caesia* (Rabate, 1938) and a 7-tetraglucoside in the leaf of the olive tree, *Olea europea* (Bockova et al., 1964).

Glycosides of most of the other naturally occurring flavones are fewer in number and many have not been fully characterised. Acacetin (4'-O-methylapigenin) is unusual in being reported to occur as the 7-rutinoside in *Linaria vulgaris*, *Buddleia variabilis* and *Robinia pseudacacia* but as a different 7-rhamnosylglucoside in *Fortunella japonica* (Matsumo, 1958). The two rhamnosylglucosides have different melting points (265° and 215° respectively) and it has been suggested that the *Fortunella* glycoside may be the neohesperidoside. Acacetin also occurs in leaves of *Robinia* as a 7-xylosylrhamnosylglucoside (Freudenberg and Hartmann, 1954). Finally, it should be mentioned that 6-hydroxylated flavones (i.e. baicalein, scutellarein etc.) are best known as the 7-glucuronides and highly active glucuronidases can be obtained from the plants (e.g. *Scutellaria*) which contain them (Marsh, 1955).

R_f data for most common flavone O-glycosides are given in Table 2·4. While most glycosides have different values, nevertheless mixtures of closely related compounds are sometimes difficult to separate on paper. One pair, the 7- and 4′-glucosides of apigenin, have already been mentioned; a second

TABLE 2·4. R_f Values and Sources of Flavone Glycosides

Pigment	R_f values ($\times$ 100) in				Source
	BAW	H$_2$O	15% HOAc	PhOH	
O-Glycosides					
APIGENIN					
7-Glucoside	65	04	25	78	*Bellis perennis* rays
7-Rutinoside	58	09	46	74	*Rhus succedanea* leaf
7-Apiosylglucoside	57	06	42	75	*Apium graveolens* seed
7-Glucuronide	57	13	29	46	*Antirrhinum majus* pet
7,4′-Diglucuronide	12	53	—	10	*Antirrhinum majus* pet
7,4′-Diglucoside	14	31	62	56	*Taxus baccata* pollen
LUTEOLIN					
7-Glucoside	44	01	15	56	*Daucus carota* leaf
7-Apiosylglucoside	42	03	23	50	*Apium graveolens* seed
7-Diglucoside	40	05	29	54	*Dahlia variabilis* petal
5-Glucoside	82	00	07	65	*Galega officinalis* seed
7-Glucuronide	24	12	25	17 ⎱	*Digitalis purpurea* leaf
7-Glucosylglucuronide	22	33	28	00 ⎰	
ACACETIN					
7-Rutinoside	61	14	60	55	*Linaria vulgaris* petal
CHRYSOERIOL					
7-Glucuronide	36	33	14	65	*Medicago sativa* leaf
TRICIN					
5-Glucoside	23	00	14	89 ⎱	*Triticum dicoccum* leaf
5-Diglucoside	08	05	36	78 ⎰	
7-Glucuronide	29	01	11	71 ⎱	*Medicago sativa* leaf
7-Diglucuronide	22	30	61	15 ⎰	
C-Glycosyl Derivatives					
APIGENIN					
8-Glucosyl (vitexin)	41	06	24	63 ⎱	*Tamarindus indica* leaf
6-Glucosyl (isovitexin)	56	16	44	79 ⎰	
Isovitexin 7-glucoside	38	33	72	60	*Saponaria officinalis* lea
8-Rhamnosylglucosyl	45	54	80	82 ⎱	*Avena sativa* leaf
6-Arabinosylglucosyl	47	57	80	80 ⎰	
6,8-Diglycosyl	15	14	33	38	*Triticum dicoccum* seed
LUTEOLIN					
8-Glucosyl (orientin)	31	02	13	43 ⎱	*Tamarindus indica* leaf
6-Glucosyl (iso-orientin)	41	09	35	51 ⎰	

pair are the 7-glucuronides of luteolin and chrysoeriol present in snapdragon petals. Flavone glycosides (apart from glucuronides and C-glycosylflavones) can be distinguished from flavonol glycosides in plant extracts by their low mobility in water. Most flavone glycosides, having the sugar attached to the 7-position, differ from common flavonol glycosides by being more resistant to acid hydrolysis. Apigenin 7-glucuronide is exceptionally resistant and is only completely hydrolysed by 4 hr heating in N HCl–EtOH (1:1) at 100° (Harborne, 1965a).

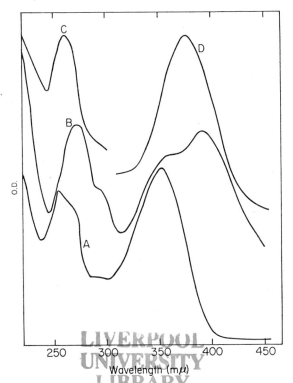

FIG. 2·2. Absorption spectrum of luteolin 7-glucoside. Curve A, 95% EtOH; curve B, EtOH–AlCl₃; curve C, EtOH–NaOAc; curve D, EtOH–H₃BO₃–NaOAc.

Most flavone glycosides have very similar spectral properties in ethanol to the corresponding aglycones. Thus, most apigenin glycosides have λ_{max} at 269 and 335 mμ and most luteolin glycosides at 255 and 350 mμ (Fig. 2·2). There is therefore no need to list the maxima of the glycosides as those of the aglycones have already been given (Chapter 1, Tables 1·1 and 1·2). It should however be noted that the addition of sugar residue may alter the relative intensity of the flavone maxima and that 4'-O-glycosylation has a hypsochromic effect on the longwave maximum.

The value of spectral data in the identification of flavones and their glycosides is considerably increased by the use of reagents such as aluminium chloride ($AlCl_3$), fused sodium acetate (NaOAc), sodium ethoxide (NaOEt) and boric acid–sodium acetate (H_3BO_3). These substances, when added in trace amounts to the solution of the flavone in ethanol, produce shifts in the maxima according to the location of the various functional groups in the molecule (Jurd, 1962) (see Fig. 2·2). In the case of flavone glycosides, a positive shift with NaOAc in band I indicates that the 7-hydroxyl is not glycosylated and likewise a positive $AlCl_3$ shift in band II shows that the 5-hydroxyl is free. The band II maximum shifts with borate only if the flavone has free hydroxyl groups in the 3'- and 4'-positions so that while the 5- and 7-glucosides of luteolin exhibit bathochromic shifts of 20 mμ, the 3'- and 4'-glucosides do not respond to this reagent. Finally the magnitude of the alkaline shift is related to the number and position of free hydroxyl groups in the molecule; for example luteolin 4'-glucoside ($\Delta\lambda + 39$ mμ) gives a smaller shift than the 7-glucoside ($\Delta\lambda + 54$ mμ). The use of these reagents for structural analysis is discussed further under flavonol glycosides (Section V.C.).

Most flavone glycosides have relatively high melting points (see e.g. Table 2·3) and form hydrates on recrystallisation from aqueous solvents. A disconcerting feature of many glycosides, especially derivatives of scutellarein and acacetin, is their remarkable insolubility. Although these pigments appear to occur in the plant in soluble form, once isolated they are difficult to dissolve in both water and organic solvents.

B. C-GLYCOSYL DERIVATIVES

The glycoflavones comprise a remarkable group of substances in which sugar is attached by a carbon–carbon bond to the flavone nucleus in the 6- or 8-position. They are not, strictly speaking, glycosides since this infers an oxygen linkage, but they are obviously closely related to the flavone O-glycosides which have just been discussed. They differ by being almost completely resistant to acid hydrolysis; all O-glycosides are hydrolysed by heating for 4 hr in 2 N HCl–ethanol (1:1) whereas C-glycosyl derivatives are largely unaffected (Harborne, 1965a). Glycoflavones can also be distinguished from O-glycosides by their mobility on water chromatograms (see Table 2·4) and by the fact that they undergo isomerisation when heated with acid and then appear as two spots in most solvent systems.

Glycoflavones are spectrally identical in the u.v. to the parent flavones from which they are derived. They differ from O-glycosides in that they undergo spectral shifts with all four of the usual reagents since their phenolic hydroxyl groups are all free. There are some differences in the infra-red spectra between the two classes of glycoside, but the most important differences occur in the nuclear magnetic resonance (n.m.r.) spectra (see below).

There has been much confusion in the past over the structure of glycoflavones, but this has been cleared up following the application of n.m.r. spectroscopy to these substances. Confusion over nomenclature still persists but the sensible proposals of Haynes (1965), which are followed here, should receive wide acceptance. The two commonest glycoflavones are vitexin, the 8-*C*-glucoside of apigenin (13), and isovitexin (formerly saponaretin), the 6-*C*-glucosyl isomer (14). They occur together in many plants, but especially

(13) Vitexin   (14) Isovitexin

in members of the Archichlamydeae (*see* Table 5·3, Chapter 5), one form usually predominating over the other. *In vitro*, they are readily interconvertible, and on heating either in acid solution, roughly equal amounts of the two compounds are produced. The luteolin analogues, orientin and iso-orientin (formerly lutexin and lutonaretin), are also widely occurring.

C-Glycosyl compounds are found almost exclusively among the flavones, but the *C*-glycosyl derivatives of isoflavones, flavanones and dihydrochalcones have also been isolated (*see* Table 2·5). The suggestions that *C*-glucosyl-

TALBE 2·5. The Known *C*-Glycosyl Flavonoids

Aglycone	8-*C*-Glucosyl deriv.	6-*C*-glucosyl deriv.
FLAVONES		
Apigenin	Vitexin	Isovitexin
Acacetin	Cytisoside	—
Genkwanin	—	Swertisin
5-Deoxyapigenin	Bayin	—
Luteolin	Orientin	Iso-orientin
Chrysoeriol	Scoparin	Isoscoparin
5-*O*-Methylluteolin	Parkinsonin-A	—
5,7-Di-*O*-methylluteolin	Parkinsonin-B	—
7-*O*-Methylluteolin	—	Swertiajaponin
ISOFLAVONE		
Daidzein	Puerarin	Isopuerarin
FLAVANONE		
Naringenin	Isohemiphloin	Hemiphloin
DIHYDROCHALCONE		
3-Hydroxyphloretin	Aspalathin	—
FLAVONOL		
7-*O*-Methylkaempferol	—	Keyakinin
7-*O*-Methylquercetin	—	Keyakinin B

kaempferol and *C*-glucosylquercetin occur in some plants of the Ranunculaceae and Iridaceae (Bate-Smith and Swain, 1960) has not been substantiated
by further investigation (Harborne, 1962b; 1965a), but the occurrence of
6-*C*-glucosyl derivatives of rhamnocitrin (7-*O*-methylkaempferol) and of
rhamnetin in the wood of *Zelkova serrata* (Ulmaceae) (cf. Funaoka, 1957) has
recently been fully established by Hillis and Horn (1966). The only widely
occurring flavonoid class still unrepresented among the *C*-glycosyl derivatives
are the anthocyanins.

6,8-Diglycosyl derivatives of apigenin (the vicenins (15)) and luteolin (the
lucenins) have recently been described by Seikel *et al.*, (1966) and Hörhammer
et al. (1965). Some probably have glucose in both the 6- and 8-positions but
others have two different sugars. Furthermore, glycoflavones occur as *O*-
glycosides in which a second sugar unit is attached either to a phenolic hydroxyl [e.g. as in isovitexin 7-*O*-glucoside, saponarin (16)] or to the alcoholic
group of the sugar already present in the 6- or 8-position (e.g. as in the vitexin
rhamnoside of *Avena*). The structure of one substance of the latter type has
already been completely elucidated by Horowitz and Gentili (1966); it is
8-*C*-(xylosyl-*β*,1→2)glucosylapigenin (17) which occurs in *Citrus sinensis*.

(15) Vicenin type   (16) Saponarin

(17) Xylosylglucosylapigenin

The identification of a new *C*-glycosylflavone today involves the use of
n.m.r. spectroscopy, since it is the only satisfactory procedure for showing
which position the sugar occupies in the flavone nucleus. It may also furnish
data on the nature of sugar–sugar linkages when more than one sugar is present.
A valuable diagnostic feature, recognised by Hillis and Horn (1965) who analysed the spectra of acetylated glycoflavones, is the fact that the acetate shows

a peak at about $\delta = 1 \cdot 75$ (instead of the more usual $2 \cdot 0$). This peak can be attributed to the proton of the 2-acetoxy group of the sugar, a group which is shielded by the aromatic moiety to which it is in particularly close proximity. None of the acetoxy groups in flavone O-glycoside acetates are so affected, all having signals at about $\delta = 2 \cdot 0$. The positions of the sugar residues in glyco-flavones are located by comparing the signals of the aromatic protons (Horowitz and Gentili, 1964). Neglecting B-ring protons (which lie at about $\delta = 8 \cdot 0$), isovitexin has signals at 6·77 and 6·56, vitexin at 6·77 and 6·29 and apigenin at 6·76, 6·24 and 6·52. The signal at 6·76 is readily assigned to the proton in the 3-position, which is present in all three substances. That at 6·5 belongs to H-8 and that at 6·24 to H-6. In the 6,8-disubstituted vicenins and lucenins, signals at $\delta = 6 \cdot 5$ and $\delta = 6 \cdot 24$ are absent.

Known glycoflavones can still be identified without recourse to n.m.r. spectroscopy, by careful u.v. and chromatographic comparison with authentic substances. R_f values for most of the common glycoflavones are therefore given in Table 2·4. Identification can be confirmed by ferric chloride oxidation, which yields the carbon-attached sugar in low yield and treatment of the glycoflavone with hydriodic acid in phenol at the boiling point yields the parent flavone.

IV. The Flavonols

A. Common Structures and Their Methylated Derivatives

Three flavonols are common in plants, kaempferol (18), quercetin (19) and myricetin (20). The first two are present in leaves and petals of over half the plants that have been surveyed and myricetin is present in a tenth. These

(18) Kaempferol   (19) Quercetin

(20) Myricetin

pale yellow pigments are the most easily recognised of all phenolic constituents (Bate-Smith, 1962), have bright yellow colours on paper in u.v. light and are clearly separated from each other and from other flavonoids on Forestal chromatograms (R_f 55, 41 and 28). Identification can be confirmed by u.v. spectroscopy and microdegradation with alkali. Their melting points and those of their acetates, are rather too high (between 280 and 357°) to be very useful. Quercetin is used in the pharmaceutical and food industries and samples of the three flavonols are normally available commercially; all three occur in "instant tea" (soluble tea powder), a convenient alternative source.

Each of the three common flavonols occurs as a range of methyl ethers (Table 2·6), most of which are of rare occurrence, the exception being

TABLE 2·6. Properties of Common Flavonols and their Methyl Ethers

Methyl ether Trivial name	Posn. of methylation	R_f (× 100) in BAW	Forestal	PhOH	Source	λ_{max} in EtOH (mμ)
KAEMPFEROL	none	83	55	58	—	268, 368
3-O-Methyl	3-	92	84	91	*Begonia* leaf	268, 352
Rhamnocitrin	7-	—	—	—	*Rhamnus* berry	— —
Kaempferide	4′-	—	—	—	*Alpinia* rhizome	266, 367
7,3′-Di-O-methyl	7,3′-	—	—	—	*Rhamnus* berries	— —
3,7-Di-O-methyl	3,7-	—	—	—	*Beyeria* leaf	— —
QUERCETIN	none	64	41	29	—	255, 374
3-O-Methyl	3-	93	84	71	*Nicotiana* calyx	257, 362
Azaleatin	5-	48	49	50	*Rhododendron* petal	254, 369
Rhamnetin	7-	72	53	66	*Rhamnus* berry	257, 371
Isorhamnetin	3′-	74	53	66	*Cheiranthus* petal	254, 369
Tamarixetin	4′-	85	—	—	*Tamarix* leaf	258, 274, 37
Caryatin	3,5-	76	80	—	*Carya* bark	253, 265,[1] 3
Rhamnazin	7,3′-	—	—	—	*Rhamnus* berry	255, 375
Ombuin	7,4′-	—	—	—	*Phytolacca* leaf	256, 369
3,3′-Di-O-methyl	3,3′-	—	—	—	*Nicotiana* calyx	256, 268, 36
Ayanin	3,7,4′-	—	—	—	*Distemonanthus* wood	254, 271, 33
MYRICETIN	none	43	28	13	—	256, 378
3-O-Methyl	3-	69	62	49	*Aegialitis* leaf	256, 366
5-O-Methyl	5-	26	33	21	*Rhododendron* petal	252, 371
Europetin	7-	51	49	55	*Plumbago* leaf	255, 379
Syringetin	3′,5′-	68	55	87	*Lathyrus* petal	254, 374
3,7,4′-Tri-O-methyl	3,7,4′-	—	—	—	*Ricinocarpus* leaf	264, 350
Combretol	3,7,3′- 4′,5′-	—	—	—	*Combretum* seed	266, 345

[1] Indicates shoulder or inflection.

[2] Typical log ϵ values for the flavonols are: quercetin 4·32 (255) and 4·34 (374), myricetin 4·21 (256) and 4·29 (378 mμ).

isorhamnetin. Six methyl ethers of kaempferol have been isolated and 10 of quercetin. No myricetin derivatives were reported in 1960 (*see* Gripenberg in Geissman, 1962) but five are now known and a sixth (the 4′-methyl ether) has been provisionally identified in black wattle leaf, *Acacia mearnsii* (A. M. Mackenzie, unpublished results).

R_f and spectral data and sources of all these methyl ethers are given in Table 2·6. Identification is based on a study of R_f values, colour reactions, melting points and spectral properties, including spectral shifts. The procedure may be illustrated by comparing the five isomeric monomethyl ethers of quercetin, which all occur naturally (Table 2·7).

Melting points are not very different and even the acetates are similar (all melt around 200°). The u.v. spectra are more useful, since methylation has a hypsochromic effect on the visible spectral band of quercetin. The 3-methyl ether is particularly easily picked out (λ_{max} at 362 instead of 374 mμ) and this is the only one to appear as a dull brown colour on paper in u.v. light. By contrast, the 5-methyl ether (azaleatin) has an intense and characteristic yellow fluorescence in the u.v. and also has a very distinct R_f in BAW. In fact, the 7- and 3'-methyl ethers are the only pair not immediately distinguishable by R_f alone. However, the 7-methyl ether does not give a sodium acetate shift whereas the 3'-methyl ether does, and conversely the 7-methyl ether responds to borate whereas isorhamnetin does not. In any case, identification is easily confirmed by reductive cleavage or alkaline fusion, since the 7-methyl ether yields phloroglucinol monomethyl ether and protocatechuic acid, whereas the 3'-methyl ether gives phloroglucinol and vanillic acid. Again, demethylation with pyridinium chloride distinguishes this pair of isomers since the 7-methyl ether is much more resistant. Finally, turning to the fifth isomer, the 4'-methyl ether, we find that this is the only one to have a long wave-length band at the same position as quercetin; like the 3'-methyl ether (from which it differs in R_f) it fails to respond to borate.

TABLE 2·7. Properties of Quercetin Monomethyl Ethers

Methyl ether	M.p.	Visible max (mμ)	Response to alk.	NaOAc	H_3BO_3	R_f (× 100) in Forestal	BAW	Colour[1] in u.v.
Parent cpd.	316–318°	374	Decomp.	+	+	41	64	bY
3-Me	273–275°	362	Mod. stable	+	+	84	93	dB
5-Me	320°	369	Decomp.	+	+	49	48	fY
7-Me	294–296°	371	Decomp.	−	+	53	72	bY
3'-Me	305–307°	369	Decomp.	+	−	53	74	bY
4'-Me	259–260°	376	Stable	+	−	—	85	dY

[1] Key: bY = bright yellow, dB = dull brown, fY = fluorescent yellow, dY = dull yellow

Less data are available on di- and trimethyl ethers but the four reported dimethyl ethers (Table 2·6) all appear to be readily distinguishable. Consider, for example, the spectral properties of the 3,5-dimethyl ether, caryatin (Fig. 2·3). The positive NaOAc shift shows that the 7-hydroxyl is free and the positive H_3BO_3 shift likewise shows that the 3',4'-dihydroxyls must be free. The stable alkaline shift indicates that the 3-hydroxyl is substituted and the lack of an $AlCl_3$ shift shows that both 3- and 5-hydroxyl groups are blocked. Its structure can be confirmed by demethylation which yields a mixture of the 3- and 5-monomethyl ethers and eventually free quercetin. Distinguishing all possible trimethyl ethers of quercetin may be more difficult but, theoretically, every isomer should be different in some respect. Separation of mixtures of closely related methyl ethers may, however, present more problems. For example, myricetin 3',5'-dimethyl ether and quercetin 3'-methyl ether, occurring together in flowers of *Lathyrus pratensis*, have very similar R_f values in most solvents and can only be separated on paper by chromatography in phenol.

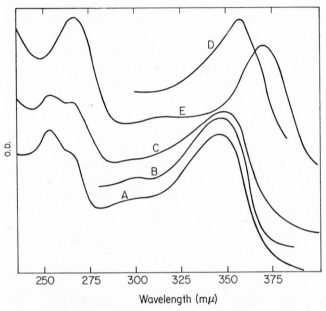

Fig. 2·3. Absorption spectrum of quercetin 3,5-dimethyl ether (caryatin). Curve A, 95% EtOH; curve B, EtOH–AlCl₃; curve C, EtOH–NaOAc; curve D, EtOH–NaOH; curve E, EtOH–H₃BO₃–NaOAc.

B. Rarer Structures

1. 6-Hydroxylated Flavonols

Of the various 6-hydroxylated flavonols known, the quercetin analogue, quercetagetin (21) is the commonest. It has been found as such (or in methylated form) in over 11 plant families and occurs especially frequently in the

(21) Quercetagetin

(22) Casticin

Leguminosae and Compositae (*see* appropriate sections in Chapters 5 and 6). Quercetagetin has recently been identified as the yellow flower pigment in the common primrose and the birdsfoot trefoil, *Lotus* (Harborne, 1965b), but the best source is still the one it was first isolated from, i.e. petals of the African marigold, *Tagetes erecta* (Latour and de la Source, 1877). It is a very distinctive substance, appearing on Forestal chromatograms as an almost black spot (in u.v. light) of low R_f (*see* Table 2·8). It has unusual spectral properties,

TABLE 2·8. R_f Values, Colours and Spectral Maxima of Various Hydroxy Derivatives of the Common Flavonols

Derivative and its name or structure		R_f (× 100) *in*			Colour in u.v. light[2]	Spectral maxima
		BAW	Forestal	PhOH		
KAEMPFEROL		83	55	58	b. yellow/green	268, 368
Deoxy	(3,7,4'-triOH)	85	70	70	f. yellow	258, 319, 357
-OH	Morin	79	73	34	v. b. yellow	263, 380
OH	Herbacetin	57	43	31	d. black	275, 327, 370
OMe	Tambuletin	38	74	93	d. brown	260, 382
QUERCETIN		64	41	29	b. yellow	255, 374
Deoxy	Fisetin	73	58	32	f. yellow	253, 315, 370
OH	Quercetagetin	31	26	12	d. black	259, 272, 364
OMe	Patuletin	68	48	56	b. yellow/green	258, 373
OH, 7-Me	Corniculatin	64	45	46	dark brown	260, 270,[1] 381
OH	Gossypetin	19	48	17	d. black	262, 278, 341, 386
MYRICETIN		40	25	07	b. yellow	256, 378
Deoxy	Robinetin	40	36	18	f. yellow	250, 319, 368

[1] Indicates shoulder or inflection.
[2] Key: b. = bright, v.b. = very bright, f. = fluorescent, d. = dull.

exhibiting as it does a double peak in the short u.v. and a very broad band (λ_{max} 364 mμ) at longer wave-lengths (Fig. 2·4). Quercetagetin, most of its derivatives and in fact all flavonols with a catechol group in the A-ring, are more easily oxidised than other flavonols and have to be handled carefully during isolation and identification.

6-Hydroxylation apparently quenches the u.v. fluorescence of flavonols but O-methylation in the 6-position restores it and patuletin, the 6-methyl ether present in *T. patula* flowers has a brighter colour than quercetin. The 7-methyl ether, which occurs in *Lotus corniculatus* petals with quercetagetin, is nearly as dark as the parent compound in the u.v. but is easily distinguished from it by other means (Harborne, 1965b). Two dimethyl ethers of quercetagetin are known: the 3,6-dimethyl ether in cocklebur, *Xanthium pennsylvanicum*, leaves (Taylor and Wong, 1965) and the 6,3'-dimethyl ether (spinacetin) in spinach, *Spinacia oleracea* (Zane and Wender, 1961). Three trimethyl ethers have been described: the 3,6,3'-trimethyl ether (centaureidin) in *Centaurea jacea* (Hörhammer *et al.*, 1965), the 3,7,4'-trimethyl ether (oxyayanin-B) in wood of *Distemonanthus* and the 6,7,3'-trimethyl ether (chrysosplenetin) from

3

Chrysosplenium japonicum (Nakaoki and Morita, 1956). Of the three tetra-methyl ethers, the structures of two are certain: the 3,6,7,4'-tetramethyl ether, casticin (22), which occurs in the seeds of *Vitex agnus-castus* (Belic *et al.*, 1961) and the 3,6,7,3'-tetramethyl ether, the lipophilic flavone of *Matricaria chamomilla* (Hansel *et al.*, 1966). The third, polycladin, was thought to have the same orientation as the *Matricaria* flavone (Marini-Bettolo *et al.*, 1957) but a recent synthesis of this derivative (Hörhammer *et al.*, 1965) has shown that polycladin must be a different isomer. Finally, a pentamethyl ether,

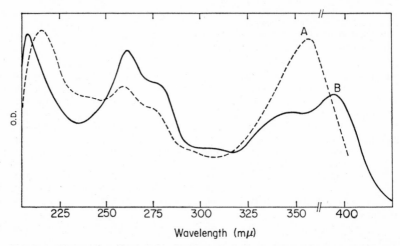

FIG. 2·4. Absorption spectra of quercetagetin (curve A) and gossypetin (curve B).

artemitin, has been isolated from the leaves of *Artemisia absinthium*; it is the 3,6,7,3',4'-pentamethyl ether (Tunmann and Isaac, 1957).

To these 10 simple methyl ethers of quercetagetin must be added three methylated derivatives which have methylenedioxy groups attached as well. These occur specifically in *Melicope* heartwood (Rutaceae) (*see* p. 178, Chapter 5, for their structures). It is not clear from the literature whether these various methyl ethers are easily differentiated; their separation from each other would almost certainly require the use of thin-layer chromatography.

6-Hydroxykaempferol, curiously enough, has not yet been completely identified in any one plant. A substance with the right R_f value, colour and u.v. spectrum has been detected in *Mimulus luteus* petals (Harborne, 1965b), and it has also been reported in petals of *Galega officinalis* (Maxyutin and Litvinenko, 1964), but in both cases further work is needed to fully establish its occurrence in nature. Two methylated derivatives are well documentated: the 5-methyl ether (vogeletin) which occurs in the seed of *Tephrosia vogelii* as the 3-arabinosylrhamnoside (Sambamurthy *et al.*, 1962), and the 3,6,7-trimethyl ether (23) (penduletin), present in *Brickelia pendula* leaves as the 4'-glucoside (Flores and Herran, 1958).

6-Hydroxymyricetin is unknown in the plant kingdom but the 3,6,7,4'-tetramethyl ether (24) has been isolated from *Eremophila fraseri* by Jefferies *et al.* (1962).

(23) Penduletin (24)

2. 8-Hydroxylated Flavonols

8-Hydroxyl derivatives of kaempferol, quercetin and myricetin (herbacetin, gossypetin (25) and hibiscetin) occur together in the same family, the Malvaceae (*Gossypium* and *Hibiscus*). They are of relatively rare occurrence themselves but many derivatives have been described. These are listed in Table 2·9, together with sources and spectral maxima; data on the five 6,8-dihydroxylated flavonols are also included.

Identification of these highly substituted flavonols is based mainly on u.v. spectral analysis, demethylation to a parent hydroxyflavonol, complete methylation and alkaline degradation. The R_f is not generally available and most have high values in the solvents used for the common flavonols; digicitrin, for example, has an R_f of 88 in BAW. The methylated compounds are probably more easily separated on silica gel than on paper. The elucidation of the structure of the *Ricinocarpus* gossypetin derivatives was based on the use of n.m.r. spectroscopy (Henrick and Jefferies, 1964, 1965). The same authors have correlated the n.m.r. spectra of a range of flavonols in order to show which signals can be relied upon for structural assignments.

3. 5-Deoxyflavonols

Flavonols lacking a 5-hydroxyl group occur characteristically in the Leguminosae and a few related families (*see* Chapter 5) as heartwood constituents. The two substances corresponding to quercetin and myricetin are called fisetin (26) and robinetin but the 3,7,4'-trihydroxyflavone has no trivial name.

(25) Gossypetin (26) Fisetin

The only methyl ether in the series is the 4'-methyl ether of fisetin, which occurs in *Schinopsis lorentzii* (Kirby and White, 1955). 5-Deoxyflavonols have

TABLE 2·9. 8-Hydroxylated and 6,8-Dihydroxylated Flavonols

Flavonol and Trivial Name	Spectral maxima	Source
8-HYDROXYKAEMPFEROLS		
Parent compound (herbacetin)	275, 327, 370	*Gossypium indicum* petals
8-Me (tambuletin)	260, 382	*Xanthoxylum acanthopodium* seeds
8,4'-diMe (prudomestin)	275, 323, 375	*Prunus domestica* heartwood
7,8,4'-triMe (tambulin)	—	*Xanthoxylum acanthopodium* seed
3,7,8-triMe	274, 331, 370	*Ricinocarpus stylosus* whole plant
3,7,8,4'-tetraMe (flindulatin)	275, 324, 362	*Flindersia maculosa* leaf
8-HYDROXYQUERCETINS		
Parent compound (gossypetin)	262, 278, 341, 386	*Gossypium herbaceum* petal
8,3'-diMe (limocitrin)	259, 273,[1] 340,[1] 378	*Citrus limon* peel
3,8-diMe	267,[1] 276, 338,[1] 373	*Ricinocarpus muricatus* leaf
3,7,8-triMe	262, 271,[1] 374	*Ricinocarpus stylosus* leaf
3,7,8,4'-tetraMe	260, 275, 366	*Ricinocarpus stylosus* leaf
3,7,8,3'-tetraMe (ternatin)	259, 273, 368	*Melicope ternata* bark
3,7,8,3',4'-pentaMe	258, 276, 340,[1] 364	*Ricinocarpus stylosus* leaf
8-HYDROXYMYRICETIN		
Parent compound (hibiscetin)	—	*Hibiscus sabdariffa* petal
6,8-DIHYDROXYKAEMPFEROLS		
3,5,7,8-tetraMe (calycopterin)	—	*Calycopteris floribunda* leaf
3,6,7,8,4'-pentaMe	—	*Citrus aurantium* peel
6,8-DIHYDROXYQUERCETINS		
6,8,3'-triMe (limocitrol)	260, 275, 350,[1] 377	*Citrus limon* peel
6,8,4'-triMe (isolimocitrol)	260, 276, 350,[1] 375	*Citrus limon* peel
3,6,7,8,3',4'-hexaMe (erianthin)	258, 280, 345	*Blumea eriantha*
6,8-DIHYDROXYMYRICETIN		
3,6,7,8,3',4'-hexaMe (digicitrin)	282, 337	*Digitalis purpurea* leaf

[1] Indicates shoulder or inflection.

a quite characteristic yellow fluorescence in u.v. light only shown otherwise by 5-*O*-methylated flavonols. Their u.v. spectra have three bands instead of two, the extra one at 315–320 mμ being absent from the related 5-hydroxy-flavonols. These substances can thus be readily identified by means of colour reactions, R_f (*see* Table 2·8) and spectra.

4. 2'-Hydroxyflavonols

2'-Hydroxyflavonols are rare and only two are known for certain in plants. Morin (27), present in the wood of *Morus tinctoria*, is well known for its dyeing and chelating properties and is used in analytical chemistry as a reagent for metals. Datiscetin, reported in *Datisca cannabina* as the rhamnoside datiscin, is 3,5,7,2'-tetrahydroxyflavone.

A third compound, oxyayanin-A, present in *Distemonanthus* wood, was thought to be 5,2',5'-trihydroxy-3,7,4'-trimethoxyflavone by King *et al.*

(27) Morin   (28) Galangin

(1953) but the flavonol has recently been synthesised and differs from oxyayanin-A (Jain *et al.*, 1965). Remarkably enough, the synthetic and natural substances agree in melting point and mixed melting point but differ in u.v. and i.r. spectra and their acetates have different melting points.

5. *Flavonols Lacking B-Ring Hydroxyls*

Galangin (28) (λ_{max} 268 and 368 mμ) is the only representative of this class. It is of rare occurrence (found mainly in *Pinus* and *Alpinia* species) but may be much more common than this as a trace constituent since, like chrysin, it is possibly an intermediate in the biosynthesis of more highly substituted flavonols. It is accompanied in the rhizomes of *Alpinia* plants by its 3- and 7-monomethyl ethers.

6. *C-Methylflavonols*

Two such substances are known: 6-*C*-methylquercetin (29) and 6-*C*-methylmyricetin. Both occur in *Pinus ponderosa* bark and have similar properties to quercetin and myricetin. They can be separated by paper chromatography (Hergert, 1962) but are best distinguished by alkaline degradation; the *C*-methylphloroglucinol they yield is easily separated from phloroglucinol (given by the common flavonols) on thin layers of silica gel.

(29) Pinoquercetin   (30) Nor-icaritin

7. *Isoprenoid Flavonols*

Three such compounds have so far been described. Nor-icaritin (30), a kaempferol derivative substituted at C-8 by a hydrated isoprenoid residue, is reported as the 7-glucoside in leaves of *Phellodendron* (Hasegawa and Shirato,

1953). Icaritin, the 4'-monomethyl ether, occurs in *Epimedium* roots and leaves (Ohta and Miyazaki, 1958); it is present with rhamnose in the 3-position and with glucose attached to the isoprenoid hydroxyl.

These two flavonols are distinguished from other aglycones by their optical activity. The third isoprenoid flavonol is sericetin (**31**), a disubstituted

(31) Sericetin

derivative of galangin, which occurs in the bark of *Mundulea sericea* (Burrows *et al.*, 1960). Flavone analogues with isoprene residues at C-6 and C-8 have been reported in *Artocarpus* (*see* p. 179, Chapter 5).

V. Flavonol Glycosides

A. GLYCOSIDIC PATTERNS

Flavonols, like anthocyanins, occur in living tissue bound to sugar and a very considerable number of different glycosides have been described. These fall into no less than 48 classes (excluding acylated glycosides), each class of glycoside differing from its neighbour in the number, nature or position of attachment of the sugars (Table 2·10). These 48 classes can be subdivided into 22 types of 3-glycosides, in which various sugars are attached to the 3-hydroxyl group of the flavonol, 16 3,7-diglycosides and 10 others, in which at least one sugar is present in the 7-, 3'- or 4'-positions.

Most of the common flavonol glycosides are similar in structure to the anthocyanins and are 3-glucosides, 3-galactosides, 3-rutinosides or 3-sophorosides. The 5-hydroxyl group in the molecule of kaempferol or quercetin is protected by hydrogen bonding with the adjacent carbonyl group and is therefore never apparently involved in glycosylation. For this reason, flavonol 3,7-diglycosides correspond to the 3,5-diglycosides in the anthocyanin series (Chapter 1). The task of listing all the known flavonol glycosides is a formidable one, partly because many incompletely characterised glycosides have been reported in the literature. Kaempferol and quercetin have each been found already in over 30 different sugar combinations and new glycosides are being discovered at an increasingly rapid pace.

In spite of the considerable complexity in glycosylation, there is obviously some pattern in the way plant enzymes control the glycosylation of flavonols. The following points may be noted.

TABLE 2·10. Glycosidic Classes of the Flavonols

3-Glycoside	Derived 3,7-diglycosides	Other derived glycosides
3-Glucoside	3,7-Diglucoside 3-Glucoside-7-rhamnoside	7-Glucoside 3'-Glucoside 4'-Glucoside 3,3'-Diglucoside 3,4'-Diglucoside 7,4'-Diglucoside
3-Galactoside	3-Galactoside-7-rhamnoside 3-Galactoside-7-xyloside	3-Digalactoside
3-Arabinoside	3-Rhamnosylarabinoside(?)-7- rhamnoside	3-Arabinoside-3'-rhamnoside
3-Xyloside	—	3-Rhamnosylxyloside
3-Rhamnoside	3,7-Dirhamnoside	3-Dirhamnoside 7-Rhamnoside 3-Rhamnoside-4'-arabino- side
3-Glucuronide	3,7-Diglucuronide	3-Glucosylglucuronide
3-Rutinoside	3-Rutinoside-7-glucoside 3-Rutinoside-7-rhamnoside 3-Rutinoside-7-glucuronide	3-(2^G-Glucosylrutinoside) 3-Rhamnodiglucoside
3-Sophoroside	3-Sophoroside-7-glucoside 3-Sophoroside-7-rhamnoside 3-Sophorotrioside-7-rhamnoside	3-Sophorotrioside
3-Gentiobioside	—	3-Gentiotrioside
3-Sambubioside	3-Sambubioside-7-glucoside	3-Sambubioside-3'-glucoside
3-Lathyroside	3-Lathyroside-7-rhamnoside	—
3-Robinobioside	3-Robinobioside-7-rhamnoside	3-Rhamnosyldigalactoside

(1) There is usually a sugar in the 3-position. If a second position is glycosylated, it is commonly the 7-position and there are many types of 3,7-diglycoside (Table 2·10). Nevertheless, flavonol 7- and 4'-glycosides are known and all but one of the five possible quercetin monoglucosides have been found in plants.

(2) Of the sugars present, glucose and rhamnose (and their combination rutinose) are by far the most common. Most di- and trisaccharides found attached to flavonols contain one, if not both, these sugars. Linkages in the di- and trisaccharides that have been reasonably fully identified are limited to three types: $\beta1 \rightarrow 6$, $\beta1 \rightarrow 2$ and $\alpha1 \rightarrow 6$.

(3) The number of sugar units is usually one, two or three. Thus, of the 48 glycosidic classes, 10 involve one sugar, 22 two and 15 three. There is only one tetraglycoside known (kaempferol 3-sophorotrioside-7-rhamnoside from *Solanum* seed) and a single partly characterised pentaglycoside (in *Equisetum palustre*) has been described.

(4) Flavonols are stored in plant cells almost always as simple sugar derivatives. Known acylated glycosides (i.e. flavonols with aromatic or aliphatic

acids attached to them via sugar, *see* Section D) are relatively few in number. The only other type of combination, which replaces glycosylation, is potassium sulphate ester formation, but this is apparently confined to plants of the genera *Polygonum* and *Oenanthe*.

B. THE SUGARS OF FLAVONOL GLYCOSIDES

A detailed discussion of the oligosaccharides found associated with flavonols in plants is outside the scope of the present work and is a subject that has recently been reviewed by the author (1964) and Pridham (1965). It is an unfortunate fact that carbohydrate chemists have generally ignored these interesting and unusual sugars, many of which are unique to the flavonoid pigments, so that their structures have rarely been established completely rigorously.

There is little difficulty in identifying monosaccharides today and it is fairly clear that six are found in association with flavonols: glucose, galactose, glucuronic acid, xylose, rhamnose and arabinose. All, except glucuronic acid, also occur attached to anthocyanidins. All probably occur in the pyranose form (except arabinose, which may occur in both furano and pyrano forms, *see* Section D) and the first four, being D-sugars, are usually β-linked to the flavonol and the last two, being L-sugars, are α-linked.

The various di- and trisaccharides that occur attached to the flavonols are given in Table 2·11 and it is at once clear that most of these carbohydrates require fuller characterisation. However, the structures of six of the 11 disaccharides seem to be reasonably well established. The linkage in rutinose and robinobiose was thought to be $\beta1\rightarrow6$ until Gorin and Perlin (1959) showed that it must be $\alpha1\rightarrow6$. It is curious that an isomer of rutinose neohesperidose ($\alpha1\rightarrow2$ link), which occurs frequently in association with flavanones (Chapter 3) has not yet been found among the flavonols. However, although rutin, quercetin 3-rutinoside, has been "identified" in a very large number of plants (*see* Tables 5·1 and 6·1 in Chapters 5 and 6), only in a very few instances has the disaccharide present been compared directly with authentic rutinose. Listing only two flavonol 3-glucosylglucosides (sophoroside and gentiobioside) in Table 2·10 may similarly be an underestimation, since many 3-diglucosides have been reported without further identification.

The trisaccharides are less well defined and only for two are the structures fairly clear: 2^G-glucosylrutinose and gentiotriose. The branched trisaccharide, which is present in the flavonols of potato flowers (Harborne, 1962c), also occurs in association with anthocyanins (Chapter 1). In the case of the potato flavonols, this trisaccharide was originally thought to be a linear structure but direct comparison with material isolated from *Begonia* anthocyanins showed that it was identical. The fact that the rhamnosyldiglucose present in tea flavonols is a different trisaccharide, i.e. isomeric with 2^G-glucosylrutinose, rests on the different R_f values of the respective kaempferol and quercetin

TABLE 2·11. Di- and Trisaccharides Attached to Flavonols

Structure	Trivial name	Occurrence
DISACCHARIDES		
6-O-α-L-Rhamnosyl-D-glucose	Rutinose	Widespread, e.g. *Ruta graveolens*
6-O-α-L-Rhamnosyl-D-galactose	Robinobiose	Common, e.g. *Robinia pseudacacia*
2-O-β-D-Glucosyl-D-glucose	Sophorose	Widespread, e.g. *Sophora japonica*
6-O-β-D-Glucosyl-D-glucose	Gentiobiose	Rare, *Primula sinensis*
2-O-β-D-Xylosyl-D-glucose	Sambubiose	Uncommon, *Aesculus hippocastanum*
2-O-β-D-Xylosyl-D-galactose	Lathyrose	Rare, *Lathyrus odoratus*
Rhamnosylarabinose	—	Rare, *Cheiranthus cheiri*
Rhamnosylrhamnose	—	Rare, *Exocarpus cupressiformis*
Galactosylgalactose	—	Rare, *Betula verrucosa*
Rhamnosylxylose	—	Rare, *Tilia argentea*
Glucosylglucuronic acid	—	Rare, *Nelumbo nucifera*
TRISACCHARIDES		
O-β-Glucosyl-(1→6)-O-β-glucosyl-(1→6)-glucose	Gentiotriose	Rare, *Primula sinensis*
O-β-Glucosyl-(1→2)-glucosyl-(1→2)-glucose(?)	Sophorotriose	Uncommon, *Pisum sativum*
Galactosyl-O-β-glucosyl-(1→2)-glucose	—	Rare, *Pisum arvense*
O-β-Glucosyl-(1→2)-O-[α-L-rhamnosyl-(1→6)]-D-glucose	2^{G}-Glucosylrutinose	Uncommon, *Solanum tuberosum*
Rhamnosylglucosylglucose	—	Rare, *Camellia sinensis*
Rhamnosylgalactosylgalactose	Rhamninose	Rare, *Rhamnus*

3*

derivatives (Table 2·13). The identification of the glucosylglucosylglucose present in *Primula* species as gentiotriose rests primarily on hydrolytic evidence (e.g. it only gives one disaccharide gentiobiose as intermediate in acid hydrolysis) and calculation of the R_f values of the respective flavonol mono-, di- and triglucosides, all of which occur together in the *Primula* plant (cf. Harborne, 1965b).

C. IDENTIFICATION

Known glycosides of the common flavonols can be rapidly identified by chromatographic means. Although most 3-glycosides appear on paper as dark brown spots in u.v. light, those of kaempferol become bright greenish-yellow when fumed with ammonia, those of quercetin yellow and those of myricetin yellow-brown. 3,7-Glycosides can be distinguished from 3-glycosides by their fluorescence in u.v. and ammonia; 7-glycosides behave like the free flavonols on paper and 4'-glycosides are duller in colour. Once the aglycone has been recognised, measurement of the R_f value in several solvent systems and consulation of R_f data (Table 2·13) will indicate which glycoside is present. Analysis of the products of acid hydrolysis will then confirm the identity of aglycone and sugar and co-chromatography and spectral analysis will confirm the identity of the glycoside.

Identification of a new glycoside requires more careful analysis but can usually be carried out on 5–10 mg of material by means of measuring R_f and u.v. spectra and studying acidic and enzymic hydrolyses. Paper chromatography is still the most reliable technique for flavonol glycosides but thin-layer chromatography may be used alternatively. Early experiments with silica gel and polyamide were largely unsuccessful (e.g. Egger, 1961) but the recent introduction of the solvent mixture, chloroform–methanol–methyl ethyl ketone (9:4:2), for use with polyamide plates is more promising. In both systems, the mobility of a glycoside is closely correlated with the number and position of substitution of the sugars. With increasing glycosylation at the 3-position, the lower the R_f is in BAW and the higher the R_f in water. 3,7-Glycosides behave in general like 3-glycosides, but move faster in water than the corresponding 3-glycosides. 7-, 3'- and 4'-glucosides have planar structures (unlike all 3-glycosides) and are barely mobile in water.

The importance of u.v. spectral measurements in characterising flavonol glycoside cannot be overemphasised. For example, all the seven known quercetin mono- and diglucosides can be distinguished by their spectral characteristics (Table 2·12 and Fig. 2·5). In neutral solution, 3-glycosides, because of the hypsochromic effect of 3-glycosylation, can be readily distinguished from all other glycosides. In alkali, glycosides which have free hydroxyl groups in the 3- and 4'-positions decompose, whereas others exhibit stable absorption maxima in the visible region. All glycosides exhibit bathochromic shifts with aluminium chloride if they have free hydroxyls in the 3- or 5-positions but only those with a 7-hydroxyl free show similar shifts with sodium

TABLE 2·12. Spectral Properties of Quercetin Glycosides

Quercetin glucoside[2]	λ_{max} in 95% EtOH (mμ)		$\Delta\lambda^{\text{NaOEt}}$	$\Delta\lambda^{\text{H}_3\text{BO}_3}$	$\Delta\lambda^{\text{AlCl}_3}$	$\Delta\lambda^{\text{NaOAc}}$
	Band I	Band II	Band II	Band II	Band II	Band I
3-Glucoside	258	364	+53	+10	+26	+6
7-Glucoside	257	375	+46	+10	+50	0
3'-Glucoside	252, 266[1]	367	decomp.	0	+62	+14
4'-Glucoside	254	367	+63	0	+56	+15
3,7-Diglucoside	258	360	+78	+22	+40	0
3,4'-Diglucoside	259,[1] 267	350	+30	0	+47	+3
7,4'-Diglucoside	256, 265[1]	371, 420[1]	+56	0	+52	0

[1] Denotes inflection or shoulder.

[2] Other quercetin glycosides generally have similar properties to the glucosides; exceptionally, the 3-rhamnoside has band II max at 352 mμ.

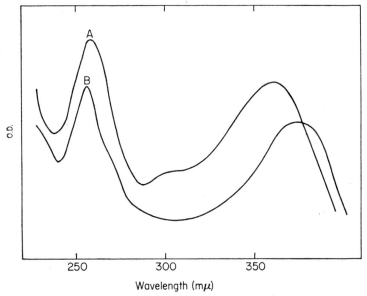

FIG. 2·5. Absorption spectra of quercetin 3-arabinoside (avicularin) (curve A) and quercetin 7-glucoside (quercimeritrin) (curve B). Solvent: 95% EtOH.

acetate. Finally, spectral shifts with boric acid are only shown by glycosides of quercetin or myricetin which have a catechol nucleus in their structure.

Infra-red spectral analysis has rarely been of much value for structural analysis of these glycosides but n.m.r. measurements show considerable promise in providing supplementary data. Mabry et al. (1965b) were able to recognise, by n.m.r. studies, the presence of a chromatographically inseparable mixture of 3-rhamnosylglucoside and 3-rhamnosylgalactoside in an isorhamnetin glycoside isolated from Opuntia. These authors overcame solubility difficulties by preparing the trimethyl silyl ethers and carrying out the measurements in carbon tetrachloride. Although the ethers of quercetin 3-rhamnoside and 3-rutinoside have different spectra, those of naringin (flavanone neohesperidoside) and hesperidin (flavanone rutinoside) have identical signals in the "sugar region". Neohesperidosides and rutinosides can, however, be distinguished by n.m.r. analysis of the acetates (Rosler *et al.*, 1965).

Much structural information can be obtained by carrying out enzymic and acid hydrolysis. As the subject has already been dealt with under anthocyanins (*see* p. 20, Chapter 1), only a few brief notes are needed here (but *see also* Harborne, 1965a). Flavonol glycosides differ considerably in their rates of acid hydrolysis and use can be made of this for determining whether a sugar is present in the 3-position (rapidly hydrolysed) or in the 7-position (slowly hydrolysed). Furthermore, sugars attached to the 3-position are lost at different rates: rhamnose and arabinose in 2–3 min, glucose and galactose in 4–6 min and glucuronic acid in 45–60 min.

β-Glucosidase attacks the 3-, 7- and 4'-O-glucosides of quercetin at the same rate, but preferentially removes glucose from the 7-hydroxyl of quercetin 3-sophoroside-7-glucoside (32) to give the sophoroside (33). Rhamnosides and

(32)   (33)

rutinosides are not attacked by β-glucosidase and require an α-rhamnosidase (present in the enzyme mixture anthocyanase) for their hydrolysis.

Finally, mention must be made of some of the classical properties of flavonol glycosides. Most crystallise as pale yellow needles or plates in one or more hydrated forms and the melting point is considerably affected by the water of crystallisation present. For characterisation, melting point is mainly of use for distinguishing closely related glycosides. All glycosides are optically active but measurement of optical rotation is only important when two forms of the same glycoside (e.g. quercetin 3-arabinosides, *see* Section D) are known to occur in plants. Finally, the classical procedure of methylation and hydrolysis for locating position of sugar substitution is valuable if there is any doubt regarding this point; the procedure has been successfully developed for use on a micro-scale (*see* e.g. Nordström and Swain, 1953).

D. INDIVIDUAL GLYCOSIDES

1. Kaempferol Glycosides

At least 31 simple glycosides have been described and R_f values and sources of some 24 are listed in Table 2·13. The seven others, which were not avail-

TABLE 2·13. Flavonol Glycosides: R_f Values and Sources

Glycoside	BAW	R_f ($\times$ 100) in H₂O	15% HOAc	PhOH	Source
KAEMPFEROL					
3-Rhamnoside	78	28	49	78	*Lathyrus odoratus* petal
3-Glucoside	70	13	43	74	*Astragalus sinicus* petal
3-Glucuronide	53	67	44	25	*Phaseolus vulgaris* leaf
3-Rutinoside	54	23	54	64	*Calystegia japonica* leaf
3-Rhamnodiglucoside	41	34	61	54	*Camellia sinensis* leaf
3-(2ᴳ-Glucosylrutinoside)	40	54	74	52	*Solanum tuberosum* petal
3-Gentiobioside	43	27	54	55	⎱ *Primula sinensis* petal
3-Gentiotrioside	31	33	51	45	⎰
3-Sophoroside	45	29	58	63	*Rosa* cv. petal
3-Sophorotrioside	30	43	72	49	*Pisum sativum* leaf
3-Xylosylglucoside	55	29	65	68	*Phaseolus vulgaris* seed
7-Glucoside	54	02	17	62	*Thespesia populnea* petal
7-Rhamnoside	75	02	18	76	*Lilium regale* petal
7-Glucuronide	53	04	18	35	From *Tulipa* petal
3,7-Dirhamnoside	56	41	50	68	*Indigofera arrecta* leaf
3-Robinobioside-7-rhamnoside	40	54	75	73	*Robinia pseudacacia* leaf
3-Sophoroside-7-rhamnoside	40	71	87	60	⎱
3-Sophorotrioside-7-rhamnoside	35	78	88	45	⎰ *Solanum tuberosum* seed
3-Rhamnoarabinoside-7-arabinoside	58	59	89	72	*Matthiola incana* petal
3-Lathyroside-7-rhamnoside	53	71	92	73	*Lathyrus odoratus* petal
3,7-Diglucoside	28	57	80	44	*Paeonia albiflora* petal
3-Sophoroside-7-glucoside	20	70	82	54	*Petunia* cv. petal
3-Rutinoside-7-glucoside	15	47	65	—	*Equisetum palustre*
3-Rutinoside-7-glucuronide	26	93	85	39	*Tulipa* cv. leaf
3-(p-Coumaroylglucoside)	83	06	31	83	*Tilia argentea* petal
3-(Feruloylsophorotrioside)	38	26	63	71	*Pisum sativum* leaf
QUERCETIN					
3-Arabinoside	70	07	31	61	*Vaccinium myrtillus* leaf
3-Xyloside	65	06	32	—	*Reynoutria japonica* leaf
3-Glucoside	58	08	37	54	*Nicotiana tabacum* leaf
3-Galactoside	55	09	35	56	*Rhododendron* cv. petal
3-Rhamnoside	72	19	49	58	*Quercus tinctoria* bark
3-Glucuronide	40	69	38	16	*Phaseolus vulgaris* leaf
3-Rutinoside	45	23	51	46	*Ruta graveolens* leaf
3-Gentiobioside	37	19	45	36	⎱
3-Gentiotrioside	23	18	41	26	⎰ *Primula sinensis* petal
3-Sophoroside	45	31	75	52	
3-Sophorotrioside	36	46	70	31	*Pisum sativum* leaf
3-Rhamnodiglucoside	36	26	54	35	*Camellia sinensis* leaf
3-(2ᴳ-Glucosylrutinoside)	36	46	71	31	*Solanum tuberosum* petal
7-Glucoside	32	00	10	40	*Helianthus annuus* ray
7-Glucuronide	26	02	10	18	From *Tulipa* leaf
4'-Glucoside	48	01	12	35	*Allium cepa* bulb
3,7-Diglucoside	30	33	63	32	*Ulex europaeus* petal
3,7-Diglucuronide	16	98	99	27	*Potentilla reptans* leaf

TABLE 2·13—*continued*

Glycoside	BAW	R_f ($\times 100$) in H₂O	15% HOAc	PhOH	Source
QUERCETIN—*continued*					
3,4′-Diglucoside	35	29	60	42	⎫ *Allium cepa* bulb
7,4′-Diglucoside	23	03	—	34	⎭
3-Robinobioside-7-rhamnoside	25	57	72	53	*Vinca minor* leaf
3-Rutinoside-7-glucuronide	20	90	78	28	*Tulipa* cv. leaf
3-Sophoroside-7-glucoside	14	61	84	37	*Petunia* cv. leaf
3-Sambubioside-7-glucoside	16	65	76	38	*Helleborus foetidus* petal
3-(Caffeoylsophoroside)-7-glucoside	21	64	75	39	*Helleborus foetidus* petal
3-(*p*-Coumaroylsophoro-trioside)	33	26	62	49	*Pisum sativum* leaf
MYRICETIN					
3-Glucoside	47	05	25	32	*Primula sinensis* petal
3-Rhamnoside	60	15	44	39	*Cercis siliquastrum* leaf
3-Rutinoside	43	16	—	29	*Solanum soukupii* petal
3-Robinobioside-7-rhamnoside	25	52	70	36	*Vinca minor* petal
AZALEATIN					
3-Rhamnoside	58	22	42	76	*Rhododendron mucronatum* petal
3-Galactoside	48	06	18	77	*Eucryphia glutinosa* leaf
ISORHAMNETIN					
3,4′-Diglucoside	38	27	62	63	*Dactylis glomerata* pollen
QUERCETAGETIN					
3-Galactoside	42	10	36	40	*Lotus corniculatus* petal
7-Glucoside	31	02	06	30	*Papaver nudicaule* petal
3-Gentiotrioside	15	40	56	30	*Primula vulgaris* petal

able in this laboratory for R_f determination, are as follows: 3-arabinoside (flowers of *Aesculus hippocastanum*); 3-galactoside (*Trifolium pratense*); 3-glucosylglucoside (*Nelumbo nucifera*); 3-rhamnoside-4′-arabinoside (*Prunus spinosa* leaf); 3-dirhamnoside (*Exocarpus cupressiformis*); 3-glucoside-7-rhamnoside (*Tilia argentea* leaf); 3-sambubioside-7-glucoside (*Helleborus niger* petal).

The most notable kaempferol glycoside is robinin (34), the 3-rhamnosyl-galactoside-7-rhamnoside, which was first isolated from flowers of *Robinia pseudacacia* in 1861 by Zwenger and Dronke. It also occurs in *Vinca minor* (Rabaté, 1933), where it is accompanied by related quercetin and myricetin derivatives. Its structure was established by Zemplen and Bognar (1941). On acid hydrolysis, it gives kaempferol 7-rhamnoside as an intermediate and on enzymic hydrolysis it yields the same compound and robinobiose (Zemplen and Gerecs, 1935). It crystallises as an octahydrate and exists in two inter-convertible forms: α-, m.p. 195–197°, and β-, m.p. 249–250°.

Plant sources of the flavonol glycosides are given in detail (with references) in Chapters 5–7. Here it is only necessary to mention convenient sources of

(34) Robinin   (35) Avicularin

the most common glycosides. The 3-rhamnoside may be crystallised from petal extracts of white sweet pea varieties (e.g. "Swan Lake") and the 3-glucoside is readily available in bracken, *Pteridium aquilinum*, a common weed in many temperate countries. The 3-galactoside conveniently occurs in clover, *Trifolium pratense*, and the 3-rutinoside in petal and leaf of the tobacco plant, *Nicotiana tabacum*.

2. Quercetin Glycosides

Twenty-four of the thirty-six glycosides known are listed in Table 2·13. The twelve not included there are the following: 3'-glucoside (Douglas Fir needles); 3,3'-diglucoside (horse chestnut); 3-sambubioside-3'-glucoside (horse chestnut); 3-rhamnosylxyloside (*Tilia argentea* leaf); 3-galactoside-7-rhamnoside (*Caltha palustris*); 3-galactoside-7-xyloside (*Caltha palustris*); 3-glucoside-7-rhamnoside (*Tilia argentea* leaf); 3-glucosylglucuronide (*Nelumbo nucifera* leaf); 3-rhamnosyldigalactoside (*Rhamnus* berry); 3-dirhamnoside (*Exocarpus cupressiformis*); 3-rutinoside-7-rhamnoside (*Viola × wittrockiana* petal); 3-rhamnosylarabinoside (*Cheiranthus cheiri* fruit).

Of the known glycosides, quercetin 3-rutinoside (rutin) is by far the most common and best known. It is easily crystallised and is available commercially. More interesting glycosides, from the carbohydrate point of view, are the arabinosides, four of which have been described. The commonest is avicularin (35), the 3-α-L-arabofuranoside, which occurs, for example, in *Polygonum aviculare*. It crystallises as a hemihydrate, m.p. 216–217° (anhydrous m.p. 222°), and shows a rotation $[\alpha]_D^{25} - 168°$ (EtOH). A second one, guaijaverin, the 3-α-L-arabopyranoside, is confined to the leaves of *Psidium guaijava*. This form crystallises as a monohydrate, m.p. 239° (anhydrous m.p. 256°), and shows a rotation $[\alpha]_D^{25} - 97°$ (EtOH). It was reported to occur with avicularin (El Khadem and Mohammed, 1958) but Seshadri and Vasishta (1965) could find no trace of a second arabinoside in *P. guaijava* leaf. The third arabinoside is polystachoside, the 3-β-L-arabinoside, which occurs in *Polygonum polystachyum* and crystallises as a dihydrate, m.p. 246–247° ($[\alpha]_D - 25·9°$ (MeOH)).

Finally, a fourth 3-arabinoside, foeniculin, apparently occurs in *Foeniculum vulgare* leaf; its only recorded property is the melting point (256°).

There are many convenient and easily accessible plant sources of quercetin glycosides. Five of the six monoglycosides are present in the skin of the "Grimes Golden" apple (Siegelman, 1955) and the sixth, the 3-glucuronide, occurs in broad bean leaf. Finally, the 4'-glucoside and 3,4'- and 7,4'-diglucosides are obtainable in quantity from onion scales.

3. Isorhamnetin Glycosides

Until recently, only three glycosides were fully characterised: the 3-arabinoside (*Taxodium distichum* leaf), the 3,4'-diglucoside (*see* Table 2·13.) and the 3-rutinoside (daffodil sepals). Following recent investigations of flavonoids in *Opuntia lindheimeri* petals (Rosler *et al.*, 1966), in pear skins (Nortjé and Koeppen, 1965), in *Peumus boldus* leaves (Krug and Borkowski, 1965) and in various crucifers (Hörhammer *et al.*, 1966) at least six other glycosides are now known. These include the 3-glucoside, 3-galactoside, 3-rhamnosylgalactoside, 7-glucoside, 3,7-diglucoside and 3-glucoside-7-rhamnoside. The 3-rutinoside and 3-rhamnosylgalactoside (possibly the robinobioside?), remarkably enough, were found to occur together both in the pear and in *Opuntia* flowers. Rosler *et al.* (1966), working with *Opuntia*, had to rely on n.m.r. spectroscopy for identification of the mixture but Nortjé and Koeppen (1965) were able to separate the pair of glycosides from pears by use of paper chromatography in phenol–water.

The identification of the 3-glucoside and the 3,7-diglucoside of isorhamnetin also deserves some comment. The 3-glucoside isolated from pear had m.p. 242–244° and $[\alpha]_D - 44°$ (pyridine) and the authors concluded that it was different from a 3-glucoside, m.p. 165–167° or 168–172°, isolated from *Argemone mexicana* (Rahman and Ilyas, 1962) or *Calendula officinalis* (Friedrich, 1962). However, there seems to be only one isorhamnetin 3-glucoside in nature, since Krishnamurti *et al.* (1965), in re-isolating it from *Argemone*, found a melting point (233–235°) for it close to that of Nortjé and Koeppen. A second glucoside in *Argemone* was described as the 7-diglucoside by Rahman and Ilyas (1962) but re-examination (Krishnamurti *et al.*, 1965) showed it to be the 3,7-diglucoside.

4. Myricetin Glycosides

To the four glycosides listed in Table 2·13 must be added the 3-digalactoside (*Betula verrucosa* leaf), the 3-rutinoside-7-rhamnoside (*Viola* petal) and the 3'-glucoside (*Cannabis indica* and *Hibiscus abelmoschus* petal). It is difficult to believe that these are the only myricetin glycosides in plants and there is little doubt that further investigation of plant groups known to contain myricetin would reveal the myricetin analogues of the many known kaempferol and quercetin derivatives.

5. Acylated Glycosides

Just as anthocyanins are found in plants in acylated form, so acylated flavonol glycosides have also been reported, but they are much fewer in number. All so far isolated appear to have a single acyl group (usually p-coumaric, ferulic or caffeic acid) attached to the flavonol by the sugar alcohol grouping. Even tiliroside (36) of lime flowers, first thought to be a 7-p-coumaroyl derivative (Hörhammer et al., 1961), is in fact the 3-(p-coumaroylglucoside) of kaempferol (Harborne, 1964a). Proof of this new structure included spectral analysis and the isolation of p-coumaroylglucose following both acid hydrolysis and hydrogen peroxide oxidation. The only other report of a different mode of attachment may similarly be in error; this is of a kaempferol 3-sophoroside in *Fagus sylvatica* buds, which is reputed to have a p-coumaric acid attached in the 4′-position (Dietrichs and Schaich, 1962).

(36) Tiliroside

The only acylated glycoside of which structure is completely known is petunoside (37), which occurs in petals of *Petunia* cultivars. The isolation of 2-feruloylglucose from its partial hydrolysis (Birkhofer et al., 1965) shows that it must have the structure shown:

(37) Petunoside

A series of p-coumaroyl and feruloyl triglucosides of kaempferol and quercetin occur in pea seedlings, but these are still waiting complete analysis. Other acylated flavonol glycosides known include a kaempferol 3-(p-coumaroylarabinoside) in *Bryophyllum daigremontianum* (Karsten, 1965) and a quercetin 3-(caffeoylsophoroside)-7-glucoside in *Helleborus foetidus* petals (Harborne, 1965a).

CHAPTER 3

THE MINOR FLAVONOIDS

I. Introduction

The major flavonoids of plants are the anthocyanins, flavones and flavonols, three very numerous and widely distributed classes of plant constituents (*see* Chapters 1 and 2). Accompanying these major flavonoids in certain plant groups are related derivatives, conveniently described as minor flavonoids. Included in this category are chalcones, aurones, flavanones, isoflavones, biflavonyls and leucoanthocyanidins. They are minor either because their contribution to plant colour is limited or because surveys have so far indicated that they are of restricted distribution. The chalcones and dihydroflavonols are almost certainly intermediates in the biosynthesis of the major flavonoids (Chapter 8) and probably occur in most plants as trace components; they have only been detected in a relatively few species as bulk constituents. Biflavonyls and isoflavones are, by contrast, end-products of synthesis but are mainly restricted in occurrence to the gymnosperms in one instance and legumes in the other. Again, the aurones are a group of yellow pigments formed late in evolutionary time and which appear frequently only in the flowers of one of the most highly developed plant families, the Compositae.

Chemically, the minor flavonoids are closely related to each other and to the major flavonoids:

(2) Naringenin

(3) Chalcononaringenin

(1) Apigenin

(4) 4,6,4′-Trihydroxyaurone

(5) Genistein

(6) Cupressoflavone

Reduction of the 2,3-double bond of the commonly occurring flavone apigenin (1), for example, gives the typical flavanone, naringenin (2). This flavanone is, in turn, in acid-base equilibrium with its chalcone, chalcononaringenin (3), which is at the same level of oxidation. Oxidation of (3) yields 4,6,4'-trihydroxyaurone (4), not yet known as a natural pigment but a typical representative of the aurone class. Returning to apigenin, isomerisation of the B-ring, i.e. transfer from the 2- to the 3-position, gives the well-known isoflavone, genistein (5). Finally, dimerisation of apigenin (1), by linking two molecules together 8- and 8'-positions, yields the biflavonyl cupressoflavone (6), one of the most recently isolated members of this group of plant substances.

Most of the reactions mentioned above can be carried out in the laboratory and all, no doubt, occur in plants (although not with the same substrates or in the same sequence as above). Certainly there are a number of plants which contain series of interrelated flavonoids, all with identical hydroxylation patterns, which must presumably be formed one from another by simple enzymic transformations. Seedlings of chana grain, *Cicer arietinum*, for example, contain 3,7,4'-trihydroxyflavone (or 7,4'-dihydroxyflavonol (7)), the related dihydroflavonol, garbanzol (8), the related flavanone liquiritigenin (9), the related chalcone isoliquiritigenin (10) and the related isoflavone daidzein (11) (Wong *et al.*, 1965). Similar series of minor flavonoids have been noted in heartwoods of *Acacia* and *Robinia* species by Roux and his co-workers (*see* Leguminosae section of Chapter 5).

(7) 3,7,4'-Trihydroxyflavone

(8) R=OH, Garbanzol
(9) R=H, Liquiritigenin

(11) Daidzein

(10) Isoliquiritigenin

Minor flavonoids, like the anthocyanins and flavones, are usually found in plant tissues as glycosides and thus they present very similar problems in isolation and identification as the major flavonoids. Most of the simple minor flavonoids can readily be identified by use of paper chromatography and absorption spectroscopy. The chalcones and aurones, being bright yellow in colour, are particularly easily recognised by means of simple colour reactions and of visible spectroscopy. On the other hand, flavanones are more difficult

TABLE 3·1. Colour Properties of Different Flavonoid Classes

Visible colour	Colour in u.v. light Alone	With ammonia	Indication
Orange Red	Dull orange, red or mauve	Blue	Anthocyanidin 3-glycosides
Mauve	Fluorescent yellow, cerise or pink	Blue	Most anthocyanidin 3,5-diglycosides
Bright yellow	Dark brown or black	Dark brown or black	6-Hydroxylated flavonols and flavones; some chalcone glycosides
		Dark red or bright orange	Most chalcones
	Bright yellow or yellow-green	Bright orange or red	Aurones
Very pale yellow	Dark brown	Bright yellow or yellow-brown	Most flavonol glycosides
		Vivid yellow-green	Flavone glycosides
		Dark brown	Biflavonyls and unusually substituted flavones
None	Dark mauve	Faint brown	Most isoflavones and flavanonols
	Faint blue	Intense blue	5-Deoxyisoflavones and 7,8-dihydroxyflavanones
	Dark mauve	Pale yellow or yellow-green	Flavanones and flavanonol 7-glycosides

General chromatographic sprays: 5% alc. $AlCl_3$—response: yellow-green fluorescence in u.v.; aqu. $FeCl_3$–$K_3Fe(CN)_6$ (1:1)—response: blue on yellow background; Folin reagent—response ($+NH_3$): blue on white background.

to deal with. They are not easily located on paper chromatograms and have u.v. absorption maxima in a region of the spectrum where impurities can seriously interfere with their characterisation.

The key to colour properties (Table 3·1) indicates broadly how to differentiate the flavonoid classes when inspecting paper chromatograms of crude plant extracts. Practically all flavonoids, when present in any quantity, can be located on paper without using chromogenic sprays. Spraying the chromatogram with ethanolic aluminium chloride (5% w/v) is sometimes useful in increasing the intensity and brilliance of the colours under u.v. light. Some minor flavonoids can be distinguished by special colour tests; flavanones, for example, are the only flavonoids to produce red colours when the paper is sprayed with sodium borohydride and then fumed with mineral acid.

Minor flavonoids can be distinguished from each other by means of u.v. spectroscopy and in Table 3·2 the principal and subsidiary maxima of each flavonoid class are given. The range of values refer mainly to the common substances in each class. There are compounds which do not conform precisely with the range shown; isoflavones and flavanones with 6,7,4′-trihydroxylation, for example, are atypical in having an intense band at about 330 mμ and could thus be mistaken for flavones. A useful technique for confirming the

TABLE 3·2. Spectral Characteristics of Main Flavonoid Classes

Principal maxima (mμ)	Subsidiary maxima (mμ) (with relative intensities)	Indication
475–560	ca 275 (55%)	Anthocyanins
390–430	240–270 (32%)	Aurones
365–390	240–260 (30%)	Chalcones
350–390 ⎱ 250–270 ⎰	ca 300 (40%)	Flavonols
330–350 ⎱ 250–270 ⎰	Absent	Flavones and biflavonyls
275–290 ⎱ ca 225 ⎰	310–330 (30%)	Flavanones and flavanonols
255–265	310–330 (25%)	Isoflavones

flavonoid class in such difficult cases is to prepare and measure the spectrum of the fully acetylated flavonoid; prior demethylation is necessary in the case of methyl ethers. Acetylation of phenolic hydroxyl groups completely nullifies their effect on absorption so that a fully acetylated hydroxy flavonoid has spectral properties practically identical to that of the appropriate unhydroxylated compound. The flavonoid class of an unknown thus becomes apparent since these parent substances differ significantly in their properties. Flavone (and acetylated hydroxyflavones), for example, have λ_{max} at 250 (4·06) and 297 (4·20) mμ whereas isoflavone (and acetylated hydroxyisoflavones) have λ_{max} at 245 (4·41) and 307 (3·82) mμ and flavanone (and acetylated hydroxyflavanones) have λ_{max} at 250 (3·86) and 320 (3·37) mμ.

In the present chapter, it is proposed to outline briefly the structural variation encountered in each class of minor flavonoid, to list some properties of the commoner substances in each group and to consider some of the difficulties that may be encountered in their identification.

II. The Chalcone Pigments

A. STRUCTURAL VARIATION

Chalcones and aurones (Section III) are together termed "anthochlor" pigments because they are readily detected in yellow-flowered plants by means of the "anthochlor" test. If they are present in a plant, the petal when fumed with ammonia, or the alkaline vapour of a cigarette, turns orange or red (Gertz, 1938). Most yellow flowers, of course, are pigmented by carotenoids and chalcone pigments have a very limited distribution in nature, occurring mainly as petal pigments in the Compositae. They are by no means restricted to petal tissue and have been found in heartwood, bark, leaf, fruit and root.

Since Shimokoriyama (1962) reviewed the naturally occurring chalcones, several new pigments have been reported, bringing the total of known aglycone structures to 19. These 19 can conveniently be divided into four groups on the basis of the parental flavone types: (1) 4-hydroxychalcones, apigenin derived; (2) 3,4-dihydroxychalcones, luteolin derived; (3) 3,4,5-trihydroxychalcones, tricetin derived; and (4) chalcones lacking B-ring hydroxyls, chrysin derived. It should be pointed out at once that the nomenclature of the chalcones differs from that of the other flavonoids in the way the aromatic nuclei are numbered:

Chalcone system   Flavone system

Thus a 5-hydroxyl in a flavone is a 6'-hydroxyl in a chalcone and a 5-deoxy-flavonol corresponds in structure to a 6'-deoxychalcone. A further complication in the chalcone series is that there is free rotation of the A-ring around the carbonyl group and the 2'- and 6'- and the 3'- and 5'-positions are equivalent.

1. 4-Hydroxychalcones

There are two simple chalcones in this class; CHALCONONARINGENIN (12) and its 6'-deoxy derivative, isoliquiritigenin (13). Chalcononaringenin only

(12)   (13)

occurs naturally in a protected form with a methyl or glucosyl group attached. The chalcone⇌flavanone equilibrium in the case of the free compound is strongly in favour of the flavanone form and (12) cannot be prepared as such even in the laboratory. It is thus found in nature mostly as the 2'-glucoside, isosalipurposide, which was first isolated from willow bark, *Salix purpurea* (Charaux and Rabate, 1931), and which has recently been found as a petal pigment in *Aeschynanthus*, *Asystasia*, *Dianthus*, *Helichrysum* and *Paeonia* species (Harborne, 1966b). The most readily accessible source of isosalipurposide is undoubtedly the yellow carnation.

Chalcononaringenin also occurs in methylated form, as the 4'-methyl ether in *Prunus* bark, as the 2',4'-dimethyl ether in *Piper methysticum* root (Hansel *et al.*, 1963), as the 4',4-dimethyl ether in the sori of the golden fern *Pityrogramma chrysophylla* and as the 2',4',4-trimethyl ether in *Xanthorrhoea* resin. Two more complex derivatives of (12) are the hop constituent xanthohumol, which has an isoprenoid side-chain (14) and carthamone (15), the red quinone colouring matter of safflower, *Carthamus tinctorius*.

(14) Xanthohumol   (15) Carthamone

Isoliquiritigenin (13), a more stable substance than (12), was first isolated from liquorice root, *Glycyrrhiza glabra*, in 1881 (Claisen and Claparede; cf. Seshadri, 1956); it is present in the root as the 4-glucoside. It occurs as the 4′-glucoside and the 4′-diglucoside in yellow varieties of *Dahlia variabilis* (Nordström and Swain, 1956) and in gorse flowers, *Ulex europaeus* (Harborne, 1962d); the latter plant also contains isoliquiritigenin 4,4′-diglucoside and 4′-diglucoside-4-glucoside.

2. 3,4-Dihydroxychalcones

The chalcone corresponding to luteolin occurs in *Helichrysum bracteatum* (Rimpler and Hansel, 1965), as the 2′-glucoside, and a dimethyl derivative olivin (16) has recently been isolated from the leaf of the olive tree by Bockova *et al.* (1964). The 6′-deoxychalcone, butein (17), is a well-known compound and occurs fairly widely in the Compositae as the 4′-glucoside, coreopsin. This

(16) Olivin   (17) Butein

4′-glucoside was first isolated by Geissman (1941) from flowers of *Coreopsis douglasii*; the related 3,4′-diglucoside occurs in petals of *Butea frondosa* of the Leguminosae. Butein is accompanied in flowers of some *Coreopsis* species by its 3′-hydroxy derivative, okanin, present as the 4′-glucoside marein in *C. maritima*, and by its 3′-methoxylated derivative, lanceolatin, present as the 4′-glucoside lanceolin in *C. lanceolata*. The 5′-hydroxy derivative, neoplathymenin (18), may occur in petals of *C. stillmanii* but is certainly present in heartwood of *Plathymenia reticulata* (King *et al.*, 1953).

(18) Neoplathymenin

(19) Robtein

3. 3,4,5-Trihydroxychalcones

The only representatives of this class are 2',4',6',3,4,5-hexahydroxychalcone which occurs as the 2'-glucoside in *Helichrysum* (Rimpler and Hansel) and the 6'-deoxy derivative, robtein (19) recently found in the heartwood of the acacia tree, *Robinia pseudacacia* (Drewes and Roux, 1963).

4. Chalcones Lacking B-ring Hydroxyls

The simple chrysin-derived 2',4',6'-trihydroxychalcone is present as the 2',4'-dimethyl ether (20) in the heartwood of pines (Mahesh and Seshadri, 1954) and in *Piper* root (Hänsel, Ranft *et al.*, 1963). Two more complex substances of this type, pedicin (21) and its dimethyl ether pedicellin, occur in the orange-red deposit on the under surface of the leaves of *Didymocarpus pedicellata* (Seshadri, 1951); two related quinones, pedicinin and methylpedicinin, are also present in the same material.

(20)

(21) Pedicin

B. IDENTIFICATION

Chalcones are easily detected by their characteristic colour reactions, spectra and R_f values. Their complete characterisation is more difficult because of their instability. On acid treatment, for example, they readily isomerise to the flavanone. In the case of the acid hydrolysis of the glucoside isosalipurposide, the free chalcononaringenin cannot be detected since it is completely and rapidly isomerised to naringenin. 6'-Deoxychalcones, e.g. butein, are more stable and form a mixture of chalcone and flavanone on acid treatment; this makes the measurement of sugar–aglycone ratios in the glycosides rather difficult. 3,4-Dihydroxychalcones are rather easily oxidised to the corresponding aurone; the okanin glycoside, marein, for example, is completely converted to the corresponding aurone during storage on chromatography paper. *Cis-trans*-isomerisation may also occur during the isolation procedure.

2',3',4',3,4-Pentahydroxychalcone, for example, was obtained by King and King (1951) from the wood of the African greenheart, *Cylicodiscus gabunensis*, as a mixture of *cis*- and *trans*-isomers; the *cis*-, iso-okanin, has m.p. 140° (as a hydrate) and the *trans*-, okanin, has m.p. 235–240°.

Paper chromatography is satisfactory for distinguishing the known chalcone aglycones and may also be used for separating the various glycosides. The four known glycosides of isoliquiritigenin, for example, have distinctive R_f values in BAW and water (Table 3·3). Thin-layer chromatography of chalcones may be carried out on silica gel, using benzene–ethyl acetate–formic acid (9:7:4).

TABLE 3·3. Chromatographic and Spectral Properties of the Naturally Occurring Chalcones

Chalcone	R_f (× 100) in BAW	Forestal	PhOH	λ_{max} in 95% EtOH (mμ)	$\Delta\lambda$ Alk. (mμ)
AGLYCONES					
2',4',4-TriOH (isoliquiritigenin)	89	73	90	235, 372	+70
2',4',3,4-TetraOH (butein)	78	51	66	263, 382	Unstable
2',4',5',3,4-PentaOH (neoplathymenin)	—	—	—	268, 320,[1] 393	Unstable
2',3',4',3,4-PentaOH (okanin)	56	38	—	260, 330,[1] 384	+50 (Unstable)
GLYCOSIDES	BAW	H₂O	PhOH		
Isoliquiritigenin					
4'-Glucoside	61	06	80	} 240, 372	+53
4'-Diglucoside	53	24	56		
4,4'-Diglucoside	34	19	70	} 237, 361	+49
4-Glucoside-4'-diglucoside	36	80	40		
Butein					
4'-Glucoside (coreopsin)	56	—	—	245, 265, 385	+65
Okanin					
4'-Glucoside (marein)	45	—	—	266, 320,[1] 383	+66
Chalcononaringenin					
2'-Glucoside (isosalipurposide)	67	04	36	240, 369	+72

[1] Indicates shoulder or inflection.

Spectral measurements clearly distinguish 4-hydroxy- from 3,4-dihydroxy-chalcones, but substitution in the A-ring has little effect on the visible maximum (Table 3·3). Spectral measurements are useful for determining the position of glycosylation; the attachment of sugars on both A- and B-ring hydroxyls has a hypsochromic effect on the spectrum in neutral solution and the intensity and position of the visible max. in alkaline solution are also affected. Correlations between chalcone structure and spectra are discussed in more detail by Jurd and Horowitz (1961).

III. The Aurone Pigments

A. STRUCTURAL VARIATION

The aurones are the most recently discovered group of flavonoids and are, as their name implies, essentially golden yellow flower pigments. Benzalcoumaranones, to give them their full title, have long been known synthetically but were only found in plants, in *Coreopsis grandiflora*, as recently as 1943 (Geissman and Heaton, 1943, 1944). It is not surprising, therefore, that only seven aurones are known to occur naturally at the present time. Their distribution is also limited, and the only families with aurone records in more than one genus are the Compositae, Leguminosae and Scrophulariaceae.

Aurones with the hydroxylation patterns of luteolin and tricetin (or of quercetin and myricetin) are known–AUREUSIDIN (22) (yellow crystals, m.p. 270–275°) and BRACTEATIN (23) (m.p. 350° with decomposition)—but the apigenin analogue has yet to be discovered. Note that the numbering of the hydroxy groups in the aurone series differs slightly from that in the flavones (cf. 22) and aurone 4- and 6-hydroxyls are equivalent to 5- and 7-hydroxyls of flavones. Aureusidin is the most widely distributed aurone flower pigment outside the Compositae. It occurs as the 6-glucoside (aureusin) in *Antirrhinum majus* (Scrophulariaceae), as the 4-glucoside (cernuoside) in *Chirita micromusa* and *Petrocosmea kerrii* (Gesneriaceae) and in *Limonium bonduellii* (Plumbaginaceae), and as a mixture of the two glucosides in *Oxalis cernua* (Oxalidaceae). In *Antirrhinum*, aureusin is accompanied by the 6-glucoside of bracteatin (23) (Harborne, 1963c), a pigment which is also present in *Helichrysum bracteatum* (Compositae) (Hänsel, Langhammer *et al.*, 1963) but as the isomeric

(22) Aureusidin   (23) Bracteatin

4-glucoside. Finally, aureusidin occurs as the 4-methyl ether, rengasin, in the heartwood of *Melanorrhoea* (Anacardiaceae) (King *et al.*, 1962; Farkas *et al.*, 1963).

The remaining four aurones all lack a hydroxyl in the 4-position (i.e. correspond in structure to 5-deoxyflavonols such as fisetin) and occur only in the Leguminosae and Compositae. The simplest, 6,4'-dihydroxyaurone (hispidol) was found recently in seedlings of *Soya hispida* (Wong, 1966). The next, sulphuretin (6,3′,4′-trihydroxyaurone), occurs as the 6-glucoside sulphurein, in several Compositae genera, but especially *Coreopsis*. It is most easily accessible, perhaps, in the flowers of the yellow dahlia, where it is also present as the 6-diglucoside (Nordström and Swain, 1956). It also occurs free in *Cotinus* wood (King and White, 1957) and as the 6,3′-diglucoside (palasitrin) in *Butea frondosa* petals (Puri and Seshadri, 1955). Sulphurein is accompanied in flowers of *Coreopsis maritima* by the 7-hydroxy derivative maritimein (24)

(24) Maritimein　　　　　　　　　　(25) Maesopsin

(aglycone: maritimetin) and in flowers of *C. grandiflora* by the 7-methoxy derivative, leptosin (aglycone:leptosidin). This latter substance was, in fact, the first naturally occurring aurone to be isolated (*see* above.)

As has already been mentioned, a plant record of the aurone related to apigenin is still lacking but it seems certain that it will be found, since the corresponding "hydrated" aurone, maeosopsin (25), is present in *Maesopsis eminii* heartwood (Janes *et al.*, 1963). "Hydrated" aurones related to sulphuretin and aureusidin have also been discovered recently (cf. Janes *et al.*, 1963). These substances are themselves colourless but are dehydrated by boiling with acid to the aurone; they can be synthesised from the corresponding dihydroflavonol by alkaline treatment.

B. Identification

Apart from bracteatin derivatives, which readily polymerise, the golden yellow aurones are quite stable substances and are readily characterised on a micro-scale by the usual procedures. Since there are so few structures, identification by means of R_f and spectral maxima presents no problems (Table 3·4). Paper chromatography may be supplemented by separations on thin layers of silica gel, buffered with sodium acetate, using the solvent benzene–ethyl acetate–formic acid (9:7:4) (Hänsel, Langhammer *et al.*, 1963).

The spectral properties of a large number of natural and synthetic aurones have been examined (Geissman and Harborne, 1956); here it is only necessary to point out that pigments can be identified by observing the visible maxima in 95% EtOH, the position and stability of the maxima in alkali and the presence (or absence) of a shift with aluminium chloride. The position of the main visible peak ranges from 388 mμ (for hispidol) to 413 mμ (for maritimetin) (Fig. 3·2). The value of measuring alkaline and aluminium chloride shifts is

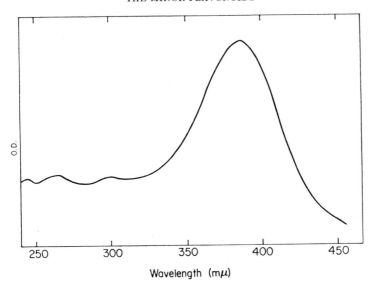

FIG. 3·1. Absorption spectrum of butein 4′-glucoside (coreopsin). Solvent: 95% EtOH.

best illustrated by reference to the isomeric 4- and 6-glucosides of aureusidin and bracteatin. Both pairs of glucosides are nearly indistinguishable by their spectra in neutral solution; however, in alkali, the 4-glucosides are unstable, whereas the 6-glucosides exhibit large stable bathochromic shifts (Table 3·4), Again, the two 6-glucosides, both having a hydroxyl group in the 4-position

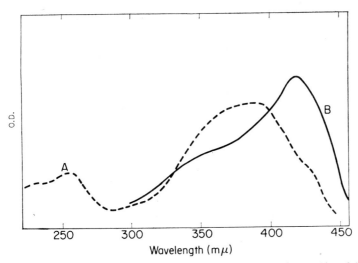

FIG. 3·2. Absorption spectra of 6,4′-dihydroxyaurone (hispidol) (curve A) and 6,7,3′,4′-tetrahydroxyaurone (maritimetin)(curve B). Solvent: 95% EtOH.

which is free to join with the adjacent carbonyl group in chelation, respond to aluminium chloride ($\Delta\lambda = +60$ to 70 mμ), whereas the 4-glucosides are unaffected by the reagent.

TABLE 3·4. Chromatographic and Spectral Properties of the Naturally Occurring Aurones

Aurone	R_f (× 100) in			λ_{max} in	$\Delta\lambda$ alk.
	BAW	30% HOAc	PhOH	95% EtOH (mμ)	(mμ)
AGLYCONES					
6,4'-DiOH (hispidol)	—	—	—	234, 254, 388	66
6,3',4'-TriOH (sulphuretin)	80	19	70	257, 270, 399	65
4,6,3',4'-TetraOH (aureusidin)[2]	57	10	29	254, 269, 399	Unst.
7-OMe,6,3',4'-TriOH (leptosidin)	76	19	80	257, 340, 406	53
6,7,3',4'-TetraOH (maritimetin)	53	10	—	270, 355,[1] 413	Unst.
4,6,3',4',5'-PentaOH (bracteatin)	32	05	13	260, 327, 403	Unst.
GLYCOSIDES					
Aureusidin					
4-Glucoside (cernuoside)	49	25	45	255, 267,[1] 405	45
6-Glucoside (aureusin)	28	16	35	272, 322, 405	85
Bracteatin					
4-Glucoside (bractein)	27	15	07	259, 330, 409	Unst.
6-Glucoside	06	09	04	263, 322, 408	62
Maritimein (6-glucoside)	42	21	—	242, 274, 330, 419	86
Leptosin (6-glucoside)	51	33	73	257, 277, 329, 411	111
Sulphurein (6-glucoside)	49	41	69	257, 277, 340[1] 404	111

[1] Indicates shoulder or inflection. Unst. = unstable.
[2] Aureusidin 4-methyl ether, rengasin, has λ_{max} 254 and 403 mμ.

IV. Flavanones

A. STRUCTURAL VARIATION

The flavanones, perhaps because they are colourless compounds, are a rather neglected group of flavonoids. They have been largely ignored in plant surveys and our knowledge of them is based on the fact that they happen to occur in quantity in several economically important plants. Most notably, they occur in the peel and juice of *Citrus* fruits and make a major contribution to the taste, being bitter substances. Over 30 flavanones (excluding glycosides) have been described but their structures display few new features since each can be related to a naturally occurring chalcone, flavone or flavonol.

The distribution of the rarer flavanones, with their structures, are mentioned in later chapters in this book. Here, the emphasis will be on the commoner flavanones, their combined forms and their identification. The known flavanones can conveniently be divided (with some overlapping) into four groups, depending on whether they are common structures or whether they are related to a chalcone, flavone or flavonol.

1. Common Structures

The flavanone analogues of apigenin and luteolin, namely naringenin (26) and eriodictyol (27), probably occur fairly widely in plants, although this has

(26) Naringenin   (27) Eriodictyol

still to be demonstrated by means of surveys. NARINGENIN (colourless plates, m.p. 248–251°) was first isolated from *Citrus* peel by Will in 1885. It is present as the 7-methyl ether, sakuranetin, in *Prunus* bark (Asahina, 1908) and as the 4′-methyl ether, isosakuranetin, in grapefruit (Hattori *et al.*, 1944). Eriodictyol (colourless needles, m.p. 262–267°), isolated first from *Eriodictyon glutinosum* leaves (Power and Tutin, 1907), occurs as the 3′-methyl ether, homoeriodictyol, in the same plant and as the 4′-methyl ether, hesperetin, in *Citrus* peel (Hoffmann, 1876).

The usual flavanone glycosidic combination is 7-rhamnosylglucoside and the most interesting fact here is that there are two series: 7-neohesperidosides, which are intensely bitter substances, and 7-rutinosides, which have no appreciable taste. The bitterness, which is of the same order as quinine, is due to the unique combination of the flavanone structure with the particular sugar, neohesperidose (rhamnosyl-α(1→2)-glucose) (Horowitz, 1964). Both types of glycoside can be isolated from *Citrus* fruits: the bitter naringin (naringenin 7-neohesperidoside) from the bitter lemon and the tasteless hesperidin (hesperetin 7-rutinoside) from the sweet Navel orange. Naringenin occurs outside *Citrus* as the 7-neohesperidoside, e.g. in flowers of *Antirrhinum majus*; other glycosidic combinations include the 5-glucoside (in *Salix*), the 7-glucoside (prunin in *Prunus*) and the C-glucosyl derivative, hemiphloin (cf. Chapter 2, Section III.B), in *Eucalyptus*.

2. Flavone-related Flavanones

The interconversion, by oxidation-reduction mechanisms, of flavones and flavanones is very facile in plants and many of the rarer flavones are accompanied in the same plant by the corresponding flavanones. For example, chrysin and its methyl derivatives in *Pinus* and *Prunus* woods occur with flavanones such as pinocembrin (dihydrochrysin) and strobopinin (the 6-*C*-methyl derivative). *C*-methyl flavanones also occur, apparently without the related flavones, in *Rhododendron* leaves and in the fern *Cyrtomium*; hence the names farrerol for 6,8-di-*C*-methylnaringenin (from *Rhododendron farrerae*) and cyrtominetin for 6,8-di-*C*-methyleriodictyol (from *Cyrtomium falcatum*).

3. Chalcone-related Flavanones

At least six flavanones are known which do not have a hydroxyl group in the 5-position and these are all clearly related to the corresponding chalcones and usually occur in association with them. In fact, it is not always clear whether flavanones so reported actually occur as such in the living plant or whether they are formed from the related chalcone during the isolation procedure. 5-Deoxyflavanones are found mainly in the Leguminosae or Compositae and have names similar to their related chalcone, e.g. 7,4'-dihydroxyflavanone is liquiritigenin, 7,3',4'-trihydroxyflavanone is butin and so on. It is interesting that, besides occurring as 7-glycosides (e.g. in *Dahlia*), these particular flava-

(28) Liquiritigenin   (29) Dihydrokaempferol

nones are known with sugars attached to the B-ring; two examples are liquiritigenin 4'-glucoside (in *Glycyrrhiza*) and butin 7,3'-diglucoside (in *Butea*).

4. Flavonol-related Flavanones

Possibly the most widely distributed of all flavanones are those related to the flavonols, i.e. the dihydroflavonols or flavanonols. The most common are dihydrokaempferol (29) (also known as aromadendrin), dihydroquercetin (taxifolin) and dihydromyricetin (ampeloptin). Other representatives include dihydrogalangin (pinobanksin), named after *Pinus banksiana*, and dihydro-robinetin, found with robinetin in *Robinia*.

Dihydroflavonols seem to accumulate mainly in heartwoods, in which they occur in the free state or occasionally as 3-glucosides or 3-rhamnosides. They are occasionally found in flower petals, where they appear as 7-glucosides (e.g. *Primula sinensis*) or as 4'-glucosides (e.g. *Petunia*). In a recent review, Pacheco

(1966) lists 28 known flavanonols, this number including aglycones, their optical isomers and their glycosides.

B. IDENTIFICATION

Flavanones, on reduction in ethanol with magnesium and hydrochloric acid, give intense cherry-red to crimson colours; flavones only react weakly so this simple test is quite useful for preliminary identification. The same test can be carried out on paper chromatograms by spraying them with sodium borohydride solution and exposing the paper subsequently to acid vapours (Horowitz, 1957). Chromatographically, the flavanone aglycones can be readily distinguished by their mobility in water (Table 3·5); practically all other flavonoid aglycones (except isoflavones, see next section) have zero R_f in this solvent.

TABLE 3·5. R_f Values and Spectral Maxima
of the Commonly Occurring Flavanones

Flavanone	R_f (× 100) in BAW	H₂O	30% HOAc	λ_{max} in EtOH (mμ)	$\Delta\lambda$ alk.[2]
AGLYCONES					
Butin	82	20	64	233, 278, 312	59
Eriodictyol	85	10	56	226, 288, 325[1]	40
Naringenin	89	16	66	224, 290, 325	37
Sakuranetin	90	09	65	232, 290, 335[1]	110
Homoeriodictyol	88	14	67	228, 288, 322[1]	39
Hesperitin	89	11	67	226, 288, 300[1]	40
Dihydrokaempferol	87	30	72	215, 292, 330[1]	26
Dihydroquercetin	78	28	67	228, 289, 325[1]	25
GLYCOSIDES					
Naringin	59	62	87	⎱ 226, 284, 330[1]	143
Naringenin 7-glucoside	64	44	80	⎰	
Sakuranetin 5-glucoside	69	48	81	228, 282, 315[1]	80
Hesperidin	48	50	85	⎱ 225, 286, 330	77
Hesperetin 7-glucoside	60	33	79	⎰	
Dihydrokaempferol 7-glucoside	57	64	84	215, 287, 325[1]	78

[1] Indicates shoulder or inflection.

[2] Alkaline shift of the main peak at 278–292 mμ measured. Some phloroglucinol-derived flavanones which are substituted in the 7-position (e.g. sakuranetin, naringenin) undergo immediate ring opening to the chalcone in alkali and thus give unusually large alkaline shifts.

Because the B-ring is not conjugated with the A-ring in a resonating system, flavanones have pronounced absorption bands only in the short u.v. They have

4

a rather characteristic intense peak at 278 to 292 mμ, with an inflection or less intense maximum at about 325 mμ (Fig. 3·3). Useful structural information can be obtained by measuring the shifts in the spectral bands produced by alkali, aluminium chloride (a positive shift indicating a free 5-hydroxyl) and

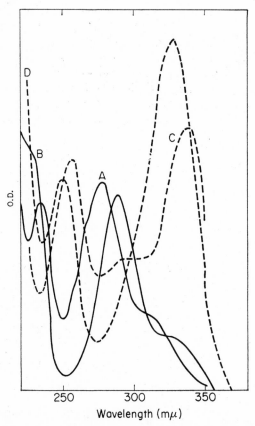

Fig. 3·3. Absorption spectra of 7,4'-dihydroxy-3'-methoxyflavanone and of 5,7,4'-trihydroxy-3'-methoxyflavanone (homoeriodictyol). Solvents: 95% EtOH (curves A and B) and EtOH–NaOH (curves C and D, respectively).

sodium acetate (a positive shift indicating a free 7-hydroxyl). Measurements in ethanol–boric acid, normally diagnostic for catechol B-rings in flavonoids, are of no value here, since flavanones such as eriodictyol fail to respond to this reagent.

Flavanones have lower melting points than flavones and can usefully be characterised by this means. They are also optically active. Most are isolated as racemates, but if care is taken, optically active forms can usually be obtained from plants. All 13 flavanones so far isolated as such are laevorotatory and it may thus be assumed they all have the same configuration as that of

hesperitin which has shown to be (30) (Whalley, 1962). Flavanonols have two
optically active centres and are therefore capable of existing in four forms. The

(30) (−)-Hesperitin

bitter taste of certain flavanone glycosides in aqueous solution has already
been mentioned; this constitutes a useful, if limited, means of identification.

V. Isoflavones

A. Structural Variation

The isoflavones have attracted the attention of organic chemists, perhaps
because of the immense structural variability encountered in this series and
their chemistry is well understood. They are isomeric with the flavones and
differ only in the position of the attachment of the aryl B-ring to the central
pyran nucleus. Common isoflavones have the usual 5,7,4'- or 5,7,3',4'-hydroxy-
lation patterns but there are many rarer isoflavones which have no known
counterpart in the flavone series. Isoflavones are found characteristically in
one sub-family (the Papilionatae) of the Leguminosae and rarely elsewhere so
they provide a character of some taxonomic importance (*see* Chapter 5). Since
excellent recent reviews of the chemistry of isoflavones are available (Ollis,
1962; Dean, 1963), a comprehensive coverage is not necessary and is, in any
case, outside the scope of the present book. The structural variation will be
outlined and methods of detection discussed.

The isoflavones corresponding to apigenin and luteolin are genistein (31)
and orobol (32), both well-known substances present in a number of legumes
Genistein (m.p. 292°) is widely distributed in *Genista* (it was first isolated

(31) Genistein   (32) Orobol

from dyer's broom, *G. tinctoria*), and also occurs in *Lupinus, Sarothamnus,
Sophora, Soya, Trifolium* and *Ulex* species. All three monomethyl ethers occur
naturally. The 7-methyl ether, prunetin, is present in *Prunus*, the 4'-methyl

ether, biochanin-A, occurs in chana *Cicer arietinum* and the 5-methyl ether was recently detected in wood of *Cytisus laburnum* and *Genista hispanica* (Paris and Fougeras, 1965; Chopin *et al.*, 1963). Genistein glycosides characterised include the 7-glucoside (in *Genista*), the 7-rutinoside (in *Baptisia sphaerocarpa*, Rosler *et al.*, 1965), the 4'-glucoside and the 4'-rhamnosylglucoside (both in *Sophora* fruits). A more unusual sugar, apiosylglucose, occurs attached to the 7-hydroxyl group of biochanin-A (the 4'-methyl ether) in the root bark of *Dalbergia lanceolata* (Malhotra *et al.*, 1965).

Orobol (m.p. 270°), the luteolin analogue, was the first isolated from roots of *Lathyrus montanus* (formerly *Orobus tuberosus*) but has also been detected in *Baptisia* leaves and flowers (Alston *et al.*, 1965). The 7-methyl ether, santal, is present in sandalwood, *Pterocarpus santalinus*, and the 4'-methyl ether has fairly recently been obtained from clover, *Trifolium pratense* (Wong, 1963).

Of the other isoflavones, the most common is the 5-deoxy derivative of genistein, namely daidzein. First isolated from soya bean, it also occurs in *Cicer, Genista, Psoralea, Trifolium* amd *Pueraria*. It occurs in this last genus as a rare *C*-glucosyl derivative, puerarin (cf. Chapter 2, Section III.B), and in *Ononis spinosa* and *Trifolium repens* as the 4'-methyl ether, formononetin. 7,3',4'-Trihydroxyisoflavone has yet to be reported as a natural constituent, but the methylenedioxy derivative, ψ-baptigenin, has long been known as a

(33) ψ-Baptigenin   (34) Irigenin

constituent of *Baptisia tinctoria* root and the trimethyl ether, cabreuvin, is present in the heartwood of *Myrocarpus fastigiatus*.

ψ-Baptigenin is accompanied in *Baptisia* root by 7,3',4',5'-tetrahydroxyisoflavone, baptigenin (Farkas *et al.*, 1963).

Of the eight 6-hydroxylated isoflavones known, four are derivatives of 6-hydroxygenistein, a substance which has not itself yet been found naturally. The 6-methyl ether, tectorigenin, was first isolated from *Iris tectorum* roots but has subsequently been found in *Baptisia* leaf (Alston *et al.*, 1965) and, accompanied by the 6,7-dimethyl ether, in *Dalbergia sisoo* flowers (Banerji *et al.*, 1963). The 5,7-dimethyl ether, muningin, is present in *Pterocarpus angolensis* heartwood; the 6,4'-dimethyl ether (irisolidone) and the 5-methyl-6,7-methylenedioxy derivative occur together in *Iris nepalensis* (Prakash *et al.*, 1965). Structurally, the most interesting isoflavone of this group is irigenin (34), because of its rare B-ring substitution pattern (3-hydroxy-4,5-dimethoxy). Irigenin is one of the most easily accessible isoflavones, occurring as it does as the major phenol in the rhizome of the garden *Iris* (*I. germanica*). It is one of the very few isoflavones which do not apparently occur in the Leguminosae.

One of the most recently discovered isoflavones is another of the 6-hydroxylated series, afrormosin, 6,4′-dimethoxy-7-hydroxyisoflavone (35). First detected in *Afrormosia elata* heartwood by McMurry and Theng (1960),[1] it

(35) Afrormosin   (36) Pomiferin

has since been found in no less than four other genera and is present as the 7-glucoside in the flowers of the common garden *Wistaria floribunda* (Shibata *et al.*, 1963). The parent 6,7,4′-trihydroxyisoflavone has been identified as the major anti-oxidant of fermented soya beans (Gyorgy *et al.*, 1964) and appears also to occur, along with genistein and daidzein, in the native bean.

The remaining known isoflavones are briefly mentioned below. Two have isoprenoid residues attached to the 6- and 8-positions; the orobol-derived pomiferin (36) and the genistein analogue, osajin. Osajin and pomiferin were originally isolated from the Moraceae, from the osage orange *Maclura pomifera* (Wolfrom and Mahan, 1942), but recently osajin, an isomer scandenone and 5-*O*-methylosajin have been found in the Leguminosae, in *Derris scandens* root (Pelter and Johnson, 1966). A further three isoflavones have 2′-hydroxy or 2′-methoxy substituents: tlatlancuayin of *Iresine* (Amarantaceae), podospicatin of *Podocarpus* (Podocarpaceae) and cavuinin of *Dalbergia* (Leguminosae). Finally, there are four legume constituents with both 2′-hydroxy and isoprenoid attachments: munetone, jamaicin, mundulone and toxicarol isoflavone. More complex derivatives containing the isoflavan nucleus are also known, the chief of these being the insecticidal rotenoids. Their phytochemical distribution is discussed in Chapter 5, along with that of the isoflavones.

B. IDENTIFICATION

The isoflavones are one of the most difficult classes of flavonoids to characterise, since they do not respond specifically to any one colour reaction. Probably the most useful diagnostic property is u.v. spectroscopy but even here occasional confusion with the flavones is possible. Unlike many other

---

[1] Afrormosin was named "afromosin" by these authors due to a misspelling of the generic name as "*Afromosia*" instead of the correct *Afrormosia*.

flavonoids, however, they are normally very easily crystallised from plant extracts and as a class lend themselves particularly well to the classical techniques of organic chemistry.

In isoflavones, the phenyl ring at position 3 is not in conjunction with the pyrone carbonyl as it is in the flavones so that the spectral properties differ significantly from those of flavones. All have intense absorption in the short u.v. (*ca* 210 mμ), an intense band at 255–275 mμ and generally a less intense band or inflection at 320–330 mμ (Table 3·6). Only in the case of 6,7,4′-tri-hydroxyisoflavone and its derivatives is this second band nearly of equal intensity to the first (log ϵ values for afrormosin are 4·49 at 260 mμ and 4·11 at

TABLE 3·6. R_f Values and Spectral Maxima of Isoflavones

Isoflavone	BAW	R_f (× 100) in H₂O	30% HOAc	MC[2]	λ_{max} in EtOH (mμ)	$\Delta\lambda$alk.[4] (mμ)
Daidzein	92	08	62	36	250, 260,[1] 302	30
Formononetin	94	07	67	—	250, 256, 300[1]	42
ψ-Baptigenin	94	07	66	66	241, 251, 296	43
Genistein	94	04	59	41	263, 325[1]	10
Prunetin[3]	94	03	67	—	263, 325[1]	25
Afrormosin	95	06	—	67	258, 320	28
Tectorigenin	90	09	70	32	268, 320	—
Irigenin	91	16	79	56	269, 335[1]	5
Pomiferin	94	00	06	00	275, 355	−10
Genistin	67	25	75	02	262, 330[1]	33
Tectiridin	67	38	83	04	268, —	—
Iridin	66	48	87	04	268, 325[1]	38

[1] Indicates shoulder or inflection.

[2] R_f in 11% MeOH in CHCl₃ on silica-gel plates; other R_f values on paper.

[3] The 7-methyl ether of genistein; the other two monomethyl ethers can be distinguished from it by spectral means: the 5-methyl ether has λ_{max} 256 mμ, $\Delta\lambda$alk. 13 mμ and the 4′-methyl ether (biochanin-A) has λ_{max} 263 mμ, $\Delta\lambda$ alk. 10 mμ.

[4] Alkaline shift of band II, i.e. the band above 300 mμ.

320 mμ) so that the overall spectrum resembles that of a simple flavone. Otherwise, the low intensity of absorption above 300 mμ of isoflavones is a valuable diagnostic feature.

Isoflavone spectra undergo bathochromic shifts in the presence of alkali and such measurements are often of value in structural studies, e.g. for distinguishing the three genistein monomethyl ethers (cf. Table 3·6 and Fig. 3·4). The presence of a free 5-hydroxyl group in an isoflavone can be detected by means of measuring the effect of ethanolic aluminium chloride on the spectrum and that of a free 7-hydroxyl by measuring the effect of adding sodium acetate to the cell solution. In both instances, the maxima undergo bathochromic shifts of 10–15 mμ.

Paper chromatography is of limited value for isoflavone separations; solvents such as water, dilute acetic acid or butanol-ammonia give useful separations (see Table 3·6), but in BAW, Forestal or phenol, isoflavones have uniformly high R_f values. The complementary use of thin-layer chromatography is recommended for purposes of separation and identification. The clover isoflavones, for example, are readily distinguished on silica gel with 11%

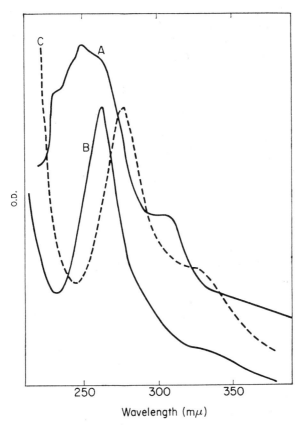

Fig. 3·4. Absorption spectra of 7,4'-dihydroxyisoflavone (daidzein) and of 5,7,4'-trihydroxyisoflavone (genistein). Solvents: 95% EtOH (curve A, daidzein; curve B, genistein) and EtOH–NaOH (curve C, genistein).

methanol in chloroform as solvent (Beck, 1964). Isoflavones can be detected on paper or silica gel with a general phenolic reagent (e.g. $FeCl_3$–$K_3Fe(CN)_6$ or Folin), but there does not seem to be a more specific spray available. However 5-deoxyisoflavones (e.g. daidzein, afrormosin) have a brilliant light blue fluorescence in u.v. light and ammonia which is quite unmistakable; genistein and other 5-hydroxyisoflavones appear as dull purple spots in u.v. light, changing to dull brown with ammonia.

VI. Biflavonyls

Biflavonyls are almost completely confined in their occurrence to the gymnosperms and so these dimeric flavones are discussed in detail under this heading in Chapter 4 (Section IV). Here it is only necessary to mention that all known substances are dimers of apigenin; they fall into three classes, depending on whether their linkage is through a carbon–carbon bond between the 8- and 3'-positions (amentoflavone, 37) or between the 8- and 8'-positions (cupressoflavone) or through a carbon–oxygen bond between the 8- and 4'-positions (hinokiflavone). Most known biflavonyls are mono-, di- or trimethyl ethers of amentoflavone (37).

(37) Amentoflavone

The problem of detecting biflavonyls on a micro-scale is not completely solved, but there is usually no trouble in differentiating them from monomeric flavones. On paper chromatograms, they appear as dull brown spots and have uniformly high R_f values in the solvent systems commonly used for mono-meric flavones such as apigenin (e.g. BAW, PhOH, Forestal, etc.). However, in butanol–ammonia they separate from each other fairly satisfactorily (Table 3·7); monomeric flavones are immobile in this solvent. Di Modica and

TABLE 3·7. R_f Values and Spectral Maxima of Biflavonyls

Biflavonyl	R_f [1]($\times$ 100) in BN (paper)	TEF (silica gel)	λ_{max} (mμ) Band I	Band II	Ratio: E band II/ E band I (as %)
Amentoflavone (parent compound)	22	27	272	342	—
Bilobetin (4'-MME)	30	35	271	341	94
Sotetsuflavone (7"-MME)	—	—	271	337	—
Ginkgetin (7,4'-DME)	66	43	272	335	97
Isoginkgetin (4',4'''-DME)	45	46	272	330	90
Sciadopitysin (7,4',4'''-TME)	75	50	272	330	94
Kayaflavone (7",4',4'''-TME)	74	49	271	329	86
Hinokiflavone	19	14	271	338	—

[1] BN = butanol–2 N NH$_4$OH (1:1); TEF = toluene–ethyl formate–formic acid (5:4:1)

Rivero (1962) have indicated that water-saturated chloroform is a good mixture for biflavonyls but in our hands it does not give reliable results. Turning to thin-layer chromatography on silica gel, the mixture toluene–ethyl formate–formic acid (5:4:1) produces reasonable separations of the known biflavonyls (see Table 3·7; Kawano et al., 1964).

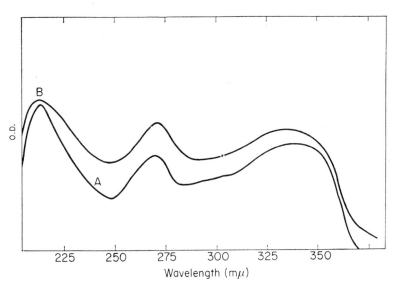

FIG. 3·5. Absorption spectra of the flavone apigenin (curve A) and the biflavonyl bilobetin (curve B). Solvent: 95% EtOH.

From the spectral point of view, biflavonyls resemble the parent flavone, apigenin, from which they are derived; isoginkgetin, with two 4′-methyl groups, not unexpectedly resembles acacetin (4′-methylapigenin). The data assembled in Table 3·7 indicate that the various biflavonyls can be distinguished from each other reasonably well by spectral measurements. Further differences appear on comparing spectra in alkaline solution (see Baker and Ollis, 1961). A spectral feature which may be of general diagnostic value for distinguishing dimers from monomers is the ratio of the extinction coefficients of the two main spectral bands. In biflavonyls, band II is uniformly less intense than band I, i.e. has 86–97% of the intensity of band I (see Table 3·7), whereas in related monomeric flavones, the situation is reversed (band II/band I ratios in apigenin 112%, acacetin 122% and 7,4′-dimethylapigenin 114% (Fig. 3·5).

VII. Leucoanthocyanidins

A brief mention of the leucoanthocyanidins is necessary here, since these substances, which are really condensed tannins, yield anthocyanidin pigments

4*

on acid treatment. They are, indeed, related biosynthetically to the antho-cyanins and flavonols and their phytochemical distribution is of considerable interest in relation to that of the flavonoids. The structural variability en-countered among the leucoanthocyanidins has already been mentioned, in terms of the anthocyanidins they yield with acid, in Chapter 1. Only a few words on chemical structure and means of identification are needed here.

Structural studies of leucoanthocyanidins have lagged behind those of simple flavonoids, mainly because of difficulties in isolation and resolution. Recently, convincing evidence has been obtained that most chromatographi-cally mobile leucoanthocyanidins are dimers, derived from one unit of flavan-3,4-diol linked to a catechin (flavan-3-ol). Structural work has centred on leucocyanidin, the most common of all leuco- substances, and three groups of workers (Geissman and Dittmar, 1965; Wienges and Freudenberg, 1965; Creasy and Swain, 1965) have shown, mainly from n.m.r. studies, that it probably has structure (38). The only question still in doubt is whether the bond joining

(38) Leucocyanidin

the two flavan units is 4-8' (as shown) or 4-6'. Polymeric leucoanthocyanidins, the true condensed tannins, are then assumed to have similar structures with a repeating flavan-3,4-diol unit and a flavan-4-ol end-unit. Most recent work has indicated that these substances do not have covalently linked sugar units in their structure; the sugars detected after acid treatment of tannins are probably free monosaccharides adsorbed onto their surfaces during isolation (Somers, 1965; but see also King, 1966).

From the point of view of identification, leucoanthocyanidins are readily distinguished from simple flavonoids. When isolated pure, they are colourless, optically active compounds, with u.v. max at 278 mμ (alkaline shift 10–20 mμ). Some are mobile in the usual solvent systems, with an R_f of about 30 in BAW and water), but the more highly polymeric substances are increasingly less mobile. As polymers, they can be separated on the grades of sephadex usually used for proteins (Somers, 1965). They are individually identified by the nature of the anthocyanidin and catechin they yield on acid treatment.

VIII. Other Phenolic Pigments

There is some overlap in chromatographic or colour properties between flavonoids and certain other phenolic pigments, notably xanthones and quinones, and errors in identification have occurred in the past. Most of these errors have been due to placing too much reliance on simple colour tests and can largely be eliminated by the use of the accurate physical methods of analysis now available. The properties of xanthones and quinones are given here, since they may occasionally be confused with flavonoids during plant surveys.

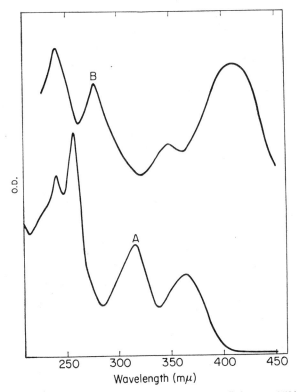

FIG. 3·6. Absorption spectrum of the xanthone mangiferin. Solvents: 95% EtOH (curve A) and EtOH–NaOH (curve B).

Xanthones are yellow phenolic pigments, similar in colour reactions and chromatographic mobility to flavones. They have quite characteristic spectral properties, exhibiting maxima at 230–245, 250–265, 305–330 and 340–400 mμ (*see* Fig. 3·6). Mangiferin (39), one of the commonest xanthones, has R_f 45 in BAW, 74 in Forestal and 12 in water, and λ_{max} at 242, 258, 316 and 364 mμ (relative intensities 82:100:50:38 respectively). On paper, it has an orange-brown colour in u.v. light, changing to fluorescent yellow-green with ammonia vapour.

Hydroxyquinones show some superficial resemblances to flavonoids, some anthraquinones having colours similar to anthocyanins. The distribution of quinones differs considerably from that of flavonoids and whereas they are a

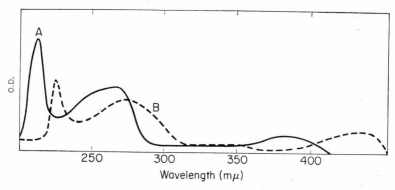

(39) Mangiferin

(40) Plumbagin

most important group of pigments in micro-organisms, their contribution to higher plant colour is negligible. Thus, they often occur in heartwood or root or are present in the plant in colourless form. From a spectral point of view, quinones are quite distinct from flavonoids, having two (or more) intense absorption bands in the u.v. region between 220 and 280 mμ, and one or two less intense bands in the visible region. Plumbagin (40), a typical naphtha-

FIG. 3·7. Absorption spectrum of the naphthoquinone plumbagin. Solvents: 95% EtOH (curve A), and EtOH–NaOH (curve B).

quinone present in roots and leaves of *Plumbago* species in colourless form, appears on paper as a yellow spot, which is dull brown in u.v., a colour unchanged by ammonia vapour. In the solvents used with flavonoids, it is either immobile (aqueous solvents) or runs to the front (e.g. in BAW). It is best detected on silica gel plates with the solvent mixture petroleum ether (b.p. 100–120°)–ethyl acetate (7:3) in which it has an R_f of 68; flavonoids are generally immobile under these conditions. In its spectrum (Fig. 3·7), plumbagin has intense peaks at 212 and 270 mμ and two weaker maxima at 404 and 420 mμ.

CHAPTER 4

GENERAL DISTRIBUTION OF FLAVONOIDS

I. Introduction

The main facts about the general distribution of flavonoids in nature are now well known. These water-soluble pigments occur almost universally in higher plants but are of restricted distribution in lower orders. The anthocyanins are particularly characteristic of the angiosperms and, apart from a few records in ferns and mosses, are not known elsewhere in nature. Flavonoids are not synthesised by animals but they do accumulate in a few insects, especially butterflies, as a result of the caterpillars' feeding habits.

A vast amount of information has been obtained through surveys and detailed chemical investigations about the occurrence and distribution of flavonoids. In the present chapter, the general aspects of the distribution of flavonoids in higher plants will be summarised; detailed accounts of the pigments in the mono- and dicotyledons are reserved for the following three chapters (5–7). Subsequent sections in this chapter will cover what is known of their occurrence in lower plants, gymnosperms and animals.

II. Flavonoids of Higher Plants

A. DISTRIBUTION OF ANTHOCYANINS

Surveys of plants for anthocyanins were carried out even before adequate methods of identification were available. Thus, Gertz in 1906 studied the histological distribution of these pigments in representative species from every plant order. Major surveys, however, were not attempted until Robinson and Robinson (1932) devised their well-known distribution and colour tests for distinguishing the common anthocyanidins and for determining the glycosidic class of the pigments. These authors, in collaboration with workers at the John Innes Institute, published seven survey papers, the results of which were collected together in the classic publication (Lawrence et al., 1939) entitled "Distribution of Anthocyanins in Flowers, Fruits and Leaves". The flower pigments of 382 species representing 240 genera and 78 families were examined, the plants being mainly either of temperate origin or glasshouse ornamentals. Data for an additional 200 species were published by Beale et al. in 1941.

The Robinson distribution tests were also used in surveying several different floras: those of Australia (Gascoigne et al., 1948), the Galapagos Islands (Taylor, 1940) and the Himalayas (Acheson, 1956). Paper chromatography, a method first applied to anthocyanins by Bate-Smith (1948), provides a more precise technique for identification and has completely replaced the distribution tests in survey work. Paper chromatography was used, for example, by Forsyth and Simmonds (1954) for studying the flora of Trinidad, by Hayashi and Abe (1955) for surveying the flora of Japan and by Reznik (1956) for studying European cultivars.

Surveys have to be based on knowledge gained by the isolation and proper identification of pigments in a reasonable sample of plants and the Robinson tests were devised for identifying the pigments known at the time. Chromatographic methods, when applied to isolation and identification in the 1950's, revealed a much wider range of pigments in plants, particularly of glycosides but also of aglycones, so that the results obtained in earlier surveys have had to be modified in the light of more recent findings. In the following paragraphs, the main results of surveys are summarised, taking this into account. All the anthocyanins that have been fully identified in higher plants are listed in Table 5·1 (monocotyledons), Tables 6·1 and 7·1 (dicotyledons) which can be found in the following chapters, pp. 127, 186 and 234.

1. The Common Anthocyanidins

Surveys have shown that six anthocyanidins are widely distributed in plants: pelargonidin, cyanidin, peonidin, delphinidin, petunidin and malvidin. Of these, petunidin is the only one about which there has been some difficulty in identification. Thus, in earlier surveys, it was not readily distinguished from cyanidin–delphinidin mixtures and again in chromatographic surveys, the similarities in the R_f values of petunidin and cyanidin have caused some confusion. Thus, Forsyth and Simmonds (1954), in their chromatographic survey of 246 tropical plants, failed to detect it in a single species. More recent work in this laboratory has shown that it occurs very widely as a trace pigment accompanying malvidin. It still does not appear to occur very frequently as a major flower pigment except in plants of the Solanaceae, but neither does the related peonidin. Thus, of the three methylated pigments, malvidin is the only one to occur really widely.

If one ignores the factor of methylation, there are only three anthocyanidin types, pelargonidin, cyanidin and delphinidin, and of these three, cyanidin is the most frequent in most floras. Thus, figures from the European survey (Lawrence *et al.*, 1939; Beale *et al.*, 1941) are pelargonidin in 23%, cyanidin in 52% and delphinidin in 50% of the sample (530 species) and for the West Indian flora (Forsyth and Simmonds, 1954) pelargonidin in 17%, cyanidin in 70% and delphinidin in 47%. Exceptionally, delphinidin (63%) occurs more frequently than cyanidin (47%) in the Australian flora, but this result is partly due to the fact that the survey was confined to wild species. Cultivation involves strong human selection for brilliant scarlet and orange-red colours and leads to some increase in the number of pelargonidin and cyanidin types recorded.

The remarkable scarcity of pelargonidin in Australian plants, only in 2% of the sample, must be due to other factors, the chief being the absence of bird pollinating mechanisms in the flowers. In the West Indies where bird pollination is well known (van der Pijl, 1961), pelargonidin is present in 17% of the sample and the figure is the same for both native and cultivated plants (Forsyth and Simmonds, 1954).

Anthocyanidins other than the six common ones are few in number (*see* Chapter 1) and are very rare; they are found regularly in only three sympetalous families (Primulaceae, Plumbaginaceae and Gesneriaceae). Other occurrences of rare pigments are mentioned in the introductory sections of the next three chapters.

2. Glycosidic Variation

In their surveys using the distribution tests, Robinson and Robinson (1931) recognised four glycosidic classes (3-monoside, 3-bioside, 3-pentoside and 3,5-dimonoside) of anthocyanidins. With the advantage of chromatographic techniques, the author has noted no less than 21 classes of anthocyanidin glycosides (Chapter 1). Details of the occurrence of these different glycosides and their acylated derivatives are given elsewhere in this volume. Here, the position may be summarised by giving the percentage occurrences by genera of monosides, biosides and triosides (i.e. anthocyanidins with one, two and three sugars attached), which are 45, 61 and 16% respectively. No anthocyanidin with four sugars attached has been fully characterised. 3-Glycosides (63%) are more frequent than 3,5-diglycosides (50%), while 3,7-diglycosides are only known in three genera (*Watsonia, Petunia* and *Papaver*). Acylated anthocyanins have been recorded in some 20 genera representing 12 families, which are equally divided between the monocotyledons, the Archichlamydeae and the Sympetalae; there is also a single record in a fern (*see* Section III.C of this chapter).

3. Organ Distribution

While anthocyanins have been found in almost every type of plant tissue, their distribution has only been surveyed extensively in flower (usually corolla), fruit and leaf. While pelargonidin and delphinidin types are common in flowers (*see* Section II.A.*1*) they are unusual or rare as leaf and fruit pigments. Thus, a survey of the transient red colours in young leaves of 200 plant species by Price and Sturgess (1938) showed that cyanidin was present in 93%. In a narrower survey, Reznik (1956) found cyanidin in nine species, delphinidin in four and peonidin in three. The situation is the same in permanent leaf pigmentation; Lawrence *et al.* (1939) found cyanidin in 80% of those examined and Reznik records cyanidin in 28 out of 36 species surveyed. The figures for autumnal colouring in leaves are even more clear-cut. Hayashi and Abe (1953) analysed 74 plants from 25 families and found cyanidin 3-glucoside in all of them; similarly, Reznik (1956) found cyanidin in the autumnal leaves of 47 of 49 plants.

In fruits, cyanidin occurs to the extent of about 70%; pelargonidin is not unknown (e.g. in passion fruit, *Passiflora*, and strawberry, *Fragaria*) and delphinidin is fairly frequent (e.g. in pomegranate, grape, *Vitis*, and aubergine, *Solanum*).

With regard to the more detailed histological distribution of anthocyanins, it may be noted that these pigments have been found in almost every part of the flower, whether it be corolla, sepal, bract, stamen, style or pollen. They have been isolated, for example, from style, stamen and pollen of *Anemone*, from sepals of *Solanum* and from bracts of *Poinsettia* and *Musa*. In leaves, anthocyanin may be confined to the upper (e.g. *Begonia*) or lower surface (e.g. *Hoffmannia ghiesbreghtii*) or to the leaf hairs (e.g. *Gynura aurantiaca*). Anthocyanin may alternatively occur just in the stem (e.g. *Solanum microdontum*). Pigmented roots are well known (e.g. the radish *Raphanus* and the deep red form of *Daucus carota*) as well as anthocyanin-coloured tubers (e.g. the potato, *Solanum tuberosum*). Finally, in the fruit, anthocyanin may occur throughout (e.g. *Rubus*) or just in the juice (e.g. in the blood orange *Citrus*) or skin (apple, *Malus*). Among the legumes, anthocyanin may colour the pod (*Pisum sativum*) or the seed coat (*Abrus* and *Phaseolus*).

4. Anthocyanins in Acyanic Plants

Surveys have shown that practically all red to blue colours in higher plants are anthocyanin-based. Thus of 832 dicotyledons in the British flora with conspicuous flowers, no less than 49% are known to have anthocyanins (Beale *et al.*, 1941). The question now remains whether plants which lack visible cyanic colour are able to make anthocyanin under any circumstances. There are at least three pieces of evidence showing that all higher plant species have the potentiality of producing anthocyanin, i.e. that visible coloration is only lacking in certain plants because environmental conditions do not favour its synthesis or because anthocyanin synthesis has no selective advantage.

First, the observation that a large number of plants, including some which are otherwise acyanic (e.g. the lettuce), produce anthocyanin in the young leaves as a "temporary flush" of colour (Price and Sturgess, 1938) indicates that anthocyanin is closely related (probably via carbohydrate) to the basic metabolism of these plants. Second, it is possible to cause plants, which are otherwise green in their leaf (e.g. the tomato), to produce anthocyanin under artificial growth conditions (nitrogen starvation, low temperatures etc.). Third, the discovery that plants, which otherwise lack anthocyanin, are able to synthesise it when cells of the plant are grown in tissue culture. A striking example here is the composite *Happlopappus gracilis*, which has green leaves and stems and yellow petals and yet which produces red cyanidin coloration in callus tissue grown on agar.

It is, of course, possible that many otherwise acyanic plants are able to store anthocyanins in their leaves or flowers in a colourless pseudo-base form (e.g. 1). This is certainly true of a number of semi-aquatic plants investigated by Reznik and Neuhausel (1959); extraction of the green leaves with cold acid produced normal anthocyanins (e.g. 2) from these plants. Anthocyanin pseudo-bases have also been found in the petals of a few plants. Certain white forms of the garden *Iris*, for example, contain the anthocyanin of the mauve form in a colourless condition.

(1) Pseudo-base   (2)

5. *Pigments of the Centrospermae*

Although anthocyanins have been found in nearly all angiosperm families surveyed, it cannot be assumed that they are ubiquitous as flower pigments in higher plants. Indeed, they are known to be replaced by a different class of vacuolar pigments, the betacyanins, in most families of the order Centrospermae. The fact that pigments in these plants, which include the beetroot, *Beta*, and showy ornamentals such as *Amaranthus* and *Bougainvillea*, differ in their colour properties and chemical stability from the anthocyanins was recognised as long ago as 1876 by Bischoff and was fully confirmed by Lawrence *et al.* (1939), who used the rather erroneous term "nitrogenous anthocyanin" for these colouring matters. The structure of betanidin (3), the aglycone of the beetroot pigment, was elucidated by Mabry *et al.* in 1962 and shows no obvious chemical relationship to the anthocyanins or any other flavonoids. Subsequently, at least 44 different betacyanins have been isolated by electrophoresis on polyamide from plants of the Centrospermae. All appear to have the same basic betanidin structure, differing from betanin, the 6-*O*-glucoside, in beetroot, only by the nature and number of sugar or other attachments (Piattelli and Minale, 1964). The only other betacyanin to be fully characterised, besides the 6-*O*-glucoside, is a 6-*O*-glucuronosylglucoside present in leaves of *Amaranthus tricolor* (Piattelli *et al.*, 1964).

(3) Betanidin   (4) Vulgaxanthin-I

Accompanying the betacyanins in flowers, leaves and roots of plants of the Centrospermae are yellow, water-soluble pigments, now called betaxanthins. These pigments were at one time thought to be related in structure to the flavones as the "nitrogenous anthocyanins" were to the anthocyanins, but in fact they have a very similar chromatophore to betanidin. They differ only in that the dopachrome moiety of (3) is replaced by an amino acid or an amine, to give, for example, vulgaxanthin-I (4) which is the proline analogue. Other known betaxanthins have the same chromophore as vulgaxanthin-I and are derivatives of hydroxyproline, methionine sulphoxide, tyramine, dopamine, glutamic acid, glutamine and aspartic acid respectively (Piattelli *et al.*, 1965). A detailed discussion of the chemistry and distribution of the betacyanins

TABLE 4·1. Properties which Differentiate Betacyanins (and Betaxanthins) from Anthocyanins

Property	Betacyanins (and betaxanthins)	Anthocyanins (including 3-Deoxyanthocyanins)
Extraction from dried plant material	Extracted by water but not methanol	Extracted by methanol but not water
Behaviour towards hot mineral acid (2N HCl at 100°)	Colour destroyed	Relatively stable
Visible spectra	$\lambda_{max}^{H_2O}$ 532–554 mμ [$\lambda_{max}^{H_2O-HCl}$ 459–471 mμ]	$\lambda_{max}^{MeOH-HCl}$ 499–545 mμ [$\lambda_{max}^{MeOH-HCl}$ 473–497 mμ]
Chromatographic mobility	Very low R_f values in butanol–acetic acid–water (0·03–0·06)	Moderate R_f values in butanol–acetic acid–water (0·09–0·50)
Electrophoretic behaviour	Mobile at pH 2·4	Immobile at pH 2·4

is outside the scope of this book and the reader is referred to the reviews of Reznik (1957), Dreiding (1961), Piattelli and Nicolaus (1965) and Mabry (1966). Mention, however, must be made of methods of differentiating these pigments from anthocyanin and also of the taxonomic significance of their occurrence in the Centrospermae in place of anthocyanins.

Some of the more important means of distinguishing betacyanins and anthocyanins are assembled in Table 4·1. Betaxanthins are included, since they have the same colour as 3-deoxyanthocyanins such as apigeninidin. It is clear from the table that the two classes of pigments are readily distinguished; there are many differences in colour reactions which are not listed (see e.g. Reznik, 1957). Electrophoresis is probably the most important method for confirming the presence of a betacyanin in a plant tissue. Although the visible spectra are superficially similar, there are differences in detail (e.g. in the effect of solvent on visible maxima, in the position and intensity of the ultra-violet bands, etc.).

The presence of betacyanins instead of anthocyanins in all but two of the families in the order Centrospermae (*sensu* Engler and Prantl) and in Cactaceae (included in the order by Wettstein, 1935) (*see* Table 4·2), and their apparent absence from all other angiosperm orders, provides one of the most striking

TABLE 4·2. Plant Genera Known to Have Betacyanin Pigments

Family	Genus[1] (No. of species in parenthesis when more than one examined)
Amaranthaceae	*Achyranthes, Aerva, Alternanthera* (6), *Amaranthus* (13), *Celosia* (4), *Iresine* (2), *Froelichia, Gomphrena* (4), *Mogiphanes, Tidestromia*
Azoiceae	*Conophytum* (17), *Fenestraria, Gibbaeum* (2), *Lampranthus* (2), *Lithops, Mesembryanthemum* (4), *Pleiospilos* (2), *Sesuvium, Tetragonia, Trichodiadema* (2), *Malephora, Dorotheanthus, Trianthema*
Basellaceae	*Anredera, Basella* (2)
Cactaceae	*Ariocarpus, Aylostera, Cereus* (3), *Chamaecereus, Cleistocactus, Gymnocalycium* (3), *Hariota, Hylocereus, Lobivia* (2), *Mammillaria* (7), *Melocactus, Monvillea, Neoporteria, Notocactus* (2), *Nopalxochia, Opuntia* (6), *Parodia* (3), *Pereskia, Phyllocactus, Rebutia* (4), *Selinocereus, Thelocactus, Zygocactus*
Chenopodiaceae	*Atriplex* (5), *Beta, Chenopodium* (7), *Corispermum* (2), *Cycloloma, Kochia, Salicornia, Salsola, Spinacia, Suaeda* (2)
Didieraceae	*Alluandia* (6), *Alluandiopsis* (2), *Decarya, Didierea* (3)
Nyctaginaceae	*Abronia* (4), *Allionia, Boerhaavia* (6), *Bougainvillea* (3), *Cryptocarpus, Cyphomeris, Mirabilis* (2), *Nyctaginia, Oxybaphus*
Phytolaccaceae	*Petiveria, Phytolacca* (3), *Rivina* (2), *Stegnosperma,*[2] *Trichostigma*
Portulacaceae	*Anacampseros, Calandrinia, Claytonia* (3), *Montia, Portulaca* (3), *Spraguea, Talinum*

[1] Data from Dreiding (1961); Mabry *et al.* (1963); Rauh and Reznik (1961).
[2] Hutchinson (1959) elevates *Stegnosperma* to family rank (Stegnospermaceae).

correlations between chemistry and morphology known at the present time. In his recent revision of Engler (1964), Melchior appears to have accepted the chemical data in part. Thus, he places the Didieraceae as an appendage to the family, following Rauh and Reznik's discovery (1961) of betacyanins in the family. The Cactaceae is not united with the order but is now placed next to the Centrospermae in an order of its own. Finally, the Caryophyllaceae, the only family without betacyanins, is separated from the other families by being placed in a sub-order of its own. Thus, the chemical data suggest that the affinities of this family, which contain anthocyanins (Chapter 5 p. 127), are not what they seem and it is to be hoped that taxonomists will take this factor into account in future studies of this anomalous family (Table 4·3).

TABLE 4·3. The Order Centrospermae

According to chemistry	According to Engler (1936)	According to Engler (1964)[1]
Chenopodiaceae	Chenopodiaceae	Phytolaccaceae
Amarantaceae	Amarantaceae	Gyrostemonaceae[2]
Nyctaginaceae	Nyctaginaceae	Achatocarpaceae[2]
Phytolaccaceae	Cynocrambaceae	Nyctaginaceae
Aizoaceae	Phytolaccaceae	Molluginaceae[2]
Portulacaceae	Aizoaceae	Aizoaceae
Basellaceae	Portulacaceae	
Cactaceae	Basellaceae	Portulacaceae
Didieraceae	Caryophyllaceae	Basellaceae
	Opuntiales	
	Cactaceae	Caryophyllaceae
		Dysphaniaceae
		Chenopodiaceae
		Amarantaceae
		Appendage:
		Didiereaceae
		Cactales
		Cactaceae

[1] The order is now divided into four sub-groups (shown by horizontal lines).
[2] These families have been split off from the Aizoaceae or Portulacaceae.

6. Evolutionary and Systematic Aspects

Lawrence et al. (1939) concluded from their surveys that cyanidin is the most primitive anthocyanidin and that delphinidin and pelargonidin were derived biogenetically from it. Similarly, they considered 3-glucoside to be more primitive than 3,5-diglucoside. These conclusions were derived from three sources of evidence.

(a) *Organ distribution of aglycones.* Cyanidin occurs in 94% of young and autumnal leaves, in 80% of permanently pigmented leaves, in 69% of fruits, but in only 50% of flowers.

(b) *Aglycone distribution in Archichlamydeae compared with Sympetalae.* A comparison by genera shows that while pelargonidin occurs to the extent of 24% in both groups, cyanidin is present in 56% of the Archichlamydeae surveyed and in 45% of the Sympetalae. The figures for delphinidin are the reverse of those for cyanidin, 57% in the Sympetalae and 41% in the Archichlamydeae.

(c) *Glycoside distribution.* A very striking association between delphinidin and 3,5-dimonoside formation is apparent from a consideration of the occurrences of 3,5-dimonosides of the different anthocyanidins: pelargonidin 30%, cyanidin 43% and delphinidin 90%. This is presumably due to the operation of selection for the greater stability of 3,5-diglycosides over the simple 3-monosides. The distribution of cyanidin glycosides according to the part of the plant analysed also indicates that cyanidin 3-monoside is the primitive type. Thus, the distributions of 3-monoside and 3,5-dimonoside respectively in autumnal leaves are 83 and 5%, in fruits 50 and 17% and in flowers 26 and 43%.

Later work (cf. Harborne, 1963a) has largely confirmed the conclusions of Lawrence *et al.* (1939). Thus, acylated 3,5-diglucosides and 3-rutinoside-5-glucosides of methylated anthocyanidins (e.g. 6) have been isolated from many members of the Tubiflorae, while simple cyanidin (5) or delphinidin 3-glucosides have been found in many primitive angiosperm stocks. There are, of course, exceptions to every generalisation and examples will be found where this simple picture does not make sense. A most striking exception is the presence of monardein, an acylated pelargonidin 3,5-diglucoside known otherwise as a labiate pigment, in the fern *Davallia divaricata*. The occurrence of peonidin 3-rutinoside-5-glucoside in flowers of *Magnolia* (Magnoliaceae) is less exceptional, since although the Magnoliaceae was at one time reckoned to be one of the most primitive angiosperm taxa (cf. Hallier, 1912), the family has more recently been recognised as being more highly specialised (Smith, 1943).

Monardein, present in *Davallia divaricata*, is an exceptional fern pigment and all other ferns so far examined have 3-deoxyanthocyanins based on apigeninidin or luteolinidin (*see* Section III); in addition, the only moss anthocyanins identified are luteolinidin derivatives. There is, thus, a suggestion that 3-deoxyanthocyanidins are more primitive than cyanidin, and the ability to add the 3-hydroxyl group to the pyrylium nucleus arose during the transition in plants from an asexual to a sexual breeding system. The need to attract insects and birds to the flowers for cross-pollination purposes is obviously the main raison d'être for anthocyanin synthesis in plants. Selection for colour has operated in two directions: towards blue colours in temperate climates for bee pollinators and towards red colours in tropical climates for bird pollinators. It is interesting that selection for red colour has led to the production of 3-desoxyanthocyanins, only known otherwise in the ferns, in the highly evolved sympetalous family, the Gesneriaceae. Thus, plants, as a result of evolutionary forces, have returned full circle to making the simplest pigments of all.

From the systematic point of view, the anthocyanins show considerable promise as being useful taxonomic markers at the generic and familial level. The glycosidic type is perhaps more useful than the anthocyanidin type, partly because it is less affected by genetic factors. Detailed surveys of the glycosidic patterns present in at least twenty genera have been carried out and the conclusion can be drawn that species within the same genus generally have the same glycosidic type. Thus, all species of *Malus* examined have 3-galactoside, of *Fritillaria* 3-rutinoside, of *Rosa* 3,5-diglucoside, of *Solanum* 3-rutinoside-5-glucoside and so on. Genera having more than one glycosidic type present are mostly those which contain an unusually large number of species (e.g. *Rhododendron* and *Begonia*) and which might be further divided when taxonomically revised. In other genera having more than one glycosidic type, anthocyanin distribution appears to follow natural divisions. For example, New World and Old World *Antirrhinum* species have different pigments (delphinidin 3,5-diglucoside and cyanidin 3-rutinoside, respectively) and blue-flowered *Streptocarpus* species all have malvidin 3-rutinoside-5-glucoside, whereas red-flowered species (e.g. *S. dunnii*) have cyanidin 3-sambubioside.

There are signs, as more extensive surveys are carried out within limited groups, that anthocyanins may be of value for supporting tribal and sub-family divisions. Thus, the tribe Vicieae in the sub-family Papilionatae of the Leguminosae, differs from the other nine tribes in the sub-family by having anthocyanidin 3-rhamnoside-5-glucosides instead of 3-glucosides and 3,5-diglucosides. In the Plumbaginaceae, one tribe, the Plumbagineae, has unusual methylated anthocyanidins which are absent from the other, the Staticeae. Again, in the Gesneriaceae, the New World sub-family has 3-deoxyanthocyanins, which have not been found in any of the Old World group. These and other examples are discussed in more detail in Chapters 6 and 7.

B. DISTRIBUTION OF FLAVONOLS

A leaf survey of 1000 angiosperms has shown that kaempferol, quercetin and myricetin occur singly or together in nearly three-quarters of them (Table 4·4). When this result is considered in conjunction with the fact that the surveys were limited to detecting only the common flavonols, the conclusion can be drawn that all higher plants have the ability to synthesise flavonols (or flavones) in their leaves. Surveys of flower petals (*see* Table 4·4) indicate a similar state of affairs and plant tissues completely lacking flavonol or flavone are obviously rare. Even cells in tissue culture synthesise flavones; *Happlo-pappus gracilis* material has yielded these substances in this laboratory.

Although flavonols are so widespread in higher plants, their distribution nevertheless definitely follows certain evolutionary trends. In particular, flavonol formation is very closely associated with lignification in plants (Bate-Smith, 1962) and flavonols, like leucoanthocyanins, are found more frequently (in the ratio 3:2, *see* Table 4·4) in woody than in herbaceous dicotyledons.

TABLE 4·4. Distribution of the Common Flavonols and Related Compounds in the Angiosperms

Flavonoid	% occurrence in leaf[1]		Total
	Monocots	Dicots	
Leucocyanidin	27	37	35
Leucodelphinidin	9	19	19
Total leucos	35	61 (woody) 19 (herb.)	40
Kaempferol	26	52	48
Quercetin	28	62	56
Myricetin	4	11	10
Total flavonols	41	90 (woody) 58 (herb.)	72
	% occurrence in flowers[2]		
	White petals	Cyanic petals	
Kaempferol	86	42	—
Quercetin	17	68	—

[1] Data from Swain and Bate-Smith (1962).
[2] Data from Reznik (1956) and Roller (1956).

The relatively low flavonol percentage in monocotyledons again reflects the same fact, i.e. that monocotyledons are predominantly herbaceous in character and contain relatively few lignified members. Although flavones have not been so widely surveyed as flavonols, there are clear indications from studies of the more advanced taxa (e.g. the Tubiflorae) of a correlation between absence of flavonol and presence of flavone, i.e. that replacement of flavonol by flavone is an evolutionary advance occurring with the changeover from woody to herbaceous character.

Turning to the hydroxylation pattern of flavonols, the main point about distribution here is that while quercetin and kaempferol are very common, myricetin (7) is definitely less frequent. The indication, in the case of its occurrence in leaf tissue, is that myricetin is a primitive character which is lost during evolutionary advancement. Its presence in only 4% of the mono-cotyledons as compared to 11% of the dicotyledons supports the view. It is clear from detailed studies of the flavonols in families such as the Iridaceae, Plumbaginaceae etc. that, with leucodelphinidin, it represents a primitive synthetic mechanism.

(7) Myricetin   (8) Isorhamnetin

The distribution of myricetin in flowers is rather different; it is found mostly in association with anthocyanins and is formed probably as a by-product of delphinidin synthesis. Since delphinidin is an advanced character in flowers, it is not surprising that myricetin is produced in flowers of some highly evolved families.

Only one flavonol methyl ether can be considered common in the angiosperms; this is the 3'-methyl ether of quercetin, isorhamnetin (8), corresponding in structure to peonidin. There are, however, a large number of methylated flavonols of limited occurrence. Their known distribution is so erratic [consider 5-O-methylquercetin, azaleatin, which is known both in the Archichlamydeae (Juglandiaceae, Eucryphiaceae) and in the Sympetalae (Ericaceae, Plumbaginaceae)] that little can be made of their systematic importance. Most of the other known flavonols, of which there are a considerable number, are closely related to one or other of the three common flavonols, differing only by the gain (or loss) of one or more nuclear substituents. They are all mentioned in the following three chapters.

Glycosidic variation in the flavonol series is even greater than with the anthocyanins. This means, inevitably, that it is difficult to perceive evolutionary trends in flavonol glycosylation. 3-Glycosides, 3-rhamnosides and 3-rutinosides are extremely common in all the main groups of plants (see Tables 5·2, 6·2, 7·2). Complex glycosides, i.e. those with three or more sugars attached and those with acylated sugars, are rarer and occur mainly in the more specialised families. One might assume they represent an evolutionary advanced character, except that a flavonol with as many as five sugars has been found in the very primitive horsetail (see Section III.C). Many of the rarer glycosidic types are of interest within the plant groups that contain them.

Plants have not been extensively surveyed for flavonoid types other than the anthocyanins and flavonols. This is either because simple and reliable methods of detection are not available (e.g. isoflavones) or because they are obviously of limited distribution (e.g. the yellow aurones). The flavones, apigenin and luteolin, are probably widely distributed; they have been detected with particular frequency in the Gramineae (as C-glycosyl derivatives), in the Compositae and families of the Tubiflorae. Simple methyl ethers of these two flavones are also present in a number of plants but the trihydroxy derivative corresponding to myricetin, i.e. 5,7,3′,4′,5′-pentahydroxyflavone has never been reported; the 3′,5′-dimethyl ether, tricin, is however a known, albeit rare, plant constituent.

Flavanones appear to be of erratic occurrence (*see* under Rutaceae, Rosaceae and Ericaceae in Chapters 5 and 6) but this may be simply because they are not readily detected. In many cases where single plant species have been exhaustively investigated (e.g. for genetic reasons as in the case of *Antirrhinum majus*), flavanone types corresponding to kaempferol and quercetin, i.e. naringenin and eriodictyol, have been obtained. The same is true, to some extent, of flavanonols which may be on the direct biosynthetic pathway to flavonols and anthocyanins. Flavanonols, apart from the occasional report of them in flowers of plants such as *Primula sinensis* and *Petunia*, are otherwise known mainly as heartwood constituents (*see* under Rosaceae). Isoflavones again are recorded almost entirely in the Leguminosae but may well be found in other plant groups, when micro-methods for their detection become available. Finally, the chalcones and aurones, because of their unmistakable colour reactions, are not easy to overlook; they almost certainly only occur in the Compositae as regular petal constituents, being otherwise of limited but erratic distribution.

III. Flavonoids in Lower Plants

A. BACTERIA, ALGAE AND FUNGI

The situation here may be summarised succinctly by stating that there are no satisfactory records of flavonoid pigments occurring in any of these three plant orders. Various reports have appeared from time to time of anthocyanins in bacteria and fungi but in no case has the relevant pigment been compared directly with authentic flavylium salts. In addition, phenazines and quinones, pigments known to occur in these organisms, are similar in colour and solubility to anthocyanins. Thus, early reports of anthocyanin-like red pigments in bacteria were dismissed by Erikson *et al.* (1938) on the grounds that they were almost certainly phenazines.

In the case of the algae, anthocyanin-like colours seem to be due to iron–tannin complexes. Alston (1958), in following up a report of anthocyanins in several genera of the Zygnemataceae by Mainx (1923), investigated the purple vacuolar pigments of *Zygogonium ericetorum* but failed to find any flavylium salt. The purple pigment was very similar in its response to a series of colour tests to a synthetic complex of iron and gallic acid. It is interesting in this connection that two groups of workers have recently reported tannin-like substances in algae.

Craigie and McLachan (1964), in investigating a yellow pigment excreted by *Fucus vesiculosus* and seven other Phaeophyta, found material with strong u.v. absorption at 275 mμ, which gave phloroglucinol on alkaline degradation. They suggested that their pigment was a flavonol or tannin, but did not have any evidence proving it was of flavonoid structure. Sieburth and Conover (1965), who examined an antibiotic which retards fouling on ships and which is produced by *Sargassum*, identified a tannin, by solubility, astringency and ability to precipitate gelatine. In absence of further evidence, this would seem to be gallic acid derivative, and not a flavan-3,4-diol, since gallotannin has been characterised in another alga, *Spirogyra arcta*, by Nakabayashi, 1957.

The reports (e.g. Moewus, 1950) of flavonoids in the alga, *Chlamydomonas eugematos*, are now known to be spurious (cf. Ryan, 1955) and do not require further comment here.

Turning to the fungi, two recent reports of anthocyanin pigments are worth mentioning. Petersen *et al.* (1961) found an anthocyanin-like pigment in a fungus, which had λ_{max} 280 and 500 mμ and melted at 280–300°, with decomposition. More recently, Avadhani and Lim (1964) reported an anthocyanidin in an unidentified fungus collected from an orchid. It had λ_{max} 225, 280 and 510 mμ and was recorded as having an R_f of 0·70 in Forestal solvent. Although both pigments superficially appear to be anthocyanin-like on the available evidence, they could well be anthraquinones. Since no anthocyanin has yet been isolated from a fungus, stronger corroboration than this is required before the presence of flavylium salts in the above fungi can be accepted.

B. Mosses

The discovery of luteolinidin in the red-coloured moss genus *Bryum* by Bendz *et al.* in 1962 constituted the first indisputable record of flavonoids in mosses. Bendz *et al.* (*see also* Bendz and Martensson, 1961) isolated and identified the 5-glucoside (9) and 5-diglucoside of luteolinidin in *Bryum cryophyllum*, *B. rutilans* and *B. weigelii*. Other red-pigmented mosses are

(9) Luteolinidin 5-glucoside   (10) Saponarin

known but no other anthocyanins have yet been fully identified. Work, however, has been going on for some time on the red coloration in the cell wall of *Sphagnum* species. Paul, as long ago as 1908, observed anthocyanin-like pigments in *S. magellanicum*, *S. rubellum* and *S. nemoreum* and in members of the Hepaticae. Recently, Rudolph (1965) described his work with these rather intractable substances. After difficult extraction and separation, he was able to note the properties of two very similar pigments in *S. magellanicum* and

S. rubellum. That in the first species, for example, has very low chromatographic mobility (R_f 0·06 in BAW, 0·04 in BuHCl and 0·33 in PhOH), λ_{max} at 288 and 546 mμ and does not give any sugar on hydrolysis. As Rudolph (1965) points out, neither pigment corresponds at all closely with any of the common anthocyanins or their glycosides. The properties recorded for the *Sphagnum* pigments do suggest that they might be highly hydroxylated 3-deoxyanthocyanidins so further work may provide another flavonoid record in the mosses; at present the possibility that these pigments are of a different structure (e.g. extended quinones of some sort) cannot be ruled out.

Flavones almost certainly occur in mosses, but the evidence for their presence needs substantiation. The early report by Molisch (1911) of the glycoflavone saponarin (10), in the moss *Madotheca platyphylla*, was followed up in this laboratory and in a survey of some 30 mosses, glycoflavones were detected in *Mnium cuspidatum, M. undulatum*, in *Plagiochila asplenoides* and in a *Cleridium* species. That the substances are glycoflavones is based on the similarity of their spectral maxima and R_f values to those of vitexin and lutexin and their resistance to acid hydrolysis.

Similar studies by Melchert and Alston (1965) have shown that eight glycoflavones occur in *Mnium affine* and that a quercetin diglycoside is present in *M. arizonicum*. No other flavonoids have been reported in mosses and there is good reason for believing that leucoanthocyanidins are absent from the group; a survey of 33 species representing 31 genera gave uniformly negative results (Bendz *et al.*, 1966).

C. Ferns

By contrast with the situation in mosses, there is abundant evidence that most of the flavonoid types found in angiosperms are present in the Pteridophyta. Anthocyanins, flavonol and flavone glycosides, leucoanthocyanidins, flavanones and chalcones have been recorded, many of these types being of widespread distribution.

Red coloration is by no means common in ferns, but a number of species are so coloured in the fronds at an early stage of development; the intensity of pigmentation appears to depend on environmental factors. The first suggestion that these colours were anthocyanin in nature was made by Price *et al.* in 1938. These authors examined some eight species, finding in most of them new or unusual pigments, resembling 6-hydroxypelargonidin, or 6-hydroxycyanidin in their colour properties. They did find ordinary pelargonidin and cyanidin

(11) Gesnerin   (12) Monardein

in one species, *Davallia divaricata*, and ordinary pigments were later reported by Hayashi and Abe (1955) in a second, *Dryopteris erythrosora*. Re-examination in this laboratory (Harborne, 1965d, 1966c) of most of the species examined by Price *et al.* and of several others established that the major anthocyanidin types in these ferns are the 3-deoxy compounds, apigeninidin and luteolinidin. The 5-glucosides (9, 11) of one or both of these aglycones were isolated from *Adiantum pedatum*, *A. veitchianum*, *Blechnum brasiliense* var. *corcovadense*, *Pteris longipinnula*, *P. quadriaurita* and *P. vittata*; unidentified glycosides of apigeninidin and luteolinidin occurred in several of these species and also in *Osmunda regalis* (for their R_f values, see Table 1·6). Re-examination of *Davallia divaricata* confirmed that ordinary anthocyanins were present and led to the identification of the main pigment as monardein (12), pelargonidin 3-(p-coumaroylglucoside)-5-glucoside. Neither fronds nor sori of *Dryopteris erythrosora* contained ordinary anthocyanins (cf. Hayashi and Abe, 1955) but unidentified glycosides of apigeninidin, luteolinidin and (possibly) tricetinidin were present in both tissues. At least three new glycosides of luteolinidin were noted in ferns (besides the known 5-glucoside) but none corresponded with the 5-diglucoside reported in the moss *Bryum* by Bendz *et al.* (1962).

Surveys have shown that luteolin, apigenin, kaempferol and quercetin are widespread in ferns, one or other of these compounds having been detected in acid-hydrolysed extracts of over 20 fern genera (Harada and Saiki, 1955; cf. Lee *et al.*, 1961). Leucodelphinidin, leucocyanidin or leucopelargonidin have been found in most ferns that have been examined for leucoanthocyanidins (Bate-Smith, 1954; Harborne, 1966c). Relatively little work has been done on the flavonol and flavone glycosides of ferns but most of those that have been isolated are types which are extremely common in the angiosperms. Thus, the 3-rhamnosides of kaempferol and quercetin occur in *Dicranopteris dichotoma* (Kishimoto, 1956), the 3-glucosides in several *Cyrtomium* species (Kishimoto, 1956), and the 3-rutinosides in bracken *Pteridium aquilinum* (Nakabayashi, 1956), and *Dryopteris oligophlebia* (Kobayashi and Hayashi, 1952). A more unusual type, kaempferol 3-arabinosylglucoside, is reported in *Phegopteris polypodioides*, which also contains the 4'-glucoside of apigenin 7-methyl ether (Ueno, 1963). Finally, vitexin, *C*-glucosylapigenin, has been isolated from *Cyathea* and *Sphenomeris* spp. (Ueno *et al.*, 1963).

More complex flavonol glucosides have been found, however, in the horsetails, Equisetaceae. Nakamura and Hukuti (1940) isolated the 7-diglucoside and 3-glucoside of kaempferol and the 7-glucoside of luteolin from the common horsetail, *Equisetum arvense*, and more recently, Beckmann and Geiger (1963) obtained kaempferol 3-rutinoside-7-glucoside and a kaempferol 3,7-diglycoside with five sugars attached from the marsh horsetail, *E. palustre*. It remains to be seen whether there is a real difference between the glycosidic pattern in the two species, since the pigments found by the Japanese workers are partial hydrolysis products of those reported by the Germans. The above work was carried out on the barren green stems of the horsetail plants; the pale brown spore-laden stems apparently contain an as yet unidentified hexahydroxyflavone (Sosa, 1949).

Turning finally to the chalcones and flavanones of the ferns, seven compounds have been isolated, all of which are structurally and phylogenetically interesting. 2',6'-Dihydroxy-4'-methoxychalcone and 2',6'-dihydroxy-4,4'-dimethoxychalcone (13) were identified by Nilsson (1961) as the yellow pigments in the sori underneath the fronds of the fern *Pityrogramma chrysophylla*; the related dihydrochalcone (14) was found in the same source. These

(13) (14)

chalcones are interesting chemically because they are the only known phloroglucinol-derived substances in which the 2'- and 6'- hydroxy groups are free and which remain in the chalcone form; the 2',4',6',4-tetrahydroxy analogue cyclises spontaneously to the flavanone form. They are interesting phylogenetically because they are biogenetically simple flavonoid types and their presence in ferns confirms the supposition that chalcones are primitive characters in plants. The same may be said of the dihydrochalcone (14) and of the four flavanones known in ferns. Two of the flavanones, matteucinol (15) and demethoxymatteucinol (no B-ring substituent), occur in leaf and rhizome of *Matteuccia orientalis* (Fujise, 1929), the other two, farrerol (16) and cyrtominetin (the 3'-hydroxy analogue of farrerol), in the leaf of *Cyrtomium*

(15) Matteucinol (16) Farrerol

falcatum (Kishimoto, 1956). All four flavanones bear *C*-methyl substituents in both the 6- and 8- position; such substances are only known otherwise in *Pinus* (Pinaceae) and in *Rhododendron* (Ericaceae), their occurrence here thus providing links between the ferns and both gymnosperms and angiosperms.

IV. Flavonoids in Gymnosperms

A. GENERAL PATTERN

The Gymnosperms comprise some 700 plants, divided into four orders and about 20 families, and, considering that it is a relatively small group, it has been fairly extensively surveyed for flavonoids. Trees are ideal plants for chemists interested in isolating large quantities of natural products to work with and it is not surprising to find that all the economically important gymnosperms, and especially the pines which have been the special study of H. Erdtman (1956), have been examined for their phenolic constituents.

The range of flavonoid types found in the gymnosperms is generally similar to that in ferns. Thus, the occurrence of flavonols, flavones, flavanones and leucoanthocyanidins is well recorded and recently a glycoflavone has been found in *Ephedra*. Another link between the ferns and the gymnosperms is *C*-methylation, since flavanones with this modification are known in both groups. A link with the angiosperms rather than with the Pteridophyta is indicated by the presence of isoflavones in *Podocarpus* (Podocarpaceae).

Chalcones are not known in the gymnosperms but anthocyanins are certainly present. Cyanidin 3-glucoside was reported in *Picea obovata* cones by Beale *et al.* (1941), who used the old colour tests, and recently the same pigment and delphinidin 3-glucoside have been detected by modern methods in the conelets of 35 species from six genera of the Pinaceae (Santamour, 1966). Although anthocyanin is responsible for cone colours, reddish bark colours in the Gymnospermae are probably quinonoid in nature.

The main chemical feature distinguishing the gymnosperms from both lower plants and the angiosperms is biflavonyl formation. Biflavonyls based on amentoflavone (17), formed from two molecules of apigenin by carbon–carbon coupling at the 8- and 3'-positions, are widely distributed in gymnosperms and have been detected in all but two families (the Pinaceae and the Ephedraceae) that have been surveyed (Sawada, 1958).

Most known biflavonyls are simple methylated derivatives of amento-flavone; as they have been given tongue-twisting names, they are listed for ease of consulation below:

(17) Amentoflavone

(18) 7″-Methyl ether: sotetsuflavone
(19) 7,4′-Dimethyl ether: ginkgetin
(20) 4′,4‴-Dimethyl ether: isoginkgetin
(21) 7,4′,4‴-Trimethyl ether: sciadopitysin
(22) 7″,4′,4‴-Trimethyl ether: kayaflavone
(23) 7,4′,7″,4‴-Tetramethyl ether: unnamed

Even assuming that the 5-hydroxyl groups are not likely to be methylated, there are still a considerable number of amentoflavone methyl ethers which may be awaiting discovery in these trees.

Two other types of biflavonyl are known; hinokiflavone (24), formed from two apigenins by carbon–oxygen coupling at the 8- and 4'-positions and cupressoflavone (25), formed by carbon–carbon coupling at the 8- and 8'-

(24) Hinokiflavone   (25) Cupressoflavone

positions. Obviously, biflavonyls with yet different linkages may well be found, as well as dimers based on flavones other than apigenin.

Taxonomically, the widespread occurrence of biflavonyls in the gymnosperms and their almost complete absence from angiosperms is of considerable interest. The fact that biflavonyls have been found twice in angiosperms, hinokiflavone in *Casuarina stricta* (Casuarinaceae) and amentoflavone in *Viburnum prunifolium* (Caprifoliaceae), does not invalidate the value of these substances as phylogenetic markers. Indeed, *Casuarina* is recognised to be a primitive relic of an earlier phase of angiosperm evolution. The discovery of biflavonyl in the relatively highly advanced *Viburnum* perhaps only emphasises our general ignorance of the origins of most angiosperm families. *Viburnum* is at least one of the most woody and hence least specialised members of the Caprifoliaceae.

The most striking fact about the distribution of biflavonyls within the gymnosperms is their absence from the Pinaceae; at least 44 species representing nine genera of the family have been searched for them (The apparent absence of biflavonyls from the Gnetales has little significance since only one species of *Ephedra* has been looked at so far). Their absence from the pines does lend support to the views of those who regard the Pinaceae as being systematically rather isolated from the other conifers. Biflavonyls are, in effect, replaced in the Pinaceae by a group of relatively unusual flavones and flavanones not known in other Gymnosperms but whether this means that the Pinaceae is more advanced than other members remains a matter of speculation.

In drawing up the more detailed account of gymnosperm flavonoids which is by family, much use has been made of Hegnauer's (1962) account of the subject and reference has been made on biflavonyls to the review of Baker and

Ollis (1961). Most biflavonyl records are due to Sawada (1958) and this may be assumed below unless another reference is given.

B. CYCADACEAE AND GINKGOACEAE

Sotetsuflavone (18) has been isolated from *Cycas revoluta*. Ginkgetin (19) and isoginkgetin (20), the only two biflavonyl dimethyl ethers known, have been obtained from the yellow autumnal leaves of *Ginkgo biloba* (Ginkgoaceae) (Baker and Ollis, 1961). The green leaves of this primitive monotypic tree contain kaempferol, quercetin and isorhamnetin 3-rutinosides (Geiger and Beckmann, 1965; Fisel, 1965), and biflavonyl formation in this plant appears to occur only in the dying leaves.

C. TAXACEAE

Sciadopitysin (21) occurs in four *Taxus* species, including the yew *Taxus baccata* (di Modica *et al.*, 1959) and kayaflavone (22) in *Torreya nucifera*. Kayaflavone is also present in two *Cephalotaxus* spp. (Cephalotaxaceae).

The pale yellow yew pollen contains a complex mixture of flavonols and flavones, according to investigations in this laboratory. The following pigments have been identified: quercetin 3-rhamnoside, myricetin 3-rhamnoside, myricetin 3-rutinoside, apigenin 7,4'-diglucoside and luteolin 7-diglucoside.

D. PODOCARPACEAE

Kayaflavone has been found in *Podocarpus macrophylla*, *P. chinensis* and *P. nagi*, and the unnamed amentoflavone 7,4',7',4'''-tetramethyl ether occurs in *Dacrydium cupressinum* (Hodges, 1965). *Podocarpus spicatus* wood has yielded quercetin, kaempferol and their dihydro derivatives and also genistein and podospicatin (Briggs *et al.*, 1959; Brewerton, 1958). The latter is an isoflavone (26), with a quite unusual substitution pattern. The presence of

(26) Podospicatin

genistein and podospicatin, isoflavones normally associated with the Leguminosae, in the Podocarpaceae is quite remarkable.

E. PINACEAE

The absence of biflavonyls from the Pinaceae (44 species of 9 genera surveyed) has already been remarked upon. A good deal is known about the flavonoids that are present and it is convenient to consider the heartwood, bark, leaf and pollen constituents separately.

The extractives of *Pinus* heartwoods have been extensively examined by Erdtman, who collected together the results of a lifetime's work in his review of 1956. *Pinus* heartwoods, besides having a range of stilbenes and terpenoids, contain a rich array of flavones and flavanones; other genera in the family (e.g. *Pseudotsuga* and *Larix*) are much simpler, having just dihydroquercetin and dihydrokaempferol. In *Pinus*, there are a series of related flavones, flavanones and flavanonols. The flavones are chrysin (27), its 7-methyl ether (28) and 6-C-methyl derivative (29); the flavanones pinocembrin (30), related methyl compounds (31) and (32) and the 8-methyl isomer of (32), crypto-

Flavones: (27) Chrysin   (28) Tectochrysin   (29) Strobochrysin
Flavanones: (30) Pinocembrin   (31) Pinostrobin   (32) Strobopinin
Flavanonols: (33) Pinobanksin   (34)   (35) Strobobanksin

strobin; the flavanonols, pinobanksin (33) and related methyl compounds (34) and (35). Two flavonols, galangin (36) and its 7-methyl ether, and the chalcone (37) have also been isolated from the same materials (Mahesh and Seshadri, 1954).

(36) Galangin   (37)

The distribution of flavones and flavanones in over half the known *Pinus* species (approx. 90) has been studied by Erdtman and the results show a very close correlation between chemistry and morphology in that the patterns of substances in the two well-recognised sub-genera, the Haploxylon and the Diploxylon, are quite different. All 17 species of the Haploxylon contain chrysin, pinocembrin and pinobanksin and the related 7-methyl ethers (28), (31) and (34); species in the subsections Strobi and Gerardianae generally have the 6- and 8-C-methylflavanones as well. By contrast, the 35 species of the Diploxylon contain only the flavanones, pinocembrin and pinobanksin. Erdtman (1956) concluded that the Diploxylon were advanced over the Haploxylon, since as a sub-genus they have lost the ability to oxidise flavanones to flavanonols or flavones. A few species in the Gerardianae (Haploxylon) were

5

close to the Diploxylon in their flavonoid pattern and appeared to represent a bridge between the two main groups of *Pinus* taxa.

The most interesting bark constituents are pinoquercetin (38) and pino-myricetin, the 6-*C*-methyl derivatives of two of the common flavonols which occur in *Pinus ponderosa* (Venkataraman, 1956). These compounds show an

(38) Pinoquercetin   (39)

obvious relationship with the *C*-methylflavones and -flavanones of the heart-wood, as does (39), a new dihydroflavonol from the stem bark of the deodar cedar, *Cedrus deodara* (also Pinaceae) (Adinarayana and Seshadri, 1965). Other pine barks contain, not unexpectedly, myricetin and dihydromyricetin, e.g. *Pinus contorta* (Hergert, 1956), and they are known to be a rich source of condensed tannins.

Little is known of pine leaf constituents, except that quercetin, dihydro-quercetin, kaempferol and myricetin are present variously in bound form in most *Pinus* species (Takahashi *et al.*, 1960). The 3-rhamnosides of kaempferol, quercetin and myricetin, together with the xylosides of the corresponding di-hydro derivatives, have been isolated from the leaf of *Chaemocyparis obtusa* (Pinaceae) by Yasue and Hasegawa (1962). Strohl and Seikel (1965) have examined the pollen of eight pines and found dihydrokaempferol, dihydro-quercetin and naringenin in them. Hisamichi (1961) recorded isorhamnetin and quercetin in the pollen of *P. densiflora* and *P. thunbergii*.

F. Taxodiaceae

Sciadopitysin, kayaflavone and sotetsuflavone have been found in five species, representing *Cryptomeria, Cunninghamia* and *Sequoia*. The most characteristic biflavonyl of the Taxodiaceae is hinokiflavone which occurs in all nine species so far examined. It is present in *Cryptomeria japonica* mainly as the 4‴-methyl ether and a dimethyl ether has also been detected (Kawano *et al.*, 1964). The absence of hinokiflavone from the Umbrella Pine, *Sciadopitys verticillata*, which has sciadopitysin, appears to confirm the separation of the Sciadopityaceae from the Taxodiaceae, to which it was once attached.

Leaves of the Taxodiaceae also contain common flavonol glycosides: quercetin and isorhamnetin 3-arabinosides are in *Taxodium distichum* and *Glyptostrobus pensilis*, quercetin 7-glucoside is in *Cryptomeria japonica* and quercetin 3-rhamnoside in *Metasequoia glyptostroboides* (Takahashi *et al.*, 1960; Kondo and Ito, 1956).

G. Cupressaceae

All 21 species that have been examined have hinokiflavone. Six *Sabina* and two *Juniperis* species have, in addition, kayaflavone and it is interesting that there has been a move afoot to separate off these two genera into their own family, the Juniperaceae. *Thujopsis dolobrata* contains sciadopitysin and sotetsoflavone and *Chaemocyparis obtusa* sotetsuflavone. *Cupressus torulosa* and *C. sempervirens* contain cupressoflavone (25) (Murti *et al.*, 1964), a biflavonyl which appears to characterise the genus.

The common flavonols and dihydroquercetin occur widely in the leaves of the Cupressaceae (Takahashi *et al.*, 1960). Quercetin 3-rhamnoside, for example, has been identified in *Chaemocyparis pisifera* and *Thuja occidentalis*.

V. Flavonoids in Animals

Phenolic compounds are, in general, rare in animals and it seems most unlikely that the animal body has the ability to synthesise the flavonoid molecule. Flavones and anthocyanins have been reported as occurring in a number of insects but their accumulation in body tissue is probably due to a failure to metabolise substances absorbed in the food rather than to any other cause. Flavonoids are well known to be rapidly metabolised in mammals (rabbits, rats, man), being broken to smaller fragments. Quercetin, for example, yields *m*-hydroxyphenylacetic acid, excreted in the urine, and carbon dioxide (*see* Williams, 1964). The only case of a flavonoid not being completely degraded is that of the isoflavone, genistein (40), a constituent of clover, which when fed to animals is converted to equol (41) (Cayen *et al.*, 1964) which can be recovered in the urine. This substance equol was for a long time thought to be

(40) Genistein   (41) Equol

a "natural constituent" of mare's urine, instead of a metabolic product.

Reports of flavonoids occurring in insects have appeared regularly since the turn of the century but since nearly all the observations were made by entomologists and were based on simple but ambiguous colour tests, their authenticity in many cases is unfortunately open to question. A major difficulty is the collection of enough insects to work with and only in recent years have chemists applied themselves to this problem and has the presence of flavones in butterfly wings at last been firmly established.

The credit for the work on butterfly flavones must go to Morris and Thomson (1963), who were able to isolate enough material from the wings of 400 Marble

White butterflies, *Melanargea galathea*, to enable them to identify the major flavones present. The pigments present are tricin (42), free and in combined form, and a glycoside of orientin (43). Flavonols are definitely absent thus proving that an earlier report of quercetin (Thomson, 1926) in *M. galathea* is incorrect. The discovery of tricin and lutexin in this butterfly essentially

(42) Tricin   (43) Orientin

proves the point that the compounds are obtained by the insect, through the caterpillar feeding on a particular food host. The larvae of the Marbled White are known to feed on *Dactylis glomerata, Poa annua* and *Festuca ovina*, just the plants which are known to contain tricin and orientin in considerable quantity. Quercetin, although the most common plant flavone generally, is very rare in grasses and so would not be expected to appear in butterflies which depend mainly on the grasses for their food.

In a parallel study on 600 Small Heath butterfly wings, *Coenonympha pamphilus*, Morris and Thomson (1964) found tricin, both free and as glycoside, and traces of other flavonoids. Orientin is absent, which is surprising since the Small Heath larvae feed on the same grass species as the Marbled White. It is probable that orientin, having a catechol nucleus in its structure, is more readily metabolised by insects than tricin; this could explain its absence from the above species.

The Small Heath and Marbled White are by no means the only butterflies reported to have flavones in their wings since Ford (1941), in a survey based on simple colour tests, found them in 38 of 328 butterfly genera and in 10 of 192 moth genera. Ford's test consists of fuming the white areas of the wings with ammonia, a yellow coloration produced indicating flavone; in some cases, he extracted the wings and tested the wings with alkali. Ford showed that the presence of wing flavone is a valuable systematic character, since its distribution closely follows the accepted morphological classification.

Ford surveyed the families Pieridae and Papilionidae in some detail. In the Pieridae, flavones are rare except in the aberrant South American sub-family Dismorphiinae (present in 12/50 species of *Dismorphia* and in all three species of *Pseudopieris*). Flavones also occur in all three known species of the palaearctic "Wood Whites" (*Leptideo*) which have been united with the Dismorphiinae on structural grounds. Outside the Dismorphiinae, flavones only occur in 2 of 233 species tested (*Eronia cleodora* and *Gandacaparina*), indicating their close association with just the one sub-family.

In the Papilionidae, flavones are present in 4 out of 16 genera. They are found in *Polydorus* but only in some American species. They occur in 82 of 92 species of *Graphium* being replaced in the 10 other species by a fluorescent yellow pigment. They are present in both species of *Lamproptera*, confirming the suggested affinity between this genus and *Graphium* and finally, they are found throughout *Parnassius*. Ford concludes his paper by saying that "in general the occurrence of anthoxanthins within the Lepidoptera supports the present classification of the order upon evidence wholly distinct from that on which it has been constructed."

Clearly, more chemical studies are urgently needed on flavones in butterflies but even more so in the case of moths, since the presence of flavones here still requires full confirmation. It is true that a "quercetin-like" substance has been isolated from the caterpillar of the silkworm *Bombyx mori* (Hayashiya, 1959), but the pigment has not yet been fully identified; quercetin does in fact occur in mulberry leaves, the silkworm's food, as the 3-glucoside. Interestingly enough, Hamamura *et al.* (1962) recognise quercetin 3-glucoside as being the "biting factor" for silkworm larvae, and is one reason (but not the only one) for their preferential selection of these leaves.

There are many other, as yet unconfirmed, reports of flavonoids in animals. The edible snail, for example, has a yellow compound in the digestive gland which has the properties of a flavonoid (Kubista, 1950) and a similar compound has been reported in the hydroid polyps of the family Sertulariidae (Payne, 1931). It is worth noting here that the yellow pigment in the chlorogosomes of the common earthworm, *Lumbricus terrestris*, which has the superficial appearance of a flavonoid (Roots and Johnston, 1966) is in fact riboflavin (Needham, 1966). Anthocyanins are reputed to occur in caterpillars of pugmoths, *Eupithecia oblongata*; those feeding on *Scabious* flowers are bluish and those on *Cirsium* pink (Habich, 1891). Anthocyanins may also be responsible for the brown-violet colour of larvae of the weevil, *Cionus aleus*, which feed on the coloured staminal hairs of *Verbascum nigrum* (Hollande, 1913). They do not, however, occur as has been claimed by Palmer and Knight (1924) as the red pigments of *Dactynotus* (Aphididae) species; quinones related to the aphin pigments are, in fact, present (Weiss and Altland, 1965).

CHAPTER 5

FLAVONOIDS OF THE DICOTYLEDONS
THE ARCHICHLAMYDEAE

I. Introduction

The Dicotyledons are usually divided by taxonomists into the Archichlamydeae and the Sympetalae (or the Metachlamydeae), the former containing the less specialised plant groups and the latter the more specialised. Because of the large number of flavonoids that have been found in the Dicotyledons, it is convenient to consider the results of the surveys under the two headings. This has the added advantage that when considering the phylogenetic development of flavonoids (*see* Chapter 4) it is possible to compare and contrast the flavonoid patterns in the two groups.

The Archichlamydeae contain a very large number of families and orders and include many plants of economic importance. Many individual genera have been studied in detail for this reason (e.g. *Vitis*, *Pisum*, *Camellia* and *Malus*) and extensive surveys have been carried out, particularly in families of the Rosales. A general summary of the flavonoids of the Archichlamydeae will be followed by brief accounts of compounds in those families in which systematic surveys have been reported.

All the anthocyanins that have been fully identified are listed in Table 5·1. All six of the common anthocyanidins are regularly represented but other aglycones are rare; apigeninidin has been detected in *Chiranthodendron* (Sterculiaceae), tricetinidin in *Camellia* (Theaceae) and aurantinidin in *Impatiens* (Balsaminaceae). 3-Glucosides and 3,5-diglucosides are very common glycosidic types, but 3-galactoside replaces 3-glucoside in *Fagus, Polygonum, Lythrum, Theobroma* and in *Malus* and *Pyrus* (both Rosaceae). The rare 3-arabinoside accompanies the 3-galactoside in *Theobroma*. 3-Rhamnoside replaces 3-glucoside only in plants of the tribe Vicieae of the Leguminosae (*see* Section VI of this chapter) and in *Ceanothus* (Rhamnaceae). Anthocyanins with unusual branched trisaccharides attached are confined to the Rosaceae, Grossulariaceae and Begoniaceae. The only other rare glycosidic type is pelargonidin 3-sophoroside-7-glucoside, which is found only in *Papaver*. Acylated anthocyanins occur with particular abundance in only one family, the Cruciferae, but are also reported in *Vitis, Viola* and *Tibouchina*.

TABLE 5·1. Distribution of Anthocyanins in the Archichlamydeae

Order, family and species	Pigments present	Organ examined	References
ALICALES			
alicaceae			
Salix fragilis	Cy 3-glucoside	Leaf gall	Blunden and Challen, 1965.
AGALES			
agaceae			
Fagus sylvatica	Pg and Cy 3-galactosides	Leaf	Harborne and Sherratt, 1957.
RTICALES			
oraceae			
Morus nigra	Cy 3-glucoside	Fruit	Harborne, 1963b.
OLYGONALES			
olygonaceae			
Polygonum hydropiper	Cy 3-galactoside	Seedling	Sugano and Hayashi, 1960.
P. persicaria	Cy 3-rutinoside	Petal	J. B. Harborne, un-published.
Rheum rhaponticum	Cy 3-glucoside Cy 3-rutinoside	Petiole	Gallop, 1965.
ENTROSPERMAE			
ryophyllaceae			
Dianthus caryophyllus	Pg and Cy 3-glucosides Pg and Cy 3,5-diglucosides	Petal	Ootani and Miura, 1961.

TABLE 5·1—*continued*

Order, family and species	Pigments present	Organ examined	References
MAGNOLIALES			
Calycanthaceae			
Calycanthus fertilis	Cy 3-glucoside	Flower	Hayashi and Nogud 1952.
Magnoliaceae			
Magnolia spp.	Cy and Pn 3-glucosides Cy and Pn 3-rutinosides Pn 3,5-diglucoside and Pn 3-rutinoside-5-glucoside	Anther and petal	Francis and Harbor 1966; Santamoua 1965.
RANUNCULALES			
Nymphaceae			
Nymphaea cv.	Dp 3-glucoside	Petal	J. B. Harborne, published.
Ranunculaceae			
Anemone coronaria	Pg 3-lathyroside	Scarlet petal	J. B. Harborne, published.
	Pg, Cy and Mv 3-glucosides	Pollen	Tappi and Monza 1955.
Delphinium ajacis	Dp 3,5-diglucoside		Harborne, 1964a.
GUTTIFERALES			
Paeoniaceae			
Paeonia spp.	Cy and Pn 3,5-diglucosides	Petal	Beale *et al.*, 1941.
Theaceae			
Camellia sinensis	Cy 3-glucoside Tricetinidin	Leaf	Roberts and Willia 1958; Chandra, published.
Eurya japonica	Cy 3-glucoside	Fruit	Shibata *et al.*, 1962.
SARRACENIALES			
Droseraceae			
Dionaea muscipula	Cy 3-glucoside	Mature leaf gland	Dipalma, 1965.
PAPAVERALES			
Cruciferae			
Brassica oleracea var. *rubra*	Cy 3-(p-coumaroylsophoro-side)-5-glucoside, Cy 3-(feruloylsophoroside)-5-glucoside, Cy 3-(diferuloyl-sophoroside)-5-glucoside	Leaf	Harborne, 1962e; Stroh, 1959.
Matthiola incana	Pg 3-glucoside Pg 3-(p-coumaroylferuloyl-sambubioside)-5-glucoside	Petal (orange)	Harborne, 1964a.
Raphanus caudatus	Mv 3,5-diglucoside	Pod	Lele, 1959.
R. sativus	Pg and Cy 3-(p-coumaroyl-sophoroside)-5-glucosides Pg and Cy 3-(feruloyl-sophoroside)-5-glucosides	Pod, root and flower	Harborne, 1964a, 19 Ishikura and Ha ashi, 1962.

Order, family and species	Pigments present	Organ examined	References
PAVERALES—*continued*			
ⱣAPAVERACEAE			
Papaver spp.	Pg 3-sophoroside-7-gluco-side Pg and Cy 3-sophorosides	Petal	J. B. Harborne, unpublished.
	Cy 3-glucoside	Petal blotch	J. B. Harborne, unpublished.
ⱣSALES			
ⱣSSULACEAE			
Kalanchoe blossfeldiana	Cy 3-glucoside Cy and Pg 3,5-diglucosides	Flower	Neyland *et al.*, 1963.
	Cy 3-glucoside Cy 3,5-diglucoside	Leaf	Neyland *et al.*, 1963
ⱣUMINOSAE			
Cicer pinnatifolia	Dp, Pt and Mv 3-rhamnoside-5-glucosides	Petal	J. B. Harborne, unpublished.
Glycine maxima	Cy 3-glucoside	Seedcoat	Kuroda and Wada, 1935.
Lathyrus odoratus	Pg, Cy, Pn, Dp and Pt 3-rhamnosides, Pg, Cy, Pn, Dp, Pt and Mv 3-rhamno-side-5-glucosides, Pg, Cy and Pn 3-galactosides, Pg and Pn 3-galactoside-5-glucosides, Pg, Cy and Pn 3-lathyrosides	Petal	Harborne, 1960a, 1963b.
L. sativus	Mv 3-rhamnoside	Petal	Harborne, 1960a.
Lathyrus spp.	Dp, Pt and Mv 3-rhamno-side-5-glucosides	Petal	Harborne, 1963b.
Lespedeza thunbergii	Mv 3,5-diglucoside	Flower	Hayashi *et al.*, 1955.
Lupinus polyphemus	Pg 3,5-diglucoside	Flower	Bayer, 1959.
Medicago sativa	Dp, Pt and Mv 3,5-digluco-sides	Petal	Cooper and Elliott, 1964.
Ononis spinosa	Dp 3-glucoside	Petal	J. B. Harborne, unpublished.
Phaseolus multiflorus cv. "Scarlet Runner"	Pg 3-sophoroside	Scarlet petal	Harborne, 1963a.
cv. "Blue Coco"	Mv 3,5-diglucoside	Pod	J. B. Harborne, unpublished.
P. vulgaris	Pg, Cy, Dp, Pt and Mv 3-glucosides Pg, Cy and Dp 3,5-digluco-sides	Seedcoat	Feenstra, 1960.
Pisum sativum	Pg, Cy, Pn, Dp, Pt and Mv 3-rhamnoside-5-glu-cosides, Dp 3-glucoside, Dp 3,5-diglucoside	Petal	Dodds and Harborne, 1964.
	Cy 3-sambubioside-5-gluco-side, Cy 3-sophoroside-5-glucoside	Pod	J. B. Harborne, unpublished.

TABLE 5·1—*continued*

Order, family and species	Pigments present	Organ examined	References
ROSALES—*continued*			
Leguminosae—*continued*			
Tamarindus indica	Cy 3-glucoside		Lewis and Johar, 1
Vicia spp.	Dp, Pt and Mv 3-rhamno-side-5-glucosides	Petal	J. B. Harborne, published.
Wistaria sp.	Dp 3,5-diglucoside		Harborne, 1963a.
Rosaceae			
Fragaria vesca *Fragaria × ananassa*	} Pg and Cy 3-glucosides	Fruit	Sondheimer and I ash, 1956.
Malus pumila	Cy 3-galactoside	Fruit	Sando, 1937.
Prunus spp.	Cy 3-glucoside Cy 3-rutinoside Cy 3-sophoroside Cy 3-(2G-glucosylrutinoside)	Fruit	Li and Wagenknee 1956, 1958; Har borne and Hall, 1964b.
Pyrus communis	Cy 3-galactoside	Bark	Harborne, 1963a.
Rosa spp.	Pg, Cy and Pn 3-glucosides Pg, Cy and Pn 3,5-digluco-sides	Petal and leaf	Harborne, 1961.
Rubus spp.	Pg and Cy 3-glucosides Pg and Cy 3-rutinosides Pg and Cy 3-sophorosides Pg and Cy 3-(2G-glucosyl-rutinosides) Cy 3-sambubioside Cy 3-(2G-xylosylrutinoside)	Fruit	Harborne and H 1964b.
Saxifragaceae			
Bergenia crassifolia	Pn 3,5-diglucoside Cy 3-glucoside		J. B. Harborne, published.
Hydrangea macrophylla	Dp 3-glucoside	Petal	Asen *et al.*, 1957.
Ribes spp.	Cy 3-(2G-glucosylrutinoside) Cy 3-(2G-xylosylrutinoside) Cy 3-sambubioside Cy 3-sophoroside Cy and Dp 3-rutinosides Cy and Dp 3-glucosides	Fruit	Harborne and H 1964b; Chandler a Harper, 1962.
GERANIALES			
Euphorbiaceae			
Daphniphyllum macropodum	Dp 3-sambubioside	Pericarp	Shibata and Ishiku 1964.
Poinsettia pulcherrima	Pg and Cy 3-glucosides Pg and Cy 3-rutinosides	Bract	Asen, 1958.
Geraniaceae			
Pelargonium spp.	Pg, Pn and Mv 3,5-digluco-sides	Petal	Harborne, 1963a.
Tropeolaceae			
Tropeolum majus	Pg 3-sophoroside	Petal	Harborne, 1962.
RUTALES			
Rutaceae			
Citrus sinensis	Cy and Dp 3-glucosides	Fruit	Chandler, 1958.

TABLE 5·1—*continued*

Order, family and species	Pigments present	Organ examined	References
PINDALES			
raceae			
cer pseudoplatinus	Cy 3-glucoside	Tissue culture	J. B. Harborne, unpublished.
cer spp.	Cy 3-glucoside	Leaf	Hattori and Hayashi, 1937.
saminaceae			
mpatiens aurantiaca	Aurantinidin glucosides	Flower	Clevenger, 1964.
. balsamina	Pg and Pn 3,5-diglucosides	Flower	Hayashi *et al.*, 1953.
LASTRALES			
uifoliaceae			
lex crenata	Cy 3-sambubioside	Fruit	Hayashi, 1942.
AMNALES			
amnaceae			
Ceanothus cv. "Autumn Blue"	Mv 3-rhamnoside	Flower	Harborne, 1963a.
aceae			
Parthenocissus tricuspidata	Cy and Mv 3,5-diglucosides Mv 3-glucoside	Callus tissue	Stanko and Bardinskaya, 1962.
Vitis vinifera	Cy, Pn, Dp, Pt, and Mv 3-glucosides		Ribereau-Gayon, 1959.
	Dp, Pt and Mv 3-(*p*-coumaroylglucosides), Pn, Dp, Pt and Mv 3,5-diglucosides Dp, Pt and Mv 3-(*p*-coumaroylglucoside)-5-glucosides Pn, Dp, Pt and Mv 3-(caffeoylglucosides)	Fruit	Albach *et al.*, 1965.
ALVALES			
lvaceae			
Abutilon insigne	Cy 3-rutinoside	Flower	J. B. Harborne, unpublished.
Hibiscus manihot	Cy and Dp 3-glucosides	Flower	Kuwada, 1964.
H. rosa-sinensis	Cy 3-sophoroside	Flower	Hayashi, 1944.
Malva silvestris	Mv 3,5-diglucoside	Flower	Willstäter and Mieg, 1915.
rculiaceae			
Chiranthodendron pentadactylon	Apigeninidin glucoside	Calyx	Pallares and Garza, 1949.
Theobroma cacao	Cy 3-arabinoside Cy 3-galactoside	Pod	Forsyth and Quesnel, 1957.
OLALES			
goniaceae			
Begonia spp.	Pg and Cy 3-sophorosides Cy 3-sambubioside Cy 3-xylosylrutinoside Cy 3-glucosylrutinoside Cy 3-glucoside	Leaf and flower	Harborne and Hall, 1964b.

TABLE 5·1—*continued*

Order, family and species	Pigments present	Organ examined	References
VIOLALES—*continued*			
Passifloraceae			
Passiflora edulis	Pg 3-glucosylglucoside	Rind	Pruthi *et al.*, 1961.
	Dp 3-glucoside	Fruit	J. B. Harborne, published.
Violaceae			
Viola × *wittrockiana*	Cy and Dp 3-(*p*-coumaroyl-rutinoside)-5-glucosides	Petal	Endo, 1959.
MYRTIFLORAE			
Lythraceae			
Cuphea ignea	Pg and Mv 3,5-diglucosides	Flower	Harborne, 1963a.
Lythrum salicaria	Mv 3,5-diglucoside	Flower	Harborne, 1963a.
	Cy 3-galactoside		Paris and Paris, 196
Melastomaceae			
Tibouchina semidecandra	Mv 3-(*p*-coumaroylgluco-side)-5-glucoside	Flower	Harborne, 1964a.
Myrtaceae			
Eucalyptus sieberiana	Cy and Dp 3-glucosides	Leaf and young bark	Hillis, 1956.
Metrosideros excelsa	Dp and Mv 3-glucosides	Flower	Cambie and Seelye 1961.
Onagraceae			
Clarkia elegans	Mv 3,5-diglucoside	Flower	Harborne, 1963a.
Epilobium hirsutum	Mv 3,5-diglucoside	Flower	J. B. Harborne, published.
Fuschia cvs.	Pn and Mv 3,5-diglucosides	Flower	Harborne, 1963a.
Godetia amoena	Cy 3-glucoside	Flower	J. B. Harborne, published.
	Mv 3,5-diglucoside		
G. cylindrica	Mv 3,5-diglucoside	Flower	J. B. Harborne, published.
G. purpurea	Mv 3,5-diglucoside	Flower	J. B. Harborne, published.
Oenothera odorata	Cy 3-glucoside	Flower	Wada, 1950.
Punicaceae			
Punica granatum	Pg 3,5-diglucoside	Petal	Harborne, 1962e.
	Dp 3,5-diglucoside	Fruit juice	Harborne, 1962e.

The complete absence of anthocyanins (and their replacement by beta-cyanins) from all but one of the families of the order Centrospermae is of considerable taxonomic interest and has already been commented on (Chapter 4). One may note here that ordinary anthocyanins have been found in at least four genera of the aberrant family, the Caryophyllaceae; in *Dianthus, Lychnis, Saponaria* and *Silene* (Lawrence *et al.*, 1939). The pigments of *Dianthus* have been studied in some detail, particularly in relation to their inheritance (Geissman and Mehlquist, 1947; Ootani and Miura, 1961). It is significant that other flavonoids, i.e. flavonol glycosides and isoflavones, occur regularly in the Centrospermae; their structures are however unusual in many respects, the most regular constituents being the isorhamnetin glycosides that have been found in *Opuntia lindheimeri* (Rosler et al., 1966) (Table 5·2). The isoflavone tlatlancuayin (1) in *Iresine celosioïdes* (Amarantaceae) (Crabbé et al., 1958) has a most remarkable structure. Other unusual methylated derivatives are the 7,4'-dimethyl ether of quercetin in *Phytolacca dioicca* (Phytolaccaceae) (Marini-Bettolo et al., 1950) and the 6-monomethyl ether and the 6,3'-dimethyl ether (2) of quercetagetin present in spinach leaf, *Spinacia oleracea* (Chenopodiaceae) (Zane and Wender, 1961).

(1) Tlatlancuayin

(2) Spinacetin

All the flavonol glycosides that have been fully identified in the Archichlamydeae are listed in Table 5·2. The glycosylation patterns are very variable, and many types are apparently confined to single genera (e.g. the 3-digalactoside to *Betula*) or to single species (e.g. 3-lathyroside-7-rhamnoside to *Lathyrus odoratus*). Other types have scattered occurrences. Thus, 3-glucuronides are found in *Populus* (Salicaceae), *Phaseolus* (Leguminosae) and *Vitis* (Vitaceae) and 3-xylosides in *Polygonum* (Polygonaceae), *Begonia* (Begoniaceae) and *Malus* (Rosaceae). Yet other types are found abundantly in a few families. Flavonol 3,7-diglycosides, for example, occur very frequently in the Ranunculaceae and in parts of the Leguminosae but not often elsewhere. Acylated flavonol glycosides are of infrequent occurrence. They have been reported in *Helleborus* (Ranunculaceae), *Pisum* (Leguminosae) and *Rosa* (Rosaceae) and in three trees, the lime *Tilia*, the plane *Platanus* and the beech *Fagus*.

TABLE 5·2. Flavonol Glycosides of the Archichlamydeae

der, family and species	Flavonols present	Organ examined	References
JGLANDALES			
glandaceae			
Carya pecan	Azaleatin	Heartwood	Sasaki and Mikami, 1963.
	Caryatin		
Engelhardtia formosana	Qu and Km 3-rhamnosides	Bark	Tominaga, 1956.
Juglans regia	Km 3-arabinoside	Leaf	Nakaoki and Morita, 1958;
	Qu 3-galactoside		Herrmann, 1955.

TABLE 5·2—*continued*

Order, family and species	Flavonols present	Organ examined	References
JUGLANDALES—*continued*			
Myricaceae			
Myrica gale } *M. nagi* } *M. rubra* }	My 3-rhamnoside	Bark or leaf	Charaux, 1924.
SALICALES			
Salicaceae			
Populus grandidentata	Qu 3-glucuronide	Leaf	Pearl and Darling, 1963
Salix triandra	Qu-3-rutinoside	Leaf	Rabate, 1928.
FAGALES			
Betulaceae			
Alnus cordata	Qu 3-sophoroside	Pollen	Sosa and Percheron, 19
Betula humilis	Qu 3-rutinoside	Leaf	Hörhammer, Wagner
B. pubescens	Qu 3-galactoside		Luck, 1957; Caspari
	My 3-galactosylgalactoside		al., 1946.
B. verrucosa	Qu 3-galactoside		
	My 3-galactosylgalactoside		
Corylus avellana	My 3-rhamnoside	Leaf	Collot and Charaux, 193
Fagaceae			
Fagus sylvatica	Qu and Km 3-glucosides	Leaf	Dietrichs and Schaich,
	Km 3-glucosylglucoside		1963.
Quercus digitata } *Q. tinctoria* } *Q. trifida* }	Qu 3-rhamnoside	Bark	Perkin and Everest, 191
URTICALES			
Moraceae			
Artocarpus integrifolia	Morin	Heartwood	Dave et al., 1962.
Cannabis indica	My 3'-glucoside		Rao and Seshadri, 1941.
Ficus carica	Qu 3-rutinoside	Leaf	Nakaoki et al., 1957.
Humulus lupulus	Qu and Km 3-glucosides		Bhandari, 1964.
	Qu 3-rhamnoside		
	Qu and Km 3-rutinosides		
M. pomifera	Morin	Wood	Barnes and Gerber, 195
Morus alba	Qu 3-glucoside	Leaf	Oku, 1934
M. bambycis } *M. tinctoria* }	Morin	Wood	Wagner, 1850.
Ulmaceae			
Zelcova serrata	6-C-glucosyl- 7-methylkaempferol	Wood	Nishida and Funacka, 19 Hillis and Horn, 1966.
Urticaceae			
Boehmeria spp.	Qu 3-rutinoside	Leaf	Nakaoki et al., 1957.
PROTEALES			
Proteaceae			
Protea concinnum	Qu 3-rutinoside	Leaf	Rapson, 1938.
SANTALALES			
Loranthaceae			
Loranthus parasiticus	Qu 3-arabinoside		Tseng and Chen, 1961.

TABLE 5·2—*continued*

der, family and species	Flavonols present	Organ examined	References
NTALALES—*continued*			
ntalacaceae			
Exocarpus cuppressiformis	Qu and Km 3-dirhamnosides Km 7-rhamnoside DihydroKm 7-rhamnoside	Leaf	Cooke and Haynes, 1960.
Osyris compressa	Qu 3-rutinoside	Leaf	Perkin, 1910.
)LYGONALES			
lygonaceae			
Fagopyrum esculentum	Qu 3-rutinoside	Leaf	Schunck, 1858.
Polygonum aviculare	Qu 3-arabinoside	Leaf	Ohta, 1940.
P. hydropiper	Qu 3-rutinoside Isorhamnetin-KSO₃	Leaf	Valentine and Wagner, 1953.
P. polystachum	Qu 3-arabinoside	Leaf	Hörhammer et al., 1955.
P. reynoutria	Qu 3-xyloside	Leaf	Nakaoki and Morita, 1956b.
Rheum spp.	Qu 3-rutinoside	Leaf and flower	Hörhammer, and Muller, 1954.
Rumex acetosa	Qu 3-galactoside	Fruit	Hörhammer and Votz, 1955; Volkonskaya and Minaeva, 1964.
NTROSPERMAE			
ryophyllaceae			
Herniaria glabra	Qu 3-rutinoside Qu 3-triglucoside Isorhamnetin 3-rutinoside Isorhamnetin 3-rutinosyl-glucoside	Leaf	Hörhammer et al., 1960.
ɲytolaccaceae			
Phytolacca decandra	Qu and Km 3-glucoside	Leaf	Ohta and Miyazaki, 1956.
ACTALES			
ɩctaceae			
Cereus grandiflorus	Isorhamnetin 3-glucoside		Hörhammer et al., 1966.
Opuntia dillenii	Qu 3-glucoside	Flower	Nair and Subramanian, 1964.
O. ficus-indica	Isorhamnetin	Flower	Arcoleo, 1962.
O. lindheimeri	Qu and isorhamnetin 3-galactosides Isorhamnetin 3-rutinoside Isorhamnetin 3-rutinosyl-galactoside	Flower	Rosler et al., 1966.
AGNOLIALES			
ɩlycanthaceae			
Chimonanthus praecox	Qu 3-diglucoside	Flower	Hayashi and Ouchi, 1946.
ɩuraceae			
Neolitsea sericea	Qu 3-rhamnoside	Young leaf	Nakabayashi, 1953.
ɩagnoliaceae			
Illicium anisatum	Qu 3-rhamnoside	Flower and bark	Nakabayashi, 1952.

TABLE 5·2—*continued*

Order, family and species	Flavonols present	Organ examined	References
MAGNOLIALES—*continued*			
Magnoliaceae—*continued*			
Magnolia obovata	Qu 3-rutinoside	Leaf	Nakaoki *et al.*, 1956.
Magnolia spp.	Qu 3-glucoside	Petal	Francis and Harborne,
	Qu and Km 3-rutinosides		1966a.
Monimiaceae			
Peumus boldus	Isorhamnetin 3-glucoside-7-rhamnoside	Leaf	Krug and Borkowski, 19∎
	Isorhamnetin 3-dirhamnoside		
	Km 3-glucoside-7-rhamnoside		
RANUNCULALES			
Ranunculaceae			
Anemone alpina	Km 3-rutinoside-7-glucoside		
	Km 3-rutinoside	Petal	Egger and Keil, 1965.
	Km 3-glucoside		
	Km 3-glucoside-7-glucoside		
Caltha palustris	Qu 3-galactoside	Petal	
	Qu 3-galactoside-7-rhamnoside		Egger and Keil, 1965.
	Qu 3-galactoside-7-xyloside	Stamen	
Helleborus foetidus	Qu 3-(caffeoylsophoroside)-7-glucoside	Petal	Harborne, 1965a.
	Qu 3-(xylosylglucoside)-7-glucoside		
H. niger	Km 3-sambubioside-7-glucoside		
	Km 3-sambubioside	Petal	
	Km 3-glucoside-7-glucoside		Egger and Keil, 1965.
	Km 3-sophoroside	Stamen	
PIPERALES			
Saururaceae			
Houttuynia cordata	Qu 3-glucoside	Leaf	Nakamura *et al.*, 1936.
	Qu 3-rhamnoside		
GUTTIFERALES			
Eucryphiaceae			
Eucryphia glutinosa	Qu 3-galactoside		
	Azaleatin 3-galactoside	Leaf	Bate-Smith *et al.*, 1966.
	Azaleatin 3-arabinosylgalactoside		
	Caryatin		
Guttiferae			
Hypericum perforatum	Qu 3-galactoside	Leaf	Jerzmanowska, 1937.
	Qu 3-glucoside		
Paeoniaceae			
Paeonia albiflora	Km 3-glucoside	Petal	Egger, 1961b.
	Km 3,7-diglucoside		Egger and Keil, 1965.
P. arborea	Km 3,7-diglucoside	Petal	Egger and Keil, 1965.

TABLE 5·2—*continued*

der, family and species	Flavonols present	Organ examined	References

TTIFERALES—*continued*

eaceae
Camellia sinensis   My, Qu and Km 3-glucosides ... Takino *et al.*, 1965; Roberts,
My, Qu and Km 3-rutino- ... 1962; Oshima and Naka-
sides ... bayashi, 1954.
Qu and Km 3-rhamnosyl- Leaf
glucosylglucosides
My and Km 3-galactosides
My and Qu 3-rhamnosides

PAVERALES

pparidaceae
Cleome chelidonii ⎫
C. spinosa ⎭   Qu 3-rutinoside   Petal   Subramanian, 1963; Roch-leder and Hlasiwetz, 1852.

uciferae

Brassica campestris	Qu 3-rutinoside	Seed	Francois and Chaix, 1961.
B. napus	Isorhamnetin 3-glucoside		Hörhammer *et al.*, 1966.
Bunias orientalis	Qu 3-rutinoside	Root	Jermstad and Jensen, 1951.
Capsella bursa-pastoris	Qu 3-rutinoside	Whole plant	Stepien and Krug, 1965.
Cheiranthus cheiri	Robinin	Fruit	Maksyntina, 1965.
	Qu 3-rhamnosylarabinoside	Fruit	Maksyntina, 1965.
	Isorhamnetin (as glycoside)	Petal	Perkin and Hummel, 1896.
Diplotaxis tenuifolia	Isorhamnetin (as glycoside)	Flower	Pacheco, 1955.
Matthiola incana	Isorhamnetin 3,4-′digluco-side	Seed	Rahman and Khan, 1962.
	Km 3-rhamnosylarabino-side-7-rhamnoside	Petal	Harborne, 1965a.
Sinapis alba	Km 3-glucoside	Leaf	Paris and Charles, 1962.
S. arvense	Isorhamnetin 3-glucoside		Hörhammer *et al.*, 1966.

oringaceae
Moringa oleifera ⎫
M. pterygosperma ⎭   Rhamnetin glycoside   Flower   Nair and Subramanian, 1962.

paveraceae

Argemone mexicana	Isorhamnetin 3-glucoside	Flower	Rahman and Ilyas, 1962.
	Isorhamnetin 3,7-digluco-side		
Eschscholtzia californica	Qu 3-rutinoside	Flower	Sando and Bartlett, 1920.
Papaver somniferum	Qu 3-glucoside	Flower	Sosa and Sosa, 1966.
	Qu 3-gentiobioside		

esedaceae
Reseda luteola   Km 3-glucoside-7-rhamno-side ... Flower ... Rzadkowska and Bodalski, 1965.
Isorhamnetin

OSALES

assulaceae
Bryophyllum daigremontianum   Km 3-(*p*-coumaroylarabin-oside)   Leaf   Karsten, 1965.

TABLE 5·2—*continued*

Order, family and species	Flavonols present	Organ examined	References
ROSALES—*continued*			
Hamamelidaceae			
Corylopsis spp.			
Distilium racemosum	My, Qu and Km 3-rhamnosides		
Hamamalis spp.		Leaf	Egger and Reznik, 1961.
Liquidambar styracifolia	My, Qu and Km 3-glucosides		
Parrotia persica			
Fothergilla monticola	Qu 3-rutinoside	Leaf	Hörhammer and Griesinge 1959.
Hamamalis japonica	Qu 4'-glucoside	Petal	Hörhammer and Griesinge 1959.
Loropetalum chinensis	Qu 3-glucoside	Petal	Nakabayashi, 1952.
Leguminosae			
Acacia cyanophylla	Qu 3-glucoside	Leaf	Paris, 1953
A. dealbata	My 3-glucoside	Pollen	Spada and Cameroni, 195
	Qu 3-rutinoside		Falco and der Vries, 1964
A. melanoxylon	Qu 3-galactoside	Flower	Nakabayashi, 1952.
Astragalus sinicus	Km 3-glucoside	Flower	
Afzelia spp.	Km 3-rhamnoside	Wood	King and Acheson, 1950.
Baptisia spp.	Km and Qu 3-glucosides		
	Km and Qu 3-rutinosides		
	Qu 7-glucoside	Leaf and	Alston *et al.*, 1965.
	Qu 3,7-diglucoside	flower	
	Qu 3-glucoside-7-rutinoside		
	Qu 3-rutinoside-7-glucoside		
	Qu 7-rutinoside		
Bauhinia reticulata	Qu 3-rhamnoside	Flower	Rabaté and Dussy, 1938.
B. tomentosa	Qu 3-glucoside	Flower	Row and Viswanadhar
	Qu 3-rutinoside		1954.
Cassia acutifolia	Isorhamnetin		Tutin, 1913.
C. angustifolia	Isorhamnetin		Tutin, 1913.
C. tora	Km 3-sophoroside	Leaf	Fukuchi and Imai, 1951.
Cercis canadensis	Qu 3-glucoside	Pod	Douglass *et al.*, 1949.
C. siliquastrum	My 3-rhamnoside	Leaf	Collot and Charaux, 1939.
Erythrophloeum africanum	Dihydro My	Bark	Hansel and Klaffenbac 1961.
Gleditschia monosperma	Robinetin	Heartwood	Brass and Krantz, 1932.
Indigofera arrecta	Km 3,7-dirhamnoside	Leaf	Hattori, 1951.
	Km 3-rhamnoside		
Lathyrus odoratus	Km 3,7-dirhamnoside	Flower	Harborne, 1960a.
	Km 3-lathyroside-7-rhamnoside		
	My, Qu and Km 3-rhamnosides		
L. vernus	Km 3-sophoroside-7-glucoside	Leaf	Harborne, 1965a.
Lespedeza cyrtobotrya	Km 3,7-dirhamnoside	Leaf	Hattori, 1951.
Lotus corniculatus	Km 3,7-dirhamnoside	Leaf	Nakaoki *et al.*, 1956.
Millettia stuhlmannii	Robinetin	Heartwood	Hawthorne and Morga 1962.

TABLE 5·2—*continued*

rder, family and species	Flavonols present	Organ examined	References
OSALES—*continued*			
eguminosae—*continued*			
Phaseolus vulgaris	My, Qu and Km 3-glucosides	Seedcoat	Feenstra, 1960.
	Km 3-xylosylglucoside		
	Qu and Km 3-glucuronides	Leaf	Harborne, 1965a.
	Qu and Km 3-rutinosides		
Pisum spp.	Qu and Km 3-glucosides	Leaf	Mumford *et al.*, 1961;
	Qu and Km 3-sophorosides		Furuya *et al.*, 1962; Har-
	Qu and Km 3-sophorotrio-		borne, 1963d.
	sides		
	Qu and Km 3-(*p*-coumaroyl		
	sophorosotriosides)		
	Qu and Km 3-(feruloylsoph-		
	orotriosides)		
Pueraria hirsuta	Robinin	Leaf	Nakaoki and Morita, 1956a.
Robinia pseudacacia	Robinin	Flower	Zemplen and Bognar, 1941.
Sophora japonica	Qu 3-rutinoside	Fruit	Suginome, 1959; Rabaté
	Km 3-sophoroside		and Dussy, 1938.
Trifolium pratense	Km 3-galactoside	Leaf	Hattori and Hasegawa, 1943.
T. repens	Qu 3-glucoside	Leaf	Hattori *et al.*, 1938.
Ulex europaeus	Qu 3,7-diglucoside		
	Qu 4'-glucoside	Flower	Harborne, 1962a.
	Qu 7-glucoside		
Vicia hirsuta	Qu 3-rhamnoside	Leaf	Nakaoki and Morita, 1956a.
Vigna angularis	Robinin	Leaf	Nakaoki and Morita, 1956.
latanaceae			
Platanus occidentalis	Km 3-(*p*-coumaroylgluco-side)	Leaf	Stambouli and Paris, 1961.
	Km 3-glucoside	Fruit and male flower	
	Qu 3-galactoside	Male flower	
	Qu 3-rutinoside	Leaf	
Rosaceae			
Amelanchier asiatica	Km 3-glucoside	Leaf	Matsuno and Shintano, 1962.
Crataegus curvisepala	Qu 3-galactoside	Leaf	Batyuk *et al.*, 1965.
C. pyracantha	Qu 3-rutinoside	Leaf	Paris and Etchepare, 1965.
Filipendula hexapetala	Qu 4'-glucoside	Petal	Casparis and Steinegger, 1945.
F. ulmaria	Qu 3-arabinoside	Leaf	Hörhammer *et al.*, 1956.
	Qu 3-galactoside		
Malus pumila	Qu 3-galactoside		
	Qu 3-rhamnoside		
	Qu 3-arabinoside	Fruit and bark	Siegelman, 1955; Williams, 1960.
	Qu 3-glucoside		
	Qu 3-xyloside		
Pourthiaea villosa	Km 3-glucoside	Flower	Matsuno *et al.*, 1963.
	Qu and Km 3-rutinosides		
Prunus cerasus	Km 3-glucoside	Leaf	Geissman, 1956.
P. emarginata	Qu 7-glucoside	Bark	Hergert, 1962.

TABLE 5·2—*continued*

Order, family and species	Flavonols present	Organ examined	References
ROSALES—*continued*			
Rosaceae—*continued*			
P. mume	Km 7-glucoside	Wood	Hasegawa, 1959.
P. persica	Km 3-glucoside	Leaf	Geissman, 1956.
	Km 3-galactoside	Flower	Ohta *et al.*, 1960.
P. salicina	Qu 3-rhamnoside	Leaf	Williams and Wender, 19
	Qu 3-arabinoside	Fruit	
P. serotina	Qu 7-glucoside	Bark	Pew, 1948.
P. spinosa	Km 3-rhamnoside-4′-arabinoside	Leaf	Hörhammer *et al.*, 195 Hörhammer and Wagn 1962.
	Qu and Km 3-arabinosides		
	Qu 3-rutinoside		
	Km 3-rhamnoside		
P. tomentosa	Qu 3-rhamnoside	Leaf	Wada, 1952.
P. virginiana	Qu 7-glucoside	Bark	Finnemore, 1910.
Pyrus communis	Qu and Km 3-glucosides	Leaf	Williams, 1960.
	Qu and Km 3-galactosides		
	Qu and Km 3-rutinosides		
	Qu and Km 3-arabinosides		
	Isorhamnetin 3-rhamnosyl galactoside	Peel	Nortje and Koeppen, 19
	Isorhamnetin 3-glucoside		
	Isorhamnetin 3-rhamnosyl-glucoside		
Rosa canina	Km 3-(*p*-coumaroylglucoside)	Fruit	Hörhammer *et al.*, 1961.
R. multiflora	Km 3-glucoside	Leaf	Aritomi, 1962.
R. rugosa	Qu 3-glucoside	Leaf	Noguchi, 1958.
Rosa spp. and cvs.	Qu and Km 4′-glucosides	Petal	Harborne, 1961.
	Qu and Km 3-rhamnosides		
	Qu and Km 3-sophorosides		
	Qu and Km 3-glucosides		
Rubus hirsutus	Km 3-galactoside		Aritomi, 1962.
	Km 3-rhamnoside		
Saxifragaceae			
Astilbe odontophylla	My 3-rhamnoside	Leaf	Ouchi, 1953.
Hydrangea macrophylla	Km 3-galactoside	Sepals	Sen *et al.*, 1957.
Ribes nigrum	Qu 3-glucoside	Fruit	Williams and Wender, 195
Ribes spp.	Qu 3-rhamnoside	Fruit	
GERANIALES			
Euphorbiaceae			
Aleurites cordata	Qu 3-rutinoside		
	Qu 3-rhamnoside	Leaf	Nakaoki *et al.*, 1957.
Sapium sebiferum	Qu 3-glucoside	Leaf	Shimokoriyama, 1949.
Tropaeolaceae			
Tropaeolum majus	Qu 3-glucoside	Leaf	Paris and Delaveau, 1961.
RUTALES			
Rutaceae			
Boenninghausenia albiflora	Qu 3-rutinoside	Leaf	Matsuno and Amano, 196
Ruta graveolens	Qu 3-rutinoside	Leaf	Weiss, 1842.

Order, family and species	Flavonols present	Organ examined	References
SAPINDALES			
Hippocastanaceae			
Aesculus hippocastanum	Km 3-arabinoside Km 3-rhamnoside Km 3-glucoside Qu and Km 3-rutinosides	Leaf and flower	Spiridonov *et al.*, 1964; Gehrmann *et al.*, 1955.
	Qu 3-sambubioside-3'-glucoside Qu 3,3'-diglucoside Qu 4'-glucoside Qu 3-rhamnoside	Seed	Wagner, 1964.
CELASTRALES			
Celastraceae			
Celastrus orbiculatus	Km 3,7-dirhamnoside Km 3-(*p*-coumaroylglucoside)	Leaf	Hattori, 1962.
RHAMNALES			
Rhamnaceae			
Rhamnus spp.	Rhamnetin 3-rhamninoside Qu 3-rhamninoside Rhamnazin 3-rhamninoside	Fruit and bark	Tanret and Tanret, 1899; Nystrom *et al.*, 1957.
Vitaceae			
Vitis vinifera	Qu 3-rhamnoside My, Qu and Km 3-glucosides Qu 3-glucuronide	Fruit	Ribereau-Gayon, 1964.
MALVALES			
Malvaceae			
Abutilon avicennae	Qu 3-rutinoside	Leaf	Nakaoki *et al.*, 1956.
Althaea rosea	Qu and Km 3-glucosides DihydroKm 3-glucoside	Flower	Obara, 1964.
Gossypium barbadense	Qu 3-sophoroside Qu 7-glucoside Qu 3-glucoside Qu and Km 3-rutinosides Km 3-galactoside Gossypetin 7-glucoside	Flower	Denliev *et al.*, 1963; Pakudina and Sodykov, 1963; Parks, 1965.
G. herbaceum	Qu and Km 7-glucosides Qu 3-glucoside Gossypetin 7-glucoside	Flower	Parks, 1965.
G. indicum	Qu 7-glucoside Gossypetin 8-glucoside		Neelakantan and Seshadri, 1939.
Gossypium spp.	Qu 3'-glucoside	Flower	Denliev *et al.*, 1963.
Hibiscus abelmoschus	My 3'-glucoside	Flower	Nair *et al.*, 1964.
H. cannabinus	Qu and Km 3-glucosides Qu and Km 3-rutinosides	Leaf	Schilcher, 1964.
	Hibiscetin 3-glucoside My 3'-glucoside	Flower	Neelakantan and Seshadri, 1939.

TABLE 5·2—*continued*

Order, family and species	Flavonols present	Organ examined	References
MALVALES—*continued*			
Malvaceae—*continued*			
H. mutabilis	Qu 7-glucoside	Flower	Subramanian and Swamy, 1964.
	Qu 3-sophoroside		
H. sabdariffa	Hibiscetin 3-glucoside	Flower	Rao and Seshadri, 1942.
H. tiliaceus	Gossypetin 7-glucoside	Flower ·	Nair et al., 1964.
	Gossypetin 3-glucoside		
Thespesia populnea	Km 7-glucoside	Flower	Neelakantan et al., 1943.
Sterculiaceae			
Firmiana platanifolia	Qu 3-rutinoside	Leaf	Nakaoki et al., 1957.
Guazuma tormentosa	Km 3-rhamnoside	Flower	Subramanian, 1963.
Tiliaceae			
Tilia argentea	Qu and Km 3-glucosides ⎫ Qu and Km 3-rhamnosides ⎪ Qu and Km 3-glucoside-7- rhamnosides, ⎬ Flower Qu 3-rhamnosylxyloside ⎪ Km 3,7-dirhamnoside ⎪ Km 3-(*p*-coumaroylgluco- ⎭ side)		Hörhammer et al., 1961; Hörhammer and Wagner 1962.
THYMELAEALES			
Elaeagnaceae			
Hippophae rhamnoides	Qu 3-rutinoside	Fruit	Bolley, 1860.
VIOLALES			
Begoniaceae			
Begonia spp.	Qu and Km 3-glucosides ⎫ Qu 3-rutinoside ⎬ Flower and Qu 3-xyloside ⎭ leaf		Harborne and Hall, 1964b.
Datiscaceae			
Datisca cannabina	Datiscetin rhamnoside		Charaux, 1925.
Violaceae			
Viola arvensis	Qu 3-rutinoside	Leaf	Kolos-Pethes, 1965.
V. × wittrockiana	Qu 3-rutinoside	Flower	Perkin, 1902.
	My and Qu 3-rutinoside-7- rhamnosides		Egger, 1961a.
CUCURBITALES			
Cucurbitaceae			
Trichosanthes cucumeroides	Km 3,7-dirhamnoside		Nakaoki and Morita, 1957.
MYRTIFLORAE			
Myrtaceae			
Eucalyptus macrorhynca	Qu 3-rutinoside	Leaf	Smith, 1898.
E. sideroxylon	Qu and Km 3-rhamnosides ⎫ Qu and Km 3-rutinosides ⎪ Qu 3-glucoside ⎬ Leaf DihydroKm 3-rhamnoside ⎭		Hillis and Isoi, 1965.

der, family and species	Flavonols present	Organ examined	References
YRTIFLORAE—*continued*			
rtaceae—continued			
Psidium guaijava	Qu 3-arabinoside	Leaf	El-Khadem and Mohammed, 1958.
MBELLIFLORAE			
rnaceae			
Cornus controversa	Qu 3-glucoside	Leaf	Nakaoki and Morita, 1958.
nbelliferae			
Bupleurum falcatum	Qu 3-rutinoside	Leaf	Rabaté, 1930.
B. multinerve	Qu 3-rutinoside		Minaeva and Volkhons-
	Isorhamnetin 3-rutinoside		kaya, 1964.
Daucus carota	Km 3-glucoside	Flower	Rahman *et al.*, 1963.
	Km 3-diglucoside		
Foeniculum vulgare	Qu 3-arabinoside	Leaf	Ohta and Miyazaki, 1959.
Oenanthe stolonifera	Qu 3-rutinoside	Flower	Matsushita and Iseda, 1965.
	Isorhamnetin-KSO₃		

Flavonols which differ from the common types by having extra O-methyl substituents or by having extra hydroxyl substituents are relatively rare and thus have considerable potentiality as taxonomic markers. They are found particularly in the Leguminosae and the Rutaceae, their occurrence otherwise being scattered. Little is known at present of their systematic distribution at the generic level. Of the known O-methyl ethers of kaempferol or quercetin, by far the commonest in the Archichlamydeae is quercetin 3'-methyl ether (isorhamnetin), the distribution of which is outlined in Table 5·2. Most of the other possible mono-methyl ethers of kaempferol and quercetin are known: the two 3-methyl ethers in *Begonia*, 5-O-methyl quercetin in *Carya pecan* bark (Juglandiaceae) and in *Eucryphia* (Eucryphiaceae), the two 7-O-methyl ethers in *Rhamnus* (Rhamnaceae) and the 4'-O-methyl ethers in *Prunus* (Rosaceae) and *Tamarix* (Tamaricaceae) respectively. No less than four dimethyl ethers are known in the Archichlamydeae: 3,5-di-O-methylquercetin (caryatin) is reported in *Carya pecan* and *Eucryphia glutinosa*, 7,3'-di-O-methylkaempferol in *Polygonum* (Polygonaceae) and *Rhamnus* species (Tatsuta, 1957), 3,7-di-O-methylkaempferol in *Beyeria* (Euphorbiaceae) (Jefferies and Payne, 1965) and 7,4'-di-O-methylquercetin in *Phytolacca*. One trimethyl ether of quercetin (the 3,7,4'-derivative) is known, in *Distemonanthus* (Leguminosae). The only methylated myricetin recorded is combretol, the 3,7,3',4',5'-pentamethyl ether, which occurs in the seeds of *Combretum quadrangulare* (Combretaceae) (Mongkolsuk *et al.*, 1966).

Flavonols with an extra hydroxyl function in the 6- or (less commonly) 8-position and their methyl ethers are found most abundantly in the Leguminosae (6 genera) and Rutaceae (4 genera). In the latter family, these flavonols usually have one, if not two, methylenedioxy groups attached. Miscellaneous occurrences of quercetagetin derivatives are *Papaver* (Papaveraceae), *Spinacia* (Chenopodiaceae) and *Chrysosplenium* (Saxifragaceae). The derivative in the latter genus is called chrysosplenetin and is the 6,7,3'-trimethyl ether of quercetagetin (3) (Nakaoki and Morita, 1956b). Mention should also be made of glycosides of gossypetin (4), which are found mainly in plants of the Malvaceae (*Gossypium*, *Hibiscus* and *Thespesia*) (Hattori, 1962). Three partly

(3) Chrysosplenetin   (4) Gossypetin

methylated derivatives of gossypetin have recently been discovered in the Euphorbiaceae, in *Ricinocarpus muricatus* and *R. stylosus* (Henrick and Jefferies, 1965); they are the 3,7-dimethyl ether, the 3,7,8-trimethyl ether and the 3,7,8,4'-tetramethyl ether.

There are fewer reports of flavones in the Archichlamydeae than of flavonols. The most remarkable point about the flavones that have been found is that they occur so frequently as *C*-glycosyl derivatives. There are, in fact, as many species with *C*-glycosylflavones in the Archichlamydeae as there are with *O*-glycosides (Table 5·3). Furthermore, *C*-glycosylation is a distinctive character of the Archichlamydeae since these substances are almost unknown in the Sympetalae (there are only three reports, in *Vitex*, in *Swertia* and in *Helichrysum*).

Flavone *O*-glycosides and *C*-glycosylflavones do not appear to co-occur with any frequency; the presence of both types in *Humulus japonicus* and *Spartium junceum* is exceptional. On the other hand, *C*-glycosylflavones do occur sometimes with *O*-substituents (e.g. in *Saponaria*). The second sugar in some glyco-flavones may be attached either to the *C*-glucosyl residue (as in *Adonis vernalis*) or directly to the A-ring to give a 6,8-di-*C*-glycosylflavone (as in *Psoralea*). Most variations in flavone glycosylation can be found in the Leguminosae; flavone 5- and 4'-*O*-glycosides are not unusual. The only other interesting family is the Umbelliferae, which has flavones present which contain the rare five-carbon sugar, apiose.

Many unusually substituted flavones are present in the Archichlamydeae. For example, the only known alkaloidal flavone occurs in the Moraceae, in *Ficus pantoniana*. Ficine is a chrysin derivative substituted in the 8-position by *N*-methylpyrrolidine (Johns *et al.*, 1965). Flavones which are

TABLE 5·3. Distribution of Flavone Glycosides in the Archichlamydeae

er, family and species	Pigments present[1]	Organ examined	References
LICALES			
icaceae			
alix bakko	Lu 7-glucosylarabinoside	Leaf	Morita and Shimizu, 1963.
. gymnolepis	Lu 7-glucoside	Leaf	Fujikawa and Nakajima 1948.
TICALES			
raceae			
Iumulus japonicus	Ap and Lu 7-glucosides	Leaf	Hattori and Matsuda, 1949.
	Vitexin	Leaf	Aritomi, 1963.
icaceae			
Boehmeria nipononivea	Ap 7-rutinoside	Leaf	Nakaoki, Morita and Yoshida, 1957.
LYGONALES			
lygonaceae			
Polygonum orientale	Vitexin	Leaf	Hörhammer et al., 1958.
	Iso-orientin		
NTROSPERMAE			
ryophyllaceae			
Saponaria officinalis	Isovitexin 7-glucoside	Leaf	Barger, 1906.
ANUNCULALES			
nunculaceae			
Adonis vernalis	Iso-orientin xyloside	Leaf	Hörhammer et al., 1960.
JTTIFERALES			
eoniaceae			
Paeonia arborea	Ap 7-glucoside	Petal	Egger and Keil, 1965.
	Ap 7-rutinoside		
APAVERALES			
uciferae			
Alliaria officinalis	Vitexin	Leaf	Paris and Delaveau, 1962.
Capsella bursa-pastoris	Diosmetin 7-rutinoside	Leaf	Oesterle and Wander, 1925;
	Luteolin 7-rutinoside		Stepien and Krug, 1965.
)SALES			
guminosae			
Amorpha fruticosa	Ap 5-glucoside	Leaf	Goto and Taki, 1938;
Aspalathus linearis	Orientin, iso-orientin, di H-chalcone corresp. to iso-orientin.		Koeppen et al., 1962; Koeppen and Roux, 1965.
Castanospermum australe	5-Deoxyvitexin	Heartwood	Eade et al., 1962.
Cytisus laburnum	4'-O-Methylvitexin	Flower	Paris, 1957.
Galega officinalis	Lu 5-Glucoside	Seed	Barger and White, 1923.
Kummerovia striata	Lu 7-glucoside	Leaf	Nakaoki et al., 1956, 1957.
Lespedeza capitata	Iso-orientin	Leaf	Paris and Charles, 1962.

TABLE 5·3—*continued*

Order, family and species	Pigments present[1]	Organ examined	References
ROSALES—*continued*			
Leguminosae—*continued*			
Microlespedeza striata	Lu 7-glucoside	Aerial portion	Nakaoki *et al.*, 1956.
Parkinsonia aculeata	Orientin, 5-*O*-methylorientin, and 5,7-di-*O*-methylorientin	Leaf and flower	Bhatia *et al.*, 1966.
Psoralea spp.	Orientin, iso-orientin, vitexin, isovitexin	Leaf	Ockendon *et al.*, 1966.
Robinia pseudacacia	Acacetin 7-glucosylrhamnosylxyloside	Leaf	Freudenberg and Hartmann, 1954.
Sarothamnus scoparius	Orientin 3'-methyl ether	Leaf	Hörhammer and Wagner 1962.
Sophora angustifolia	Lu 7-glucoside	Leaf and flower	Hattori and Matsuda, 1954
Spartium junceum	Lu 4'-glucoside	Petal	Spada and Cameroni, 1958
	Lu 7-glucoside	Flower	Hörhammer, Wagner and Dhingra, 1959.
	Orientin	Flower	Hörhammer *et al.*, 1959.
Tamarindus indica	Vitexin	Leaf	Lewis and Neelakantan 1964.
	Orientin		
Vicia hirsuta	Ap 7-apiosylglucoside	Leaf	Nakaoki and Morita, 1955.
Rosaceae			
Crataegus monogyna *C. oxycantha*	} Vitexin rhamnoside	Leaf	Fiedler, 1955.
Pyrus bretschneideri	Ap and Lu 7-glucosides	Leaf	Williams, 1964.
P. ussuriensis	Lu 7-glucoside Lu 4'-glucoside	Leaf	Williams, 1964.
GERANIALES			
Euphorbiaceae			
Euphorbia supina *E. thymifolia*	} Ap 7-glucoside	Leaf and stem	Nagasi, 1942.
Oxalidaceae			
Oxalis acetosella *O. cernua*	} Orientin	Leaf	Shimokoriyama and Geissman, 1962.
RUTALES			
Rutaceae			
Citrus aurantium	Ap 7-rutinoside	Fruit	Hattori *et al.*, 1952.
C. limon	Lu 7-rutinoside	Peel	Chopin *et al.*, 1964.
Citrus spp.	Vitexin xyloside	Fruit	Horowitz and Gentili, 196
Fortunella japonica *F. marginata* *F. obovata*	} Acacetin 7-rutinoside	Fruit and petal	Matsuno, 1958.
SAPINDALES			
Anacardiaceae			
Rhus succedanea	Ap 7-rutinoside	—	Hattori and Matsuda, 195
Aceraceae			
Acer cissifolium	Lu 4'-glucoside	Leaf	Aritomi, 1965
A. palmatum	Vitexin, isovitexin, iso-orientin	Leaf	Aritomi, 1963.

TABLE 5·3—*continued*

ter, family and species	Pigments present[1]	Organ examined	References
LVALES			
lvaceae			
Hibiscus syriacus	Vitexin, isovitexin	Leaf	Nakaoki, 1944.
iaceae			
Tilia japonica	Acacetin 7-glucoside	Leaf	Morita, 1960.
RTIFLORAE			
mbretaceae			
Combretum micranthum	Vitexin, isovitexin	Leaf	Jentzch *et al.*, 1962.
IBELLIFLORAE			
nbelliferae			
Ammi visnaga	Acacetin	—	Ralha, 1954.
Apium graveolens	Lu 7-apiosylglucoside	Seed	Farooq *et. al.*, 1953.
Conium maculatum	Diosmetin 7-rutinoside	Leaf	Oesterle and Wander, 1925.
Cuminum cyminum	Ap 7-apiosylglucoside	Seed	Sachindrak and Chakraborti, 1958.
Daucus carota	Lu 7-glucoside	Leaf	Nakaoki and Morita, unpublished.
Petroselinum crispum	Ap and Lu 7-apiosylglucosides	Seed and leaf	Nordström and Swain, 1953; Kaizmarck and Ostrowska, 1962.

[1] Abbreviations: Ap, apigenin; Lu, luteolin.

unusual in being completely O-methylated or in having 2′-hydroxy substituents are found almost entirely in the Rutaceae and are discussed in detail under this family. Other unusual flavones have recently been found in the Lauraceae and the Myrtaceae. The root of *Lindera lucida* (Lauraceae) contains 5,6,7,8-tetramethoxyflavone, 5,7-dihydroxy-6,8-dimethoxy-3′,4′-methylenedioxyflavone (lucidin, 5) and its 5,7-dimethyl ether (Lee and Tan, 1965). Again, the leaves of several *Eucalyptus* and *Angophora* species (Myrtaceae) contain two 6,8-C-dimethylflavones, sideroxylin (6) and its 4′-methyl ether eucalyptin together with 5-hydroxy-7,4′-dimethoxy-6-C-methylflavone (Lamberton, 1964; Hillis and Isoi, 1965). How far the occurrence of these compounds is related to taxonomy is not yet clear. It is worth noting that flavonoids with methylenedioxy groups are also found in the Rutaceae and Piperaceae but that those with C-methyl groups are not apparently present elsewhere in the Archichlamydeae.

(5) Lucidin

(6) Sideroxylin

Other flavonoids are of scattered occurrence throughout the Archichlamydeae and will be discussed in more detail under individual families. Isoflavones, for example, are mainly legume constituents but have been recorded once in the Amarantaceae (*Iresine*), once in the Moraceae (*Maclura pomifera*) and once in the Rosaceae (*Prunus*). Chalcones and flavanones are also common in the Leguminosae but rare elsewhere. Chalcones are also present in the Cannabinaceae (*Humulus*) and Caryophyllaceae (*Dianthus*), Ranunculaceae (*Paeonia*), Rosaceae (*Prunus*) and in the Piperaceae (*Piper*), and flavanones in the Rosaceae (*Prunus*) and Rutaceae (*Citrus*). Finally, aurones, which are characteristic constituents of a few sympetalous families, have been found in the Leguminosae (*Soya hispida*), in the Oxalidaceae (*Oxalis cernua*) and in the Anacardiaceae (*Cotinus coggyria* and *Melanorrhoea*).

II. Flavonoids of the Ranunculaceae and Paeoniaceae

The Ranunculaceae and Paeoniaceae, along with the Nymphaceae and Magnoliaceae, are regarded as typical "primitive" angiosperm families by most evolutionists. Are there signs in the flavonoid pattern of the first two families to indicate their primitiveness? In general, the answer is in the affirmative. Thus, the families are abundant in flavonols [recorded by Egger (1959) as being present in 34 of 51 species surveyed] and rare in flavones (recorded only in *Adonis* and *Paeonia*, see Table 5·3). Again, the anthocyanidin glycoside patterns are simple, being either 3-glucoside or 3,5-diglucoside (Table 5·1). Furthermore, a chalcone has been found in *Paeonia* and this is a class of flavonoid which is primitive on the basis of both its biogenesis and distribution. Finally, no unusually substituted flavonoids have been recorded in the Ranunculaceae or Paeoniaceae; such substitutions, if present, would represent "gain" mutations in evolutionary terms.

As mentioned above, the anthocyanins in the Ranunculaceae are now known to be simple in structure, although at one time a complex pigment was thought to be present in *Delphinium*, the genus which gave its name to the anthocyanidin, delphinidin. The pigment, called delphinin, was isolated from *D. consolida* by Willstäter and Mieg in 1915. According to these workers, it gave on hydrolysis delphinidin, glucose and *p*-hydroxybenzoic acid in the ratio of 1:2:2 and thus appeared to be a diacylated derivative of delphinidin 3,5-diglucoside (delphin). Re-investigation in this laboratory (Harborne, 1964a) of delphinin showed that the pigment was very difficult to purify. However, the best sample that could be obtained gave only delphinidin, delphinidin 3- and 5-glucosides and glucose on acid hydrolysis. No *p*-hydroxybenzoic acid was detected in the alkaline and acid hydrolysates and H_2O_2 oxidation also failed to yield a *p*-hydroxybenzoyl glucose ester. Thus, the major pigment of *Delphinium* appears to be simply delphinidin 3,5-diglucoside. The same pigment probably also occurs in *Aquilegia* (cf. Beale *et al.*, 1941) and has been provisionally identified in *Clematis*.

Paeonia contains the corresponding peonidin glycoside, the 3,5-diglucoside, peonin (7). It was first isolated from *P. albiflora* (Hayashi, 1939) and was

(7) Peonin

subsequently shown to occur in seven other species (*P. anomala, P. arietina, P. broteri, P. mollis, P. officinalis, P. triternata* and *P. veitchii*) by Beale *et al.*, 1941. A minor component in these flowers was reported as peonidin 3-glucoside but re-investigation by paper chromatography in this laboratory showed that the major pigment, peonin, was accompanied, not by the 3-glucoside, but by cyanin. *Paeonia* petal remains the best plant source of peonin, which has not been recorded as a major constituent in any other plant genus.

Anthocyanins occur in both petal and pollen of the garden anemone, *Anemone coronaria*, but relatively little is known about the pigments. Tappi and Monzani (1955) identified three monoglucosides in the pollen (Table 5·1), together with a partly characterised delphinidin arabinoside. More recently, the pigment of scarlet-petalled forms has been examined in this laboratory. It appears to be mainly pelargonidin 3-lathyroside, a type only known otherwise in *Lathyrus*, but it may be mixed with the related 3-sambubioside. Further work is needed on these pigments particularly since blue forms apparently have a delphinidin 3,5-diglycoside (Lawrence *et al.*, 1939).

The flavonol glycosides of these families were not seriously studied until Egger in 1959 reported the presence in acid hydrolysates of flowers and leaves of two unusual flavonols related in structure to kaempferol and quercetin. These substances were at first thought to be C-glycoflavonols (Bate-Smith and Swain, 1960) but were subsequently identified as kaempferol and quercetin 7-glucosides (Harborne, 1962b, 1965a), formed by incomplete hydrolysis of the flavonol 3,7-diglucoside present naturally in the plants. 3,7-Diglucosides of kaempferol and quercetin have subsequently been isolated and identified in *Anemone, Caltha, Helleborus* and *Paeonia* (Table 5·2). Egger's (1959) survey indicates that flavonol 3,7-diglucosides are widespread in the family, being present also in *Actaea, Aquilegia, Clematis, Delphinium, Pulsatilla, Ranunculus* and *Thalictrum*. They thus provide a very characteristic flavonoid character of the two families, one which is not widespread as far as is known, in any other dicotyledonous group.

III. Flavonoids of the Papaveraceae

Anthocyanins are important pigments in poppy plants, providing as they do the scarlet or crimson of the petal, the purple-black of the basal blotch and the blue-black of the stamen. Their systematic distribution in the genus *Papaver* shows some interesting variations. The first pigment to be isolated from the corn poppy, *P. rhoeas*, was a cyanidin glycoside, called mecocyanin (Willstäter and Weil, 1917). Mecocyanin was identified shortly afterwards as the 3-gentio-bioside by Robinson *et al.* (1934) who compared it directly with synthetic material. This structure for the poppy pigment was accepted for a long time but the discovery in 1957 of a cyanidin glycoside in *Primula sinensis* (Harborne and Sherratt), which gave gentiobiose on hydrolysis and was chromatographi-cally different from the *Papaver* pigment, threw doubt on the original identifi-cation. Re-examination of mecocyanin showed that its disaccharide, obtainable in small amounts from acid hydrolysis, was quite different from gentiobiose but very similar to cellobiose or sophorose in chromatographic behaviour. Direct comparison of this disaccharide by chromatography and electrophoresis with authentic sugars showed that it was not cellobiose but was definitely sophorose and thus that mecocyanin was cyanidin 3-sophoroside (8). At the same time, the pelargonidin glycoside accompanying mecocyanin in the corn poppy was

(8) Mecocyanin   (9) Orientalin

isolated and identified as the 3-sophoroside (Harborne and Sherratt, 1961; Harborne, 1963b).

Another novel pelargonidin glycoside was found in this investigation. This pigment, called orientalin, was first detected in the Oriental poppy, *Papaver orientale*, and later isolated from the Iceland poppy, *P. nudicaule*, as well. It differs from other pelargonidin glycosides in being distinctly more orange-yellow than orange-red in colour and was eventually identified as pelargonidin 3-sophoroside-7-glucoside (9). It remains one of the known anthocyanins with sugars attached at the 3- and 7-positions. It has, however, been found in one other unrelated plant genus, the monocotyledonous *Watsonia* (Iridaceae).

A third pigment of unusual interest also occurs in association with these anthocyanins in *Papaver*. This is a water-soluble yellow substance named nudicaulin. It was first isolated by Price *et al.* (1939) from *P. nudicaule* and *Meconopsis cambrica*, who found it was unusual in containing nitrogen in its structure. Its other properties, including its presence as a glucoside, indicated to these authors that it was a flavonoid and they proposed that its structure was related to the "nitrogenous anthocyanins" of the Centrospermae. A re-investigation in this laboratory (Harborne, 1965b) of nudicaulin showed that its spectral properties ($\lambda_{max}^{EtOH-HCl}$ 258, 330 and 467 mμ) were not consistent with it being a flavonoid. The most likely structure for it, assuming that the pure pigment does contain nitrogen, is that it is an alkaloid. Further work on its structure is in progress.

In carrying out studies of the distribution of these anthocyanins in *Papaver* Acheson *et al.* (1962) did not fully identify the pigments present. However, by combining their results with chemical data obtained the author's laboratory, it is possible to draw up the table as shown (5·4). The results, in general, support Fedde's division (1909) of the genus into six sections. The following

TABLE 5·4. Distribution of Anthocyanins and of Nudicaulin in the genus *Papaver*

Papaver Section	Species	Cyanidin 3-Sophoroside	Pelargonidin 3-Sophoroside 7-Glucoside	Pelargonidin 3-Sophoroside	Nudicaulin
thorhoeades	*rhoeas*	+	−	+	−
	dubium	+	−	+	−
	lecoquii	+	−	+	−
gemonerhoeades	*argemone*	+	+	+	−
	hybridum	+	−	−	−
	apulum	+	+	+	−
	pavovinum	−	−	+	−
	polonicum	+	−	+	−
cones	*glaucum*	+	−	+	−
	somniferum	+	−	+	−
osa	*pilosum*	−	+	+	−
	heldreichii	−	−	−	+
crantha	*bracteatum*[1]	+	−	−	+
	orientale	+	+	−	−
apiflora	*alboroseum*	−	−	−	+
	radicatum	−	−	−	+
	rhaeticum	−	−	−	+
	nudicaule	−	(+)	(+)	+

Data from Acheson *et al.* (1962) and Harborne (1963a, b); pigments are in petals (excluding basal blotch) and/or in stamens. Basal blotch in species that have been examined consists of cyanidin 3-glucoside.

[1] Also contains pelargonidin 3-glucoside.

points may be noted: (1) Species in section *pilosa* lack cyanidin glycosides. (2) Species in section *scapiflora* lack anthocyanins altogether. The pelargonidin glycosides in *P. nudicaule* only occur in cultivated forms, the wild species having yellow petals. (3) Species in sections *pilosa, macrantha* and *scapiflora* are characterised by the presence of nudicaulin. This yellow pigment also occurs in *Meconopsis*, a genus closely related to *Papaver* and included in the same sub-family, Papaveroideae. (4) There are differences in distribution of anthocyanin within species. For example, in *P. dubium* and *P. lecoquii*, cyanidin 3-sophorosode occurs in the petals whereas pelargonidin 3-sophoroside is confined to the stamens. In the third species in the same section, *P. rhoeas*, both glycosides are found in the petals.

While the above data solve no taxonomic problems, they provide a means of identifying hybrids, an aid in classifying newly discovered species and characters for relating *Papaver* to other genera in the Papaveraceae.

Preliminary information on the flavonol glycosides present in the family indicates that these are quite variable. Quercetagetin 7-glucoside has been found in *P. nudicaule* petal (Harborne, 1965b), isorhamnetin 3-glucoside and 3,7-diglucoside in *Argemone mexicana* seed (Krishnamurti *et al.*, 1965) and kaempferol in *Meconopsis integrifolia* petal. Rhamnetin (as the 3-arabinoside-3'-rhamnoside) has been isolated from leaves of *Peumus boldus* (Krug and Borkowski, 1965) and quercetin has been found as the 3-glucoside and 3-gentiobioside in petals of the opium poppy (Sosa and Sosa, 1966).

IV. Flavonoids of the Cruciferae

The Cruciferae are a large family, including such plants as the cabbage, mustard and garden stock and are most notable chemically for the sulphur compounds which are present (Kjaer, 1963). The flavonoid pattern, as far as it is known, is a distinctive one. The anthocyanins that have been isolated are unusually complex and occur associated with one or more cinnamic acids. Also, there are indications that the flavonol isorhamnetin is a characteristic constituent.

The first anthocyanins to be studied were those in the red cabbage, a plant which is an extremely rich source of pigment. The major constituent, called rubrobrassicin, was isolated in 1936 by Chmielewska, who later (1955) proposed structure 10 for it. Unusual features in this structure include a methyl group in the 3-position, a disaccharide in the 5-position and sinapic acid as an acyl group in the 7-position. Subsequent work showed that none of these unusual features is in fact present in the red cabbage pigment. For example, Stroh, in 1959, showed that the aglycone moiety was cyanidin, and not the 3-*O*-methyl ether, and that the pigment was a 3-triglucoside, not a 5-diglucoside.

(10) Rubrobrassicin: Chmielewska's structure   (11) Rubrobrassicin: Harborne's structure

Paper chromatographic studies in this laboratory indicated that a mixture of several acylated pigments were, in fact, present in *Brassica oleracea* var. *rubra* and that many impurities, including sulphur compounds and the sugar ester sinapoylglucose, were closely associated with the anthocyanins.

Thorough purification yielded three pigments all of which gave on deacylation the same cyanidin glycoside, identified as the 3-sophoroside-5-glucoside (Harborne, 1963b and 1964a). The acylated anthocyanins were subsequently found to have p-coumaric acid or ferulic acid as acyl groups and these were attached to the sugar residues in the 3-position; the major pigment was shown to have the structure (11) (Harborne, 1964a). In recent publications, Stroh and Siedel (1965) have corrected their earlier error regarding the position of the sugars, confirming the results obtained in this laboratory in most respects, but still claim sinapic acid as an acyl group. This latter claim is probably due to contamination of their material with sinapoylglucose or other simple sinapoyl esters which are abundant in cabbage. In our experience, repeated paper chromatography is necessary to produce rubrobrassicin free from impurities. It is, of course, possible that different strains of red cabbage have different types of acylated pigments.

An unusual feature of the red cabbage pigment is its ability to act as a wide-range indicator, appearing red in acid solution, blue at pH 7 and green or yellow in alkali (Wolf, 1956). Other anthocyanins do not provide such a wide range of colours, particularly in the alkaline range, and it is likely that, since work has only been done on the crude pigment, rubrobrassicin forms a loose but stable complex in solution with some other cabbage leaf constituent (possibly a sulphur compound).

The anthocyanins of the petals of the garden stock *Matthiola incana*, like those of the red cabbage leaf, are acylated. Investigations by Seyffert (1960) revealed extremely complex mixtures of anthocyanins in the various colour forms. He reported several series of acylated glycosides of pelargonidin and cyanidin in which the sugars were 3,5-triglucoside or 3-diglucoside and the acyl groups p-coumaric, caffeic, ferulic and sinapic acid in nearly all possible combinations. As many as nine anthocyanins were reputed to be present in a single phenotype. An examination of the garden stock in this laboratory showed that the situation is a much simpler one. Seyffert (1960) did not purify his anthocyanins at all and his misleading results are directly attributable to contamination with simple cinnamic esters known to be present in these flowers. In scarlet colour forms, only two pigments were in fact detected (Harborne, 1964a). One was identified as pelargonidin 3-glucoside, the other as a (p-coumaroylferuloyl) derivative of pelargonidin 3-sambubioside-5-glucoside. Mauve and blue forms contained but three cyanidin glycosides, two of which are probably analogous in structure to the pelargonidin glycosides.

The anthocyanins of a third crucifer, the garden radish *Raphanus sativus*, have also been studied in detail and shown to be acylated. Four pigments have been described: two pelargonidin derivatives in the roots of garden forms such as "Scarlet Globe" and two related cyanidin derivatives in the roots of the cross between "Scarlet Globe" and the white "Icicle". Genetic studies (Harborne and Paxman, 1964) showed that white varieties are genotypically purple (i.e. are dominant for cyanidin production) but that the expression of self colour is

6

inhibited at a second locus. The glycosidic type is uniformly the same (3-sophoroside-5-glucoside) as in the red cabbage. According to work in this laboratory (Harborne and Sherratt, 1957; Harborne, 1963b, 1964a) only two acyl groups are involved (*p*-coumaric and ferulic) and these are attached, as monoacyl derivatives, to the sophorose in the 3-position. Parallel work on the radish pigments has been carried out in Japan and Ishikura and Hayashi (1965), although agreeing in most respects with the author's work, claim that three other acylated pigments are present in both red- and purple-skinned roots, bringing the total number of pigments to ten. Their other three acylated derivatives include one with a caffeoyl residue, one with *p*-coumaroyl and feruloyl residues and one with all three cinnamic acids attached. Although they purified their pigments by column chromatography, paper electrophoresis and thin-layer chromatography, it is possible they may not have freed their material from contaminating cinnamic acid derivatives, which are not attached covalently to the anthocyanin. Further work is clearly required to settle this discrepancy.

The flavonols of the Cruciferae have not been studied systematically but the few reports that are available (*see* Table 5·2) indicate that isorhamnetin glycosides are remarkably frequent, being present in five of the six genera studied, i.e., *Brassica, Cheiranthus, Diplotaxis, Matthiola* and *Sinapis*. According to Perkin and Hummell (1896), the fruit of *Cheiranthus cheiri* also contain rhamnetin. A further indication that methylated flavonols are frequent in the family is the survey of Bate-Smith (1962) who reported flavonols different from kaempferol and quercetin in three of the nine plants studied; these were from the genera *Alyssum* and *Crambe*. Some interesting flavonol glycosidic patterns have been detected in *Cheiranthus* and *Matthiola*. The discovery of robinin (kaempferol 3-rhamnogalactoside-7-rhamnoside) in *Cheiranthus* is of taxonomic interest, since this is a rare type first found in the Leguminosae. In *Matthiola incana*, while isorhamnetin 3,4'-diglucoside occurs in the seeds, quite a different flavonol is present in the petals. This is a labile kaempferol derivative containing only pentose sugars in its structure. It has provisionally been identified as the 3-rhamnosylarabinoside-7-rhamnoside, but the order of the sugar units in the disaccharide in the 3-position still requires confirming. The above order is the one preferred on phytochemical grounds; an analogous quercetin 3-dipentoside isolated from the related *Cheiranthus cheiri* has been shown to be the 3-rhamnosylarabinoside, following the isolation of the 3-arabinoside as a result of controlled acid hydrolysis (Maksyntina, 1965).

V. Flavonoids of the Rosaceae

A. GENERAL ASPECTS

The Rosaceae is a very large family, comprising 100 genera and some 2000 species. It includes many food plants—the apple, pear, plum, cherry, peach, strawberry and raspberry—and also the most popular of ornamental plants, the rose. The family is usually divided into six sub-families: Spiraeoideae, Pomoideae, Rosoideae, Prunoideae, Neuradoideae and Chrysobalanoideae. The latter two sub-families are given family status in the recent revision of the Syllabus (Engler, 1964). A considerable amount of information is available on the flavonoids in the family, and plants of five of the six sub-families (the exception being Neuradoideae) have been extensively surveyed for leaf phenolics by Bate-Smith (1961, 1962). Chemical links with other Rosalian families have been observed, particularly with the Saxifragaceae, Leguminosae and Hamamelidaceae.

B. ANTHOCYANINS

The anthocyanin pattern of the Rosaceae is distinct in a number of respects, the most striking being the complete absence of delphinidin or its derivatives (Table 5·1). Cyanidin is a very common pigment and pelargonidin is relatively rare, appearing only in mutants of the rose and raspberry. The glycosidic patterns vary considerably and the occurrence of anthocyanins with branched trisaccharides in *Prunus* and *Rubus* is particularly noteworthy. The distribution of glycosidic patterns at the generic level is shown in Table 5·5. The

TABLE 5·5. Glycosidic Patterns of the Anthocyanins of the Rosaceae

Sub-family and genus	No. of species surveyed	Glycosidic pattern	Anthocyanidins[1]
POMOIDEAE			
Malus (apple)	14	3-Galactoside[2]	Cy
Pyrus (pear)	5	3-Galactoside	Cy
ROSOIDEAE			
Fragaria (strawberry)	2	3-Glucoside	Pg, Cy
Rosa (rose)	34	3-Glucoside and 3,5-diglucoside	Cy, Pn, Pg
Rubus (raspberry)	14	3-Glucoside, 3-rutinoside, 3-(2^Gglucosylrutinoside) and 3-(2^Gxylosylrutinoside)	Cy, Pg
PRUNOIDEAE			
Prunus (cherry, plum)	6	3-Glucoside, 3-rutinoside, 3-sophoroside and 3-(2^G-glucosylrutinoside)	Cy, Pn

[1] Abbreviations: Pg = pelargonidin, Cy = cyanidin, Pn = peonidin.
[2] In a survey of 14 species, 5 spp. hybrids and 10 cultivars, 3-galactoside was uniformly present. A cyanidin 3-pentoside was provisionally identified in two species.

most interesting point is that the two Pomoideae, *Malus* and *Pyrus*, both have cyanidin 3-galactoside, which is absent from the other sub-families. The earlier report of pelargonidin 3-galactoside in strawberries (Robinson, 1934) has been shown to be incorrect; only the 3-glucosides of pelargonidin and cyanidin are present (Sondheimer and Karash, 1956).

Although the rose is so popular as a garden shrub and breeders are continually trying to produce new colour varieties, the chemistry of petal pigmentation has not been studied extensively until recently. Cyanin, it is true, was isolated from *Rosa gallica* by Willstäter and Nolan in 1915 but the pelargonidin 3,5-dimonoside noted by Scott-Moncrieff (1936) in scarlet polyantha forms was not fully identified until 1961. Peonin was discovered at the same time (Harborne, 1961) in *R. rugosa* and related hybrids and the presence of the 3-glucosides accompanying the 3,5-diglucosides in many varieties was noted. A careful search of rose varieties, particularly mauve and purple shades, failed to reveal any delphinidin-containing forms. Arisumi (1963) surveyed 34 wild *Rosa* species and found cyanin in 97% of them, peonin in 56%, but no evidence of either delphin or pelargonin. Delphinidin is thus completely absent from the genus, and blue roses can only be produced by breeding for varieties in which the cyanin is strongly co-pigmented. Pelargonidin is produced in *Rosa* as a result of recent mutation in garden forms. It was found in diploid polyanthas, e.g. "Paul Crampel", in the 1930's but not detected in tetraploid hybrid tea roses, e.g. "Gloria Mundi", until 1953 (Rowley, 1957).

The only other work on anthocyanins in the Rosaceae has been on fruits, not petals. Cyanidin 3-galactoside was isolated from red apple skin in 1937 by Sando but work on the pigments in soft fruits such as raspberry, cherry and strawberry is of more recent origin. The pigments appeared to be simple cyanidin glycosides (*see* e.g. Li and Wagenknecht, 1956) until Harborne and Hall (1964b) discovered the presence of anthocyanins having one of two branched trisaccharides, 2^G-glucosylrutinose and 2^G-xylosylrutinose, attached in the 3-position. The systematic distribution of these pigments (e.g. 12 and 13) which were found in *Prunus* and *Rubus* and in the related *Ribes* (Saxifragaceae), was also studied and some significant correlations with taxonomy were noted.

(12)   (13)

The results of the survey (Harborne and Hall, 1964b) are summarised in Table 5·6 and the following points may be observed:

In *Rubus*, the 3-triglycosides (12) and (13) occur exclusively in raspberry types (section Ideobatus) but are completely absent from the blackberry types (section Rubus). Branched trisaccharide pigments are present in most raspberry cultivars, and especially in raspberry–blackberry hybrids. The black raspberry, *R. occidentalis*, is characterised by having xylose-containing glycosides e.g. (13) and the F1 with *R. ideaus* does not contain them, so that synthesis of sambubioside and xylosylrutinoside is a recessive character.

In *Prunus*, cyanidin 3-(2^G-glucosylrutinoside) (12) is present in sour cherry *P. cerasus* (in six of seven varieties examined) but uniformly absent from sweet cherries (*P. avium*). It is also absent from the sloe, *P. spinosa*, and the plum, *P. domestica*, but these latter fruits are different from cherries in containing the 3-glucoside and the 3-rutinoside of peonidin. The fruit of the peach contains simply cyanidin 3-glucoside (Hsia *et al.*, 1965).

TABLE 5·6. Distribution of Cyanidin 3-Glycosides in *Prunus*, *Rubus* and *Ribes*

Glycosidic pattern	Rubus	Distribution in Prunus	Ribes
osylrutinoside osylrutinoside noroside bubioside inoside oside	Absent	Absent	C.v. "Earliest of the Fourlands", *R. davidii*, *R. warszewiczii*, and 3 spp. hybrids
osylrutinoside noroside inoside oside	Raspberry cvs. e.g. "September"[1], raspberry hybrids, *R. strigosus*, *R. gracilis*, *R. frondosus*	Sour cherries[2] (6 cvs.), e.g. "Wye Morello" and "Late Duke"	*R. multiflorum*
osylrutinoside bubioside inoside oside	*R. occidentalis*	Absent	Cv. "Fay's Prolific", *R. sativum*, *R. rubrum* and 2 spp. hybrids
noroside oside	Three raspberry cvs., e.g. "October"	Absent	Cv. "Red Lake"
inoside oside	All blackberries and 8 spp. including *R. odoratus*[1]	Sweet cherries (3 cvs.), *P. cerasifera*, *P. spinosa*, *P. domestica*	Cv. "Rondom", *R. nigra*, blackcurrants[3], 14 spp. and 5 spp. Hybrids, *R. uvacrispa*, and gooseberries
oside	*R. parviflorus*[1]	*P. persica*	Absent

[1] These species also contain the corresponding pelargonidin glycosides; in *R. odoratus* the cyanidin glycosides are completely replaced by them.

[2] One variety "Morello A", classified as a sour cherry, lacks glycosylrutinoside.

[3] Blackcurrants also contain corresponding delphinidin glycosides.

In *Ribes*, cyanidin triglycosides are present in eleven of 29 species and four cultivars of currants studied. Significantly, no less than eight of the ten species having triglycosides are included in the same section (*Ribes*) and subsection (*Ribes*) of the genus. Black currants, not unexpectedly, differ from red currants in containing delphinidin glycosides in addition to cyanidin glycosides. Most red currant species have only the 3-glucoside and 3-rutinoside of cyanidin, which are also in the wild gooseberry and its cultivated forms.

C. FLAVONOLS

The most striking fact about the flavonoid pattern of the Rosaceae is the almost complete absence of substances with $3',4',5'$-trihydroxylic substitution. The absence of delphinidin has already been remarked upon and it is clear from the extensive surveys of Bate-Smith (1961, 1962) that myricetin is also usually absent. Leaf surveys of 200 species drawn from five of the six sub-families (and including no less than 27 *Prunus* spp. and 47 *Potentilla* spp.) showed that kaempferol and quercetin were ubiquitous in the family, but that myricetin was present only in three species, *Chrysobalanus pellocarpus*, *Licania rigida* and *Potentilla anserina*. The first two of these species are in the Chrysobalanoideae, the third is in the Rosoideae. The distribution of leucoanthocyanins in the leaves of the family follows precisely the same pattern as the flavonols; leucocyanidin is universal and leucodelphinidin is confined to the same three species mentioned above.

The exceptional occurrence of $3',4',5'$-trihydroxylated flavonoids in two species (of three studied) of the Chrysobalanoideae confirms the view of many taxonomists that this sub-family is distinct from the remainder of the Rosaceae. With regard to the occurrence of myricetin and leucodelphinidin in *Potentilla*, it is worth noting that ellagic acid (which has the gallic acid grouping present in these flavonoids) is restricted in its distribution to the same sub-family, i.e. the Rosoideae. In the Rosoideae, it is present in six tribes but is absent from the seventh, the Kerrieae. This result suggests that the affinities of the species in the Kerrieae tribe should be re-examined.

Myricetin is also apparently absent from petals as well as leaves of Rosaceae, since no myricetin glycosides have been recorded (Table 5·2). The report of Gupta *et al.* (1957) that myricetin occurs in the petals of some twenty Hybrid Tea roses could not be confirmed in this laboratory; a careful search of a number of garden varieties and *Rosa* species showed that only kaempferol and quercetin were present. Three other flavonols have been reported from single species in the Rosaceae: isorhamnetin from *Pyrus communis* (Nortje and Koeppen, 1965), kaempferol 4'-methyl ether from *Prunus mume* where it occurs as the 7-glucoside (Hasegawa, 1959) and herbacetin 8,4'-dimethyl ether in the heartwood of *P. domestica* (Nagarajan and Seshadri, 1964).

Turning to the flavonol glycosides, these have been fully identified in some eight genera and it is clear (Table 5·2) that the glycosidic pattern is very variable. The most unusual glycoside is, perhaps, the 3-rhamnoside-4'-arabinoside of kaempferol in the leaves of the blackthorn, *P. spinosa*. The most extensive range of kaempferol and quercetin glycosides are present in the apple; 3-rhamnoside, 3-glucoside, 3-xyloside, 3-arabinoside, 3-rutinoside and 3-galactoside occur together in the same tissue. An unusual collection of isorhamnetin glycosides have recently been isolated from the pear (Nortjé and Koeppen, 1965). Of the flavonol glycosides of systematic interest one must mention especially the distribution of the 4'-glucosides and the 3,7-diglucosides.

Quercetin 4'-glucoside (spiraeoside) (14) was first isolated from *Spiraea* (formerly *Filipendula*) in 1945 and has since been found in three related genera in the Archichlamydeae. It occurs in *Rosa* (Harborne, 1961), in the closely related *Hamamelis* (Hamamelidaceae) and in the not too distantly related *Ulex* (Leguminosae) (Harborne, 1962d). The only other report of its occurrence in nature is in the monocotyledon genus *Allium* (*see* Chapter 7). In *Spiraea*, quercetin 4'-glucoside is present in two species, *S. ulmaria* and *S. hexapetala*, formerly classified in a separate genus, (*Filipendula*) and is absent from 58 other species (Hörhammer *et al.*, 1956b). These other species have an interesting but unidentified quercetin dipentoside (sugars, arabinose and xylose) as a regular constituent of the leaf. In *Rosa*, quercetin and kaempferol 4'-glucosides occur almost exclusively in one group of old Scotch roses (*R. foetida*, *R. spinossisima*), which have yellow petals, and in their modern derivatives (the cultivars "Cafe" and "Lilac Time"). The 3-glycosides are, by contrast, distributed

(14) Spiraeoside   (15) Kaempferol 3,7-diglucuronide

more or less at random. The 3-sophorosides found in rose petals and in leaves of *Sorbus* do provide a chemical link between the Rosaceae and Leguminosae, since kaempferol 3-sophoroside was first found in *Sophora* and has since been found in many other legumes (*Cassia*, *Lathyrus*, *Pisum* etc.).

Flavonol 3,7-diglycosides are apparently of restricted occurrence in the Rosaceae. Bate-Smith (1961) detected an unidentified flavonol derivative GK only in the Rosoideae, in the genus *Potentilla* and in three related genera, *Geum*, *Walsteinia* and *Dryas*. In *Potentilla*, GK occurs in all the subsections of the section Gymnocarpae (13 species of 38 examined) but it is absent from all subsections of the other section Trichocarpae (9 species examined). GK was first thought to be a *C*-glycoflavonol, but examination in the author's laboratory indicates that it is kaempferol 7-glucuronide. The flavonol glycosides in *Potentilla reptans* were chosen for closer analysis, and the main component, although purified with difficulty, was identified as kaempferol 3,7-diglucuronide (15). On acid hydrolysis, the sugar at the 3-position is lost, leaving kaempferol 7-glucuronide which is only slowly hydrolysed to aglycone and sugar. Thus, the detection by Bate-Smith (1961) of GK in his leaf extracts strongly indicates that flavonol 3,7-diglycosides (in which the 7-sugar is probably glucuronic acid) are present in all the plants mentioned above.

D. FLAVONES AND OTHER FLAVONOIDS

The Rosaceae is one of the many families in the Archichlamydeae which have both *C*-glycosylflavones and flavone *O*-glycosides (Table 5·3). By far the richest source of flavones without C-glycosyl substituents is apparently the genus *Prunus*. Thus, Hasegawa (1958) has detected chrysin (16), tectochrysin (7-methoxy-5-hydroxyflavone) and apigenin 7-methyl ether in the heartwood of various *Prunus* species; the first two flavones, interestingly enough, are restricted to the section *Cerasus*. In addition, a series of flavanones are present: naringenin, eriodictyol, sakuranetin (7-O-methylnaringenin), isosakuranetin (5-O-methylnaringenin) and pinocembrin (5,7-dihydroxyflavanone). Further-

(16) Chrysin   (17) Genistein

more the two other unusual constituents of *Prunus* are the isoflavones, genistein (17) and its 7-methyl ether, prunetin. *Prunus*, thus, has a number of flavone-derived heartwood constituents, which not only distinguish it from other Rosaceous plants but also link it with the Leguminosae.

Leaf flavones have been studied, with interesting systematic results, in the pear *Pyrus* (Williams, 1964). While Western Asian and European species (including the domestic pear, *Pyrus communis*) have glycosides of kaempferol and quercetin, Eastern Asian species have, instead, glycosides of apigenin, luteolin and chrysoeriol. Glycosides so far identified include the 7-glucosides of apigenin and luteolin in *P. bretschneideri* and the 7- and 4'-glucosides of luteolin in *P. ussuriensis*. A parallel example of a relationship between flavonoid chemistry and geography is in the closely related genus, *Malus*. Western Asian species have a dihydrochalcone phloridzin (18), while Eastern Asian species have a glucoside of 3-hydroxyphloretin (19). Dihydrochalcones are,

(18) Phloridzin   (19)

unfortunately, not present in *Pyrus*, being replaced in effect by the simple quinol glucoside, arbutin, which also occurs in the allied *Docynia*. However, the chemical data available in both *Malus* and *Pyrus* would indicate that the genera have evolved in an Eastward direction, the Chinese and Japanese species being the more advanced plants.

VI. Flavonoids of the Leguminosae

A. INTRODUCTION

The Leguminosae have 600 genera and 12,000 species and are second only to the grasses in economic importance. The seeds of many plants (beans, peas, etc.) are used as foodstuffs and other species provide fodder (lucerne), timber (*Pterocarpus*) or a source of tannins (*Acacia*). The family are extremely rich in flavonoids, both common and unusual types, and are alone among the families of the dicotyledons in containing representatives of all classes of flavonoid (Bate-Smith, 1962). This may be partly due to size (the third largest of all flowering plants) but must also be because the Leguminosae occupy a central position in the Archichlamydeae and have affinities with many other families. These include the Rosaceae, the Anacardiaceae and the Celastraceae.

The family are usually divided into three sub-families, although a few taxonomists (e.g. Hutchinson, 1964) insist on raising these to the familial level. They are the mainly tropical Mimosoideae and Caesalpinioideae and the more temperate Papilionatae. Each sub-family tends to have certain characteristic flavonoids or related phenolics. The Mimosoideae are rich in catechins, leuco-anthocyanidins and other tannins. The Caesalpinioideae have ellagitannins (in *Caesalpinia*) and the rare heartwood pigments haematoxylin and brazilin. Finally, the Papilionatae are characterised by having isoflavones and rotenoids but are also rich in anthocyanins and flavonol glycosides. Unusual flavonoids which appear in all three sub-families are 6-hydroxylated flavonols (querce-tagetin derivatives) and flavonoids lacking a 5-hydroxyl group (fisetin, robinetin etc).

B. ANTHOCYANINS

Although anthocyanins have been fully identified in 12 legume genera (Table 5·1), detailed distribution studies have only been carried out in *Lathyrus*, *Pisum*, *Vicia* and *Phaseolus*. Interest in *Lathyrus* pigments came through studies of the inheritance of flower colour in the sweet pea, *L. odoratus* (Beale *et al.*, 1939), although Karrer and Widmer (1929) did isolate a partly character-ised delphinidin rhamnoside from the same species earlier. The pigments of the sweet pea were not fully identified until 1960, when it was discovered that the glycosidic patterns are very unusual, the two main being 3-rhamnoside and 3-rhamnoside-5-glucoside. In all, 19 pigments were identified (Harborne, 1960a, 1963b). Other glycosidic types (3-galactoside-5-glucoside, 3-galactoside and 3-xylosylgalactoside (or lathyroside)) are also present but are restricted to scarlet and crimson mutant forms. The origin of these types will be discussed later in Chapter 8. For the present taxonomic purpose, it is the pigments of the wild type *L. odoratus* which are important and these are the same as those in the cultivated mauve and purple forms, i.e. are either 3-rhamnosides or

6*

3-rhamnoside-5-glucosides. The same glycosidic types are present throughout the genus, having been detected in all of 19 species studied (Pecket, 1960; Harborne, 1963b). Nearly all species have mixtures of delphinidin, petunidin and malvidin 3-rhamnoside-5-glucosides (e.g. 20); *L. sativus* is unusual in having simply the 3-rhamnoside of malvidin (21).

(20)   (21)

These two rare glycosidic types, resulting from the replacement of glucose by rhamnose in the 3-position of anthocyanins, have only been found together in one other plant family, the sympetalous Plumbaginaceae (*see* Chapter 6). They do, however, occur in other genera in the tribe Vicieae; in fact, present evidence indicates that they are confined to this tribe and characterise it. The same types have been found in 10 of 12 *Vicia* species, in all wild *Pisum* material and in the one *Cicer* species available for study. In *Pisum*, other anthocyanins are present, but, as in *Lathyrus*, they are confined to mutant forms (3-gluco-side and 3,5-diglucoside) or vegetative organs (3-sophoroside-5-glucoside and 3-sambubioside-5-glucoside in the pod).

By contrast with the situation in the Vicieae, the anthocyanins of all other tribes of the Papilionatae that have been studied are common 3-glucosides or 3,5-diglucosides. The tribes so far studied are the Genisteae (*Lupinus*), Tri-folieae (*Ononis* and *Medicago*), Galegeae (*Wistaria*), Hedysareae (*Lespedeza*) and Phaseoleae (*Glycine* and *Phaseolus*). The only anthocyanin in the Caesal-pinioideae studied is also 3-glucoside (*Tamarindus* in the Amherstieae). The anthocyanins of *Phaseolus* have been studied in connection with genetical work (Feenstra, 1960) and will be discussed in detail later. While this survey of the Papilionatae anthocyanins is far from complete, the results so far strongly indicate that the tribe Vicieae differs markedly in the pattern from the other nine tribes. The results might profitably be applied to *Abrus*, a genus which is sometimes included in the Vicieae and at other times placed in the Phaseoleae.

C. Flavonols

1. Common Flavonols and their Glycosides

In a survey of the leaves of 52 species representing 42 genera from all three sub-families, Bate-Smith (1962) found myricetin in 8, quercetin in 24 and kaempferol in 21 species. These flavonols are regularly present, though they may be replaced by the flavones luteolin and apigenin in more highly advanced genera (*Phaseolus, Wistaria*). They are also frequent in petals, kaempferol and quercetin being found, for example, in nearly all *Lathyrus* species that have been studied (Pecket, 1960).

The glycosidic patterns show much variability. They range from myricetin 3-rhamnoside in the leaf of the Judas tree, *Cercis siliquastrum*, and quercetin 3-glucuronide in the leaf of the French bean, *Phaseolus vulgaris*, to quercetin 4'-glucoside in the petals of gorse, *Ulex europaeus*. 3,7-Diglycosides are particularly common in genera of the Papilionatae. 3,7-Dirhamnosides, for example, occur in *Lotus* (Loteae), *Indigofera* (Galageae), *Lespedeza* (Hedysareae) and *Lathyrus* (Vicieae) but are only known otherwise in *Celastrus* (Celastraceae). The rare 3-rhamnosylgalactoside-7-rhamnoside of kaempferol, robinin, was first isolated from *Robinia pseudacacia* (Galageae) and has been found in *Pueraria* and *Vigna* (both Phaseoleae); elsewhere, it is only known in the distant *Vinca* (Apocynaceae) and *Cheiranthus* (Cruciferae).

Flavonol glycosides which have glucose in the 7-position rather than rhamnose are restricted, so far, to *Lathyrus* and *Baptisia*. *Lathyrus vernus* differs from all other *Lathyrus* species in having kaempferol 3-sophoroside-7-glucoside (Harborne, 1965a). More characteristic of the genus are kaempferol 3,7-dirhamnoside, as mentioned above, and kaempferol 3-xylosylgalactoside-7-rhamnoside, found in *Lathyrus odoratus*. The disaccharide in this latter pigment is a new one and has been identified as xylosyl-$\beta1\rightarrow2$-glucose, lathyrose. In *Baptisia*, flavonol 3,7-diglycosides occur mainly in white-flowered species (*B. alba, B. pendula*, etc.) and are replaced by apigenin and luteolin 7-glucosides in *B. leucophaea* and related species and by flavone 7-rutinosides in *B. sphaerocarpa* and related species (Alston *et al.*, 1965). If one assumes that flavonols are more primitive than flavones, it is possible to place these three groups of *Baptisia* taxa into an evolutionary sequence *alba* type→*leucophaea* type→ *sphaerocarpa* type.

A study of the flavonol glycosides in the genus *Pisum* has brought to light a very similar evolutionary sequence (Harborne, 1966a). The only available wild pea, *P. fulvum*, has quercetin 3-glucoside as the main leaf flavonol, primitive Asian and African cultivars (e.g. *P. nepalensis*) have kaempferol and quercetin 3-sophorosides and the garden pea (*P. sativum*) kaempferol and quercetin 3-(*p*-coumaroylglucosylsophoroside); the closely related field pea, *P. arvense*, has kaempferol 3-(feruloylgalactosylsophoroside). These studies thus indicate that as a result of cultivation and development the pea plant accumulates more complex glycosides and the three groups fall into a natural sequence: flavonol monoglucoside→flavonol diglucoside→flavonol (acylated) triglucoside. The occurrence of a different triglycoside in the field pea is interesting because it is often considered as a subspecies or variety of the garden pea, *P. sativum*: the present results show that there are significant chemical differences between the two taxa. The trisaccharides present in the *Pisum* flavonols have not been fully identified, but they appear to be linear in structure and one is probably sophorotriose. The *Pisum* flavonols have aroused much interest recently because of their growth-regulating properties, a subject which will be considered in more detail in Chapter 9.

2. Methylated Flavonols and Quercetagetin Derivatives

Apart from a report of isorhamnetin in *Cassia* leaves (Tutin, 1913), methyl ethers of the common flavonols are only known in *Lathyrus* and *Distemonanthus*. *Lathyrus pratensis*, the meadow pea, contains glycosides of syringetin (22)

(22) Syringetin

(23) Ayanin

and isorhamnetin in leaf and petal. The former pigment, although related to the common anthocyanidin malvidin, has not previously been found in nature (Harborne, 1965b). Heartwood of *Distemonanthus benthamianus* contains quercetin 3,7,4'-trimethyl ether (23) (King *et al.*, 1952) and the corresponding quercetagetin derivative, oxyayanin-B as well. *Distemonanthus* is in the Caesalpinioideae (Cassieae) so it is worth noting that 4'-methyl ether of apigenin (acacetin) occurs in *Robinia* of the Papilionatae.

Quercetagetin, which is present in the Caesalpinioideae only as a trimethyl ether, occurs both methylated and unmethylated in the other two sub-families. Thus, quercetagetin itself has been detected in *Acacia catechu* wood (Hathway and Seakins, 1957) and *Leucaena glauca* petals (Nair and Subramanian, 1962), both of the Mimosoideae, and in *Coronilla glauca* and *Medicago sativa* petal (Harborne, 1965b) both of the Papilionatae. It is accompanied by the 6-methyl ether (patuletin) in *Leucaena* and by the 7-methyl ether in *Medicago*. Patuletin 7-glucoside has also been isolated from the flowers of *Prosopis spicigera* (Mimosoideae) (Sharma *et al.*, 1964). The related 6-hydroxykaempferol has been reported in the flowers of *Galega officinalis*, but a study of this plant in this laboratory failed to show the presence of a substance with the right chromatographic properties. However, its 5-methyl ether, vogeletin, has been isolated from the seed of *Tephrosia vogelii* as the 3-arabinosylrhamnoside vogelin (24) (Rangaswami and Rao, 1959; Sambamurthy *et al.*, 1962).

(24) Vogelin

(25) Robinetin

3. Flavonols Lacking a 5-Hydroxyl

Flavonols lacking a 5-hydroxyl group, such as robinetin (25), are found in all three sub-families of the Leguminosae. The simplest substances of this type are 7,4'-dihydroxy- and 7,3',4'-trihydroxyflavone, which have recently been isolated from *Trifolium repens* leaf; the former was also obtained from lucerne, *Medicago sativa* (Livingston and Bickoff, 1964; Bickoff *et al.*, 1965). The isomeric 3,7,4'-trihydroxyflavone occurs in *Acacia*, *Butea*, *Millettia* and in *Ulex* (as the 7-glucoside) and accompanies fisetin (3,7,3',4'-tetrahydroxyflavone) in the first three plants. A fisetin derivative, 3,7-dimethoxy-3',4'-methylenedioxyflavone, has been found in *Pongamia glabra* root by Mittal and Seshadri (1956). Robinetin (25) occurs in *Acacia*, *Butea* and *Gleditschia* wood.

Flavonols of this type provide a link with the Anacardiaceae and Celastraceae since fisetin has been found in *Rhus* spp. and *Schinopsis lorentsii* of the first family and in *Celastrus* of the second. Fisetin 4'-methyl ether and 3,7,4'-trihydroxyflavone have also been isolated from *Schinopsis lorentsii* (Kirby and White, 1955).

D. CHALCONES AND LEUCOANTHOCYANIDINS

Leucoanthocyanidins and chalcones occur in rich profusion in certain legume species; most of them correspond in structure to fisetin and robinetin in that they lack a hydroxy group in the 5- (or 6'-) position. Leucorobinetinidin, leucofisetinidin and melacacidin (26) occur, for example, in the Mimosoideae, in the wood of many *Acacia* species and of *Pithecolobium dulce* (Rayndu and Rajadurai, 1965). 7,4'-Dihydroxyflavan-3,4-diol is found in the heartwood

(26) Melacacidin   (27) Robtein

of *Guibourtia coleosperma* (Roux and Bruyn, 1963), being replaced by leucofisetinidin in other species investigated (*G. tessmannii* and *G. demensii*).

Chalcones and leucoanthocyanidins co-occur in the heartwood of *Robinia pseudacacia* (Papilionatae) and in the wattle bark of *Acacia mearnsii*. In fact, an almost identical array of different flavonoid types based on fisetin and robinetin (Table 5·7) is present in both these plants (Drewes and Roux, 1963; Roux and Paulus, 1962). Besides having butein and, in one case, isoliquiritigenin (2',4',4-trihydroxychalcone) both plants have robtein (27), which is the only chalcone known with a gallic acid derived B-ring.

Isoliquiritigenin is also present in young seedlings of *Cicer arietinum*, free and as the 4'-glucoside (Wong *et al.*, 1965). It is accompanied by an impressive range of related flavonoids, including the corresponding flavanone (liquiriti-genin), flavanonol (garbanzol), flavonol and isoflavone (daidzein). Four other isoflavones are also present: formononetin, biochanin-A, biochanin-A 7-glucoside and pratensein.

TABLE 5·7. Flavonoids of *Acacia mearnsii* and *Robinia pseudacacia*

Flavonoid type	7,3′,4′,5′-Tetrahydroxy series	7,3′,4′-Trihydroxy series
Chalcone	Robtein	Butein
Flavanone	Robtin	Butin
Flavonol	Robinetin	Fisetin
Flavanonol	Dihydrorobinetin	Fustin
Flavan-3-ol	Robinetinidol	Fisetinidol
Flavan-3,4-diol	Leucorobinetinidin	Leucofisetinidin

Acacia mearnsii (wattle bark) also contains gallotannin, catechin, leucodelphinidin, quercetin 3-rhamnoside and myricetin 3-rhamnoside.

Robinia pseudacacia (heartwood) lacks the flavan-3-ols present in *A. mearnsii* and contains, in addition, 4,2′,4′-trihydroxychalcone.

Isoliquiritigenin or butein (2′,4′,3,4-tetrahydroxychalcone) are found in *Butea frondosa* (petals), *Glycyrrhiza glabra* and *G. uralensis* (root) and *Ulex europaeus* and *U. gallii* (petals). In *Glycyrrhiza*, isoliquiritigenin is present as the 4-glucoside and the 4-apiosylglucoside (Litvinenko, 1964); in *Ulex*, it occurs as a series of related glucosides: the 4′-glucoside, the 4,4′-diglucoside, the 4′-glucosylglucoside and the 4′-glucosylglucoside-4-glucoside (Harborne, 1962d). The related chalcone with an extra hydroxyl in the 3′-position (okanin) was isolated from *Cylicodiscus gabunensis* heartwood (King and King, 1951) and that with an extra hydroxyl in the 5-′position (neoplathymenin) from *Plathymenia reticulata* heartwood (King *et al.*, 1953).

Hispidol (6,4′-dihydroxyaurone) and its 6-glucoside have recently been found (together with isoliquiritigenin) in soya bean seedlings, *Soya hispida*, by Wong (1966a). Aurone formation thus provides yet another link (compare fisetin above) between the Leguminosae and the Anacardiaceae. Aurone records in the Anacardiaceae are sulphuretin in the heartwood of *Cotinus coggygria* (King and White, 1957) and 4-O-methylaureusidin (rengasin) in *Melanorrhea* (Farkas *et al.*, 1963b).

E. ISOFLAVONES

Isoflavones are isomeric with the flavones and are formed biosynthetically by what is presumably a single-enzyme controlled aryl migration from the same chalcone precursor. They occur quite characteristically in the sub-family Papilionatae, and are rarely found elsewhere in nature. In structure, the

isoflavones of the Papilionatae range from simple compounds like daidzein
(28) to complex molecules such as toxicarol isoflavone (29). With the isoflavones
are included a number of related compounds in which the 2,3-double bond is
reduced: the isoflavanones [e.g. ferreirin (30)], the rotenoids [e.g. rotenone
(31)] and the coumarino-chromans [e.g. pisatin (32)].

(28) Daidzein

(29) Toxicarol isoflavone

(30) Ferreirin

(31) Rotenone

(32) Pisatin

(33) Muningin

Not unexpectedly, many of the characteristic structural features of the
legume flavonols and flavones are found also among the isoflavones. The
following six examples must be mentioned:

(1) *Lack of 5-hydroxyl group*, present in the flavonol, robinetin (22) of
Robinia and the chalcone isoliquiritegenin of *Ulex*. Isoflavone examples:
daidzein (28) of *Glycine hispida* and formononetin, its 4′-methyl ether of
Ononis spinosa. (2) *6-Hydroxylation*, present in the flavonol quercetagetin, of
Coronilla glauca. Isoflavone example: muningin (33) of *Pterocarpus angolensis*.
(3) *C-glycosylation*, present in the flavone, orientin, in *Spartium junceum*.
Isoflavone example: puerarin (8-C-glucosyldaidzein) of *Pueraria thunbergiana*.

(4) *2'-Hydroxylation*, present in the flavonol, oxyayanin-A (5,2',5'-trihydroxy-3,7,4'-trimethoxyflavone)[1] of *Distemonanthus*. Isoflavone example: toxicarol isoflavone (29) of *Derris malaccensis*. (5) *Methylenedioxy groups*, present in the flavones of *Pongamia pinnata* root (*see* p. 43). Isoflavone example: pseudo-baptigenin (7-hydroxy-3'4'-methylenedioxyflavone) of *Baptisia* species and pisatin (32) of *Pisum sativum*. (6) *Furano-attachment*. Furanoflavones (e.g. karanjin, pinnatin) are almost completely restricted to the Papilionatae (*Pongamia, Tephrosia*). Isoflavone example: munetone (furan ring in the 7,8-position and methoxyl in the 2'-position).

TABLE 5·8. Records of Isoflavones and Other Isoflavonoids in the Papilionatae

Tribe	Species
1. Sophoreae	*Afrormosia elata, Baphia nitida, Castanospermum australe, Ferreirea spectabilis, Maackia amurensis,*[1] *Myrocarpus fastigiatus, Myroxylon balsamum* and *Sophora japonica.*
2. Podalyrieae	*Baptisia tinctoria*[2] and *Piptanthus nanus.*[2]
3. Genisteae	*Adenocarpus complicatus, Cytisus laburnum, Genista tinctoria, G. hispanica, Lupinus polyphyllus, Sarothamnus scoparius* and *Ulex nanus.*
4. Trifolieae	*Medicago sativa, Ononis spinosa, Trifolium pratense* and *T. subterraneum.*
5. Loteae	No records.
6. Galegeae	*Millettia pachycarpa,*[1] *Mundulea suberosa, M. sericea, Tephrosia maxima* and *Wistaria floribunda.*
7. Hedysareae	*Ougenia dalbergioides.*
8. Dalbergieae	*Andira inermis,*[1] *Dalbergia nigra, D. sissoo, Derris malaccensis, Lonchocarpus utilis, Piscidia erythrina, Pterocarpus angolensis* and *P. santalinus.*
9. Vicieae	*Cicer arietinum, Lathyrus montanus* and *Pisum sativum.*[1]
10. Phaseoleae	*Glycine hispida, Pachyrrhizus erosus, Phaseolus vulgaris,*[1] *Pueraria thunbergiana, Neorautanenia pseudopachyrrhiza*[1] and *Spatholobus roxburghii.*

[1] Indicates that only rotenoids or coumarinochromans have been found in these species. All other records refer to isoflavones, though these do co-occur with rotenoids in some species (e.g. *Derris*).

[2] Unpublished works of Alston *et al.* indicate that isoflavones are widespread in the genus *Baptisia* (present in at least 11 of 18 species examined). A recent survey of the Podalyrieae in this laboratory indicates that isoflavones such as formononetin occur in *Thermopsis* (all five species tested positive), in *Pickeringia montana* and in *Piptanthus nepalensis.*

The distribution of isoflavonoids in the Papilionatae is shown in Table 5·8. The most remarkable point is that in spite of the fact that no systematic

---

[1] The structure of oxyayanin-A has recently been shown to require some modification following the discovery that the natural and synthetic substances have different properties (*see* Chapter 2, p. 61).

surveys have been carried out in the sub-family, isoflavonoids have been found by chemists in no less than nine of the ten tribes. The apparent absence of these from the Loteae is probably accidental; a careful search of its members would no doubt yield positive results. The only claims that isoflavonoids occur outside the Papilionatae but still in the Leguminosae are those of Harper *et al.* (1965), who isolated 7-hydroxy-4'-methoxypterocarpan (34) and five other related substances from the heartwood of *Swartzia madagascariensis* (Swartzieae), and those of Bevan *et al.* (1966), who isolated afrormosin from the heartwood of *Amphimas pterocarpoides*. In fact, the positions of the Swartzieae and of *Amphimas* are still disputed. Although Engler and Melchior (1964), as in earlier editions of the syllabus, place these taxa in the Caesalpinioideae, Hutchinson (1964), following Bentham and Hooker, prefers to consider them as part of the Papilionatae. The present chemical evidence certainly supports this latter view, since isoflavonoids are so well represented in the Papilionatae but are not known otherwise in the Caesalpinioideae or, for that matter, in the Mimosoideae.

(34)   (35) Afrormosin

Within the Papilionatae, the distribution of isoflavonoids does not follow tribal divisions to any great extent. Afrormosin (35), which has the unusual 6,7,4'-trihydroxylation pattern occurs in *Afrormosia, Castanospermum, Myrocarpus* and *Myroxylon*, all in the Sophorieae, but has also been isolated from the Galegeae (*Wistaria*) (Shibata *et al.*, 1963) and the Phaseoleae (*Glycine*) (Gyorgy *et al.*, 1964). Similarly, rotenoids such as rotenone (31) were first isolated from *Derris* root, but have since been detected in the Galegeae (*Millettia, Mundulea* and *Tephrosia*) and the Phaseoleae (*Pachyrrhizus, Neorautanenia*) as well as in the Dalbergieae (*Derris, Piscidia, Andira, Lonchocarpus*). Rotenoids are not completely specific to the Leguminosae, as they have been reported in two quite unrelated plants, in *Verbascum thapsis* (Scrophulariaceae) (Obdulio and Lobete, 1944) and *Myrica nagi* (Myricaceae) (Krishnamurty *et al.*, 1963). Rotenoids are accompanied in *Derris scandens* root by two coumarins, scandenin (36) and lonchocarpic acid, which are closely related biogenetically to the isoflavonoids, differing only in the state of oxidation of the central pyran ring from the isoflavones. Lonchocarpic acid is an isomer of (36), in which the 7-hydroxyl group is combined with the 6- (not the 8-) isoprene residue (Johnson *et al.*, 1966).

(36) Scandenin

Few studies at the generic level have been carried out, but there are indications that species differ in their isoflavonoid content. In *Pterocarpus*, for example, all species examined have the coumarino-chromans pterocarpin and/or homopterocarpin but there are differences in the minor constituents. The majority of species have pterostilbene in addition, but in *P. angolensis* this is replaced by the isoflavones prunetin and muningin. By contrast, Asian species (e.g. *P. dalbergioides*) have the chalcone, isoliquiritigenin, and the corresponding flavanone as minor constituents (Sawhney and Seshadri, 1956). Crombie and Whiting (1962) have found that *Neorautanenia pseudopachyrrhiza*, *N. edulis* and *Pachyrrhizus erosus*, three closely related species, can be distinguished from each other by their isoflavonoid constituents; nine substances are present, only two being common to all three species and one common to two.

Also occurring in the Dalbergieae are a series of compounds, called neoflavonoids (Eyton *et al.*, 1965), which bear a close structural relationship to the isoflavonoids. To this class belong quinones, such as 4-methoxydalbergione (37), and 4-phenylcoumarins such as dalbergin (38). These neoflavonoids have been found exclusively in species of *Dalbergia* and the closely related *Machaerium schleroxylon* (Eyton *et al.*, 1965; Balakrishna *et al.*, 1962). Too little is known yet of their distribution to assign to them any taxonomic significance.

(37)   (38)

VII. Flavonoids of the Vitaceae

More attention has been devoted to the anthocyanins in the grape than to those of any other plant material. In spite of this, the precise structures of a few of the pigments still require clarification. Nevertheless, the main facts about the pigments are now quite clear (Ribereau-Gayon, 1964). The grape contains the five anthocyanidins, cyanidin, peonidin, delphinidin, petunidin and malvidin, present as 3-glucosides or 3,5-diglucosides with or without one of two different acyl groups attached (*see* Table 5·1). Most of the difficulties in identification have been because these pigments occur in such complex mixtures. No known wild or cultivated grape contains less than five antho-cyanins, most have about 10 and one wild species, *Vitis lincecumii*, has no less than 17.

Pioneers in studying grape pigments were Willstäter and Zollinger (1915), who isolated malvidin 3-glucoside (oenin) from *V. vinifera*, and Shriner and Anderson (1928), who found a petunidin glycoside acylated with *p*-coumaric acid in muscadine grape. Further progress was held up until paper chromato-graphy was applied to the separation and identification of the grape-skin pigments. Most of our present knowledge rests on the work of J. and P. Ribereau-Gayon, father and son, of Bordeaux; in recent years, food scientists in the United States have also been active in this field.

Considerable controversy has raged over differences in the anthocyanin types between the *V. vinifera* grapes (from which species most European varieties are derived) and the American grape varieties, which are mainly of hybrid origin from *V. riparia* and *V. rupestris*. This is far from being an academic point, since the French government, who keep a strict control of wine pro-duction, are naturally keen to prevent the adulteration of European wines by those from California. The dispute has been settled by Ribereau-Gayon (1964), who showed that the *V. vinifera* grape pigments (apart from other structural features) are 3-glucosides whereas American grapes contain significant amounts of 3,5-diglucosides in addition. A simple paper chromatographic procedure which differentiates between mono and diglucoside was devised by the Ribereau-Gayons (see Ribereau-Gayon, 1964) and is now in official use in France.

Much of the confusion was due to the report by Bockian *et al.* (1955) that delphinidin, petunidin and malvidin 3,5-diglucosides were the main pigments of the "Cabernet Sauvignon" grape, a *vinifera* variety. That these identifi-cations were erroneous has recently been proved by Somaatmadja and Powers (1963), who could only find the corresponding 3-glucosides in this variety. The fact that *V. vinifera* grapes do not have 3,5-diglucosides has been invoked by Koeppen and Basson (1966) to establish that the "Barlinka" variety, the most important table grape in South Africa and previously of unknown origin, is *vinifera* derived; its main pigments are 3-glucosides of delphinidin, petunidin and malvidin.

The only structural features of the grape anthocyanins remaining to be settled are the type and linkage of the acyl groups in those pigments which are acylated. The 3,5-diglucosides are acylated with *p*-coumaric acid and are, presumably, 3-(*p*-coumaroylglucoside)-5-glucosides, analogous in structure with acylated pigments known in other plants. However, according to Ribereau-Gayon (1959), the 3-monoglucosides occur as pairs of isomeric mono-*p*-coumaroyl derivatives. The simplest explanation for two isomeric forms is that the *p*-coumaroyl residue is attached to different hydroxyl groups of the anthocyanin sugar, but this has still to be proved. Albach *et al.* (1965b) have recently obtained evidence, from gas chromatography of the methylated sugar produced from the acylated 3-glucosides of the variety "Tinta Pinheira", that the *p*-coumaric acid is attached to the sugar at the 4-hydroxyl group. They have not, however, been able to eliminate the possibility that the acyl group may have migrated during the methylation.

p-Coumaric acid is not the only acyl group of grape anthocyanins, since Albach *et al.* have detected caffeic acid as acylating the 3-glucosides of delphinidin, petunidin and peonidin in the variety "Tinta Pinheira". However, the reports that malic and chlorogenic acids are attached to the anthocyanins of grapes are almost certainly the results of work on impure pigments. The earlier claim by Bockian *et al.* (1955) that an anthocyanin of the "Cabernet Sauvignon" grape is acylated with malic acid has not been substantiated by any later work. This acid is, of course, one of the main free organic acids of grape juice and might well contaminate an impure anthocyanin preparation. The more recent claim by Somaatmadja and Powers (1963) that an anthocyanin from the same grape variety is acylated with chlorogenic acid also cannot be accepted. Saponification of an acylated pigment with alkali or acid is bound to hydrolyse chlorogenic acid, if it were present as an acyl group, to caffeic and quinic acid, so that with present methods it would not be possible to prove acylation with this acid. Furthermore, chlorogenic acid is not a grape constituent, although there are erroneous reports to this effect (e.g. Sondheimer, 1958; Masquelier and Ricci, 1962). Ribereau-Gayon has shown that

TABLE 5·9. Distribution of Anthocyanins in *Vitis*

Section and species	No. of pigments	Glycosidic type	Anthocyanin acylation	Characteristic aglycones[1]
MUSCADINIA				
rotundifolia	5	3,5-Diglucoside only	None	Cy, Pn and Mv
EUVITIS				
American species (e.g. *rupestris*, *riparia*)	6–17	3-Glucoside and 3,5-diglucoside	Present	Cy, Pn, Dp, Pt and Mv
Asiatic species (*amurensis*)	5	3-Glucoside and 3,5-diglucoside	None	Pn, Pt and Mv
European species (*vinifera*)	5–9	3-Glucoside only	Present	Cy, Pn, Dp, Pt and Mv

[1] Abbreviations: Cy, cyanidin; Pn, peonidin; Dp, delphinidin; Pt, petunidin; Mv, malvidi.

caffeic acid, and the other cinnamic acids, occur in *Vitis* combined to tartaric and not quinic acid.

The distribution of anthocyanins in *Vitis* species has been studied and the results are summarised in Table 5·9 (from Ribereau-Gayon, 1964). In general, the distribution fits in with the taxonomy and it is interesting to find that species within the section *Euvitis* differ in their anthocyanins according to their geographical distribution. A more general point about the *Vitis* anthocyanins may be observed: the absence of any control of hydroxylation (although pelargonidin has, it is true, not been detected) and of methylation (mixtures of methylated and unmethylated pigments occur in all species). By contrast, there is genetic control of glycosylation; the fact that hybrids between American and European varieties contain 3,5-diglucosides indicates that 5-glucosylation is controlled by a dominant gene.

Few other flavonoids have been reported in the Vitaceae. Leucoanthocyanins, catechins and flavonol 3-glucosides have been found in the grape (Ribereau-Gayon, 1964); the only distinctive compound present is quercetin 3-glucuronide.

VIII. Flavonoids of the Theaceae

In the Theaceae, as in the Vitaceae, the flavonoids of only one genus have been explored in detail, this is the genus *Camellia*, to which the tea plant, *C. sinensis*, belongs. Certain of the properties of the commercial product of tea depend on the flavonoids present, many of which have been modified in structure during processing. A full account of the flavonoids present in tea has been given by Roberts (1962); only a brief summary can be given here.

Anthocyanins are of restricted occurrence in fresh leaf—cyanidin 3-glucoside has been identified in an Indian tea (Chanda and J. B. Harborne, unpublished). Roberts and Williams (1958) did find the unusual 3-deoxyanthocyanidin tricetinidin (39) in processed "black tea" and concluded that it had arisen as an artifact by autoxidation of (−)-epigallocatechin gallate (40), one of the major flavonoid constituents of fresh tea leaf.

(40) (−)-Epigallocatechin gallate

(39) Tricetinidin

The question whether 3-deoxyanthocyanidins occur in fresh tea, however, must be kept an open one, since Vuataz *et al.* (1959) did report two anthocyanins in fresh Ceylon tea, the spectral data of which ($\lambda_{max}^{EtOH-HCl}$ 495 and 510 mμ) indicated that they might be luteolinidin and tricetinidin glycosides respectively.

A total of 12 different flavonol glycosides have been identified in tea leaf (Table 5·2); more, no doubt, remain to be studied. The structure of the trisaccharide present in the triglycoside remains to be elucidated. The glycosides have different R_f values from the kaempferol and quercetin 3-(2^G-glucosylrutinosides) present in the *Solanum* genus, which rules out one possible structure for it. More important flavonoid constituents of fresh tea leaf, however, are six catechins: (+)-catechin, (−)-epicatechin, (+)-gallocatechin, (−)-epigallocatechin, and the 3-gallates of (−)-epicatechin and(−)-epigallocatechin. These substances undergo fermentation during the processing of the tea leaf and are converted to the theaflavins and thearubigins, which are responsible for the colour and briskness of tea as a drink (Roberts, 1962). The structure of theaflavin (41) proposed by Roberts has been modified by Takino *et al.* (1965) and Brown *et al.* (1966) but still includes the unusual seven-membered ring of the tropolones.

The flavonol glycosides and other phenolics in fresh leaf and shoot of 14 *Camellia* species have been surveyed by paper chromatography (Roberts *et al.*, 1956). The authors were able to make the following three points from their results: (1) The four taxa within the section Thea (*C. sinensis*, *C. sinensis* var. *assamica*, *C. taliensis* and *C. irrawadiensis*) are closely similar in pattern and differ significantly in flavonoid content from *Camellias* in the other four sections. (2) The triglycosides of kaempferol and quercetin occur only in *C. sinensis*, the classical tea plant (China variety), and distinguish it from all other tea taxa. (3) The flavonoid patterns are useful at the species level. Thus, *C. reticulata* and *C. pitardii* are identical in pattern and are also morphologically very similar. Similarly, *C. taliensis* and *C. sinensis* are identical, which casts some doubt on the validity of *C. taliensis* as a separate species. By contrast, *C. irrawadiensis* is very distinct (it has only flavonol 3-glucosides), which supports the specific rank it was given by Barua in 1956.

(41) Theaflavin

IX. Flavonoids of the Rutaceae

The Rutaceae, as a family, is extremely versatile in its synthetic capacity and produces a wide range of unusual, highly substituted flavonoid constituents.

The Rutaceae are also remarkable for the richly diverse alkaloid structures present and for the coumarins and essential oils occurring in many members (Price, 1963). Unfortunately, the flavonoids have only been studied in relatively few genera, belonging to only four of the seven sub-families. They are *Melicope* and *Xanthoxylum* (Rutoideae), *Flindersia* (Flindersioideae), *Cassimiroa* and *Phellodendron* (Toddalioideae) and *Citrus* (Aurantioideae). The flavonoid results, as far as they go, fully support Price (1963) who, after analysing the alkaloid data, concluded that the family is a homogeneous entity and does not require the dividing up, by raising sub-families to familial status, that some taxonomists at one time wished for. Thus, the flavonoids, in all genera studied, are characterised by a high degree of methylation, by extra substitution in the 6-, 8- or 2'-position, by presence of methylenedioxy groups and by attachment of isoprenoid side-chains.

Most work has been done on *Citrus* flavonoids, so that these will be considered before those of other genera.

A. FLAVONOIDS OF CITRUS

Anthocyanins are rare in the genus *Citrus*. The only report is of cyanidin and delphinidin 3-glucosides as the pigments in the juice and skin of the blood orange (*C. sinensis*). The pigments of colour mutations which occur in the fruit of other *Citrus* species (e.g. in the grapefruit) are more frequently carotenoid in nature.

The common citrus fruits of commerce are particularly rich in flavones

(42) Limocitrol   (43) Isolimocitrol

and flavanones. The flavones present fall into three classes: (1) fully methylated derivatives and the corresponding 5-demethyl compounds, (2) partially methylated flavonols such as limocitrol (42) and isolimocitrol (43) and (3) common flavones such as apigenin, luteolin, diosmetin and chrysoeriol. The flavanones, corresponding in structure to the common flavones mentioned above, are also very well represented. They occur either as 7-rutinosides or 7-neohesperidosides, and when present in the latter combination, provide much of the bitter taste of the fruit in question (Horowitz, 1964).

A remarkable and recurring facet of these flavonoids is the co-occurrence of pairs of isomeric compounds, such as limocitrol and isolimocitrol, in which a methyl substituent alternates between the 3'- and 4'-hydroxyl group. Another

TABLE 5·10. The Flavonoids of *Citrus* Fruits

Species[1]	Common name	Methylated[4] flavones	Flavanone[5] glycosides	Bitter (+) or sweet (−)
Citrus				
aurantium[3]	Seville orange	Nobiletin + 5-demethyl derivative / Auranetin + 5-demethyl derivative / 5-Hydroxyauranetin	Hesperidin, neohesperidin and naringenin	+
deliciosa (formerly *reticulata*)	Tangerine	Tangeretin / Nobiletin	Hesperidin	−
grandis	Pummelo	Not known	Naringenin	+
jambhiri				
v. "Nagaland"	—	Tangeretin + 5-demethyl derivative	Hesperidin	?
v. "Assam"	—	Tangeretin	Hesperidin and neohesperidin	?
limon[2]	Lemon	Limocitrol, limocitrin and isolimocitrol	Eriocitrin and hesperidin	−
f. ponderosa	—	Not known	Neohesperidin	+
medica	Citron	Not known	Hesperidin	−
mitis	—	Not known	Citromitin, 5-demethylcitromitin and hesperidin	
paradisi	Grapefruit	3,5,6,7,8,3',4'-Heptamethoxyflavone	Naringin, poncirin, hesperidin and neohesperidin	+
poonensis	—	Tangeretin	—	?
sinensis	Valencia orange	Sinensetin, tangeretin, nobiletin, 5,6,7,4'-tetramethoxyflavone and 3,5,6,7,8,3',4'-heptamethoxyflavone	Hesperidin, isosakuranetin 7-rutinoside and naringenin 7-rutinoside	−
sudachi	—	Sudachitin and 3'-demethyl sudachitin	—	?
tankan	—	Nobiletin	—	
Poncirus trifoliata	Poncirus lemon	—	Neohesperidin, poncirin and naringenin	+

[1] It has not been possible to check the validity of the following species names: *jambhiri, mitis, poonensis, tankan*.
[2] Also contains diosmin and glycosides of apigenin, luteolin, chrysoeriol, quercetin and isorhamnetin.
[3] Also contains apigenin 7-rutinoside.
[4] Key: +nobiletin (5,6,7,8,3',4'-hexamethoxy-), auranetin (3,6,7,8,4'-pentamethoxy-), tangeretin (5,6,7,8,4'-pentamethoxy-), sinensetin (5,6,7,3',4'-pentamethoxy-), and sudachitin (6,7,3'-trimethoxy-5,8,4'-trihydroxyflavone).
[5] Hesperidin = hesperetin 7-rutinoside, neohesperidin = hesperetin 7-neohesperidoside, poncirin = isosakuranetin 7-neohesperidoside, erio-citrin = eriodictyol 7-rutinoside, citromitin (5,6,7,8,3',4'-hexamethoxyflavone).

pair is diosmetin and chrysoeriol, and it seems unlikely that a careful search would not reveal the isomers of the flavanone hesperetin (3'-OH, 4'-OMe), the flavonol isorhamnetin (3'-OMe, 4'-OH) and the glycoflavone, 8-C-glucosyl-chrysoeriol, compounds already known to occur in Citrus fruits. *Citrus* species apparently contain several glycoflavones; vitexin and a xylosylvitexin, in which the xylose is attached $\beta1\rightarrow2$ to the 8-sugar of the apigenin molecule, have also been reported in the Valencia orange (Horowitz and Gentili, 1966).

The distribution of these flavonoids in the genus *Citrus*, as far as is at present known, is shown in Table 5·10. The data are still far from being complete, but a few points of general interest may be noted. (1) The correlation between presence of flavanone 7-neohesperidosides and bitterness. Only one common fruit, the Seville orange, has a mixture of 7-neohesperidoside and 7-rutinoside; otherwise the correlation is complete. (2) The richness of flavonoids in the lemon skin probably only reflects the fact that it has been exhaustively extracted for its phenolics by Horowitz and Gentili. The other new flavonoid present, besides limocitrol and isolimocitrol, is limocitrin (8,3'-dimethoxy-3,5,7,4'-tetrahydroxyflavone); all three of these compounds occur as the 3-glucoside. (3) Most species can be distinguished by their flavonoid content. In the case of *C. jambhiri*, two geographical forms have different constituents.

Considering the extensive work that has been carried out on the flavonoids of lemon peel, it is not surprising that there have been some cases of misidenti-fication. Thus, Horowitz and Gentili (1960) found that the citrinin in the ponderosa lemon, reported as having the structure (44) by Yamamoto and Oshima (1931), is probably not present; they could only isolate neohesperidin from this fruit. Again, the aureusidin 6-rhamnosylglucoside isolated by Chopin *et al.* (1963) from lemon peel is probably an artifact of the isolation

(44)

procedure, formed from eriodictyol 7-rhamnosylglucoside by ring opening to the chalcone followed by oxidation; Horowitz and Gentili, in their extensive studies on the lemon, did not observe any aurone in their material.

B. OTHER FLAVONOIDS

Hesperidin, a major flavanone in *Citrus* (Aurantioideae), has also been iso-lated from *Xanthoxylum* (Rutoideae) (Arthur *et al.*, 1956). The flavonoids found in the other sub-families of the Rutaceae otherwise fall into three categories: quercetagetin derivatives, 2'-hydroxyflavones and isoprenoid-substituted flavonoids.

1. Quercetagetin Derivatives

A remarkable mixture of four highly methylated or methylenedioxy-substituted flavonols have been found by Briggs and Locker (1950) in the bark of *Melicope simplex* and *M. ternata*. The simplest is ternatin, a gossypetin derivative (45), the most complex meliternatin (46) which is derived from

(45) Ternatin   (46) Meliternatin

quercetagetin. The other two compounds are 3',4'-methylenedioxy derivatives of quercetagetin: the 3,6,7-trimethyl ether (melisimplin) and the 3,5,6,7-tetramethyl ether (melisimplexin). A fifth compound, melibentin (3,5,6,7,8-pentamethoxy-3',4'-methylenedioxyflavone), occurs specifically in *M. broadbentiana*, but it is accompanied by meliternatin, melisimplin and melisimplexin (Ritchie *et al.*, 1965).

Derivatives of herbacetin are found in other rutaceous members. The 8-methyl ether, tambuletin (47), occurs in the seed of *Xanthoxylum acanthopodium* (Balakrishna and Seshadri, 1947) and the 3,7,8,4'-tetramethyl ether, flindulatin (48), in the leaves of *Flindersia maculosa*. Flindulatin is possibly accompanied by the 7,8,4'-trimethyl ether in this plant (Bose and Bose, 1939).

(47) Tambuletin   (48) Flindulatin

2. 2'-Hydroxyflavones

A series of 2'-methylated flavones are apparently present in the root bark of *Casimiroa edulis* (Sondheimer and Meisels, 1960). They include 5,6,2'-trimethoxyflavone (49), 5,6,7,2'-tetramethoxyflavone (zapotin) and 5-hydroxy-6,7,2'-trimethoxyflavone (zapotinin); the simple 5,6-dimethoxy-flavone is also present. The structures of zaponinin and zapotin were established mainly by degradative experiments; recent synthetic studies by Farkas and Nogradi (1965) have thrown doubt on the correctness of these identifications.

It is interesting that another 2'-hydroxyflavone, 5,7,2',4'-tetrahydroxy-flavone (50), has had a chequered career as a natural product. It was first

(49)

(50) Norartocarpetin

reported (Dunstan and Henry, 1901) as "lotoflavin", the aglycone of a cyanogenetic glucoside in *Lotus arabicus* (Leguminosae) leaves but a later investigation (Doporto *et al.*, 1955) showed "lotoflavin" to be a mixture of quercetin and kaempferol. However, this 2'-hydroxyflavone has now been reported (Radhakrishnan *et al.*, 1965) as a heartwood constituent of *Artocarpus heterophyllus* (Moraceae), has been named norartocarpetin (50) and found to occur in conjunction with a related flavone with a C_5-isoprene side-chain in the 6-position. The 7-methyl ether of (50) is present in the wood of *A. integrifolia*. Although *Lotus* does not appear to have this 2'-hydroxyflavone, a related compound, oxyayanin-A (5,2',5'-trihydroxy-3,7,4'-trimethoxyflavone) has been found in the Leguminosae in *Distemonanthus benthamianus* by King *et al.* (1954) (but *see* footnote on p. 168).

Assuming that 2'-hydroxyflavones do occur in the Rutaceae (and there is no reason to doubt that 5,6,2'-trimethoxyflavone was correctly identified in *Casimiroa*), then this means that the Rutaceae joins the select group of families (all but two are in the Archichlamydeae) which make 2'-hydroxyflavonoids. The others are those already mentioned (the Moraceae and the Leguminosae) together with the Anacardiaceae, the Datiscaceae and the Amarantaceae. The relevant constituents in the latter families are: morin (3,5,7,2',4'-penta-hydroxyflavone) in *Lannea coromandelica* (Anacardiaceae) (Nair *et al.*, 1963), datiscetin (3,5,7,2'-tetrahydroxyflavone) in *Datisca cannabina* (Datiscaceae) and the isoflavone tlatlancuayin (1) in *Iresine* (Amarantaceae).

3. Isoprenoid-substituted Flavonoids

Several substances of this type have been reported. The flavanones (51) and (52) were found in the bark of *Melicope sarcococca* by Geissman (1958) and Brune and Geissman (1965) and amurensin (53) was isolated from *Phello-dendron amurense* by Hasegawa and Shirato (1953).

Substitution by isoprenyl groups is a rare occurrence in flavonoid bio-synthesis and the only other known cases are in the Moraceae (mentioned above), in the Berberidaceae [icariin (54) was isolated from *Epimedium macranthum* by Akai, 1955] and in the Leguminosae [sericetin (55) was iso-lated from *Mundulea seracea* by Burrows *et al.*, 1960]. How significant these

(51)

(52)

occurrences are, taxonomically, remains to be seen. It is, however, worth noting that the Rutaceae and Moraceae are already linked through 2′-hydroxyflavones and the Rutaceae and Berberidaceae by the presence in both families of the alkaloid, berberine.

(53) Amurensin

(54) Icariin

(55) Sericetin

XI. Flavonoids of the Umbelliferae

The Umbelliferae, a family of some 250 genera and 3000 species, is a natural taxon, its members being distinguished by the umbel-shape of the flower and the character of the fruit. Several umbellifers are of economic importance: vegetables such as carrot and parsnip and plants used in flavouring (anise, caraway) and in pharmacy (*Ammi*). Chemical studies, until recently, have largely been devoted to furocoumarins (*Peucedanum*) and furochromans (*Ammi*), to terpenoids (*Foeniculum*) and to alkaloids (*Conium*). Flavones or flavonols (Tables 5·2, 5·3) have been identified in some 10 species but no anthocyanin has been fully characterised, although the pigment in the black carrot of India (a variety of *Daucus carota*) is known to be a cyanidin derivative (Krishnamoorthy and Seshadri, 1962). The data on the flavone and flavonol glycosides is fragmentary but of some phytochemical interest. While the common rutin occurs in *Bupleurum* (Table 5·2), the rarer isorhamnetin potassium sulphate is present in the water hemlock *Oenanthe stolonifera* and the unusual disaccharide, apiosylglucose is present, attached to luteolin and apigenin, in *Apium*. Very recently, the flavone and flavonol aglycones in leaves of 285 species, representing 52% of the genera of the Umbelliferae, have been surveyed in the author's laboratory and the results are worth reporting for their general phytochemical interest.

The survey was carried out largely on herbarium leaf material, some specimens dating from 1840 or earlier. The majority (*ca* 80%) gave a positive result for flavonoids and species that were negative (most *Angelica* and *Seseli* spp.) generally had large amounts of other phenolics, particularly coumarins, in their leaves. The herbaria results were checked with those obtained on fresh leaf material for at least 30 species and there were no differences. The most significant discovery was that nearly all species had *either* flavonols *or* flavones, but not both; *Daucus* and *Laserpitium* species were the rare exceptions. Furthermore, flavones were found almost entirely in the more specialised taxa, flavonols being characteristic of the more primitive genera. Thus, the replacement of flavonol by flavone appears to be a character of phylogenetic significance within the family, supporting the picture already obtained from the general angiosperm survey (Bate-Smith, 1962) already discussed in Chapter 4 (*see* also under Tubiflorae in Chapter 6).

The flavonols found in the Umbelliferae were kaempferol and quercetin, with isorhamnetin being occasionally present. Myricetin, leucoanthocyanidins and ellagic acid were all absent, a result in line with the predominantly herbaceous character of the family. The flavones detected were luteolin and apigenin but there was evidence of methylated derivatives in certain species; 4'-methylapigenin (acacetin) and 4'-methylluteolin (diosmetin) were reported earlier in *Ammi* and *Conium* respectively (Table 5·3). The distribution of the common flavonols and flavones in the 12 tribes of the three sub-families of the Umbelliferae is outlined in Table 5·11; data for the closely related Araliaceae and Cornaceae are included for comparative purposes. Flavonols are, in fact, uniformly present in these families and in the two less specialised sub-families of the Umbelliferae, the Hydrocotyloideae and the Saniculoideae. By contrast, flavones occur in no less than six of the eight tribes in the Apioideae. The distribution within the Apioideae is clearly of evolutionary significance; compare the Peucedaneae with one flavone (*Ferula*) and 45 flavonol records with the highly specialised Scandiceae which has 27 flavone and only three flavonol records (*Molopospermum*). Flavones predominate over flavonols in only two tribes, the Scandiceae and the Dauceae, which are both closely related; flavones and flavonols occur with equal abundance in the Laserpiteae, a taxon particularly close to the Dauceae in morphology.

The taxonomy of the Umbelliferae at the generic and tribal level is still a matter of some dispute, so that clear-cut correlations between chemistry and taxonomy at the tribal level would hardly be expected. The chemical results now obtained may, indeed, be of help in future taxonomic revision. In Table

TABLE 5·11 Distribution of Flavonols and Flavones in the Umbelliferae, Araliaceae and Cornaceae

Family, sub-family and tribe	No. of spp. with flavonol	No. of spp. with flavone	No. of spp. surveyed	No. of genera surveyed	Approx. no. of genera described
CORNACEAE	4	0	4	2	15
ARALIACEAE	8	0	8	5	55
UMBELLIFERAE	162	61	293[1]	105	241[2]
Hydrocotylioideae					
1. Hydrocotyleae	11	0	11	5	} 34
2. Mulineae	7	0	8	8	
Saniculoideae					
1. Saniculeae	26	0	26	5	} 9
2. Lagoeaceae	2	0	2	2	
Apioideae					
1. Echinophoreae	2	0	2	2	5
2. Scandiceae	3	27	30	17	21
3. Coriandreae	3	0	3	2	5
4. Smirnieae	25	5	30	21	29
5. Apieae	76	19	98	52	85
6. Peucedaneae	45	1	49	23	41
7. Laserpiteae	9	11	17	5	8
8. Dauceae	4	5	9	3	4

[1] Total number of species in the Umbelliferae is about 2700 so this survey represents a 10% coverage.

[2] This figure represents a reasonably conservative estimate of the present taxonomic situation. It lies between the estimate of Engler and Melchior (1964) of 300 genera and those of Lawrence (1951) of 125 and Willis (1960) of 200.

5·11, the plants are laid out according to Engler and Melchior (1964); the Bentham and Hooker classification (1883) differs from this in several respects. The figures of 76 flavonol-containing species to 19 flavone-containing species in the Apieae indicates that the tribe as a whole contains mainly less specialised members. Some flavone-containing species may, in fact, not be properly classified. For example, *Cuminum cyminum* differs from most Apieae in having luteolin and is a monotypic genus which is sometimes included by taxonomists in the Scandiceae. Other flavone records in the Apieae may indicate the operation of evolutionary advancement at the generic level. Thus, in *Oenanthe*, 10 of the 11 species surveyed have flavonols but the water dropwort, *O. fistulosa*, one of the most specialised species in terms of floral morphology, has luteolin.

The flavone/flavonol data fit in reasonably well with what is known of the distribution of other chemical constituents. Thus, three predominantly flavonol-containing tribes (Apieae, Peucedanieae and Smirnieae) are all noted for their furocoumarins and acetylenic compounds occur in the latter two tribes and in the Araliaceae (flavonols in all eight species examined). The distribution of the glycosidic patterns of the flavones and flavonols have still to be studied but the present survey indicates that *C*-glycosyl derivatives are rare. *C*-glycosylflavones were only detected in *Opopanax* species (Peucedanieae) and the related *C*-glucosylxanthone, mangiferin, interestingly enough, was also once recorded, in *Colladonia triquetra* (Smirnieae).

CHAPTER 6

THE FLAVONOIDS OF THE SYMPETALAE

I. Introduction

The Englerian Sympetalae (cf. Engler, 1964) contain a smaller number of families (64) than the Archichlamydeae (over 200) and are divided into 11 orders. The largest group of families, 26 in number, are in one order, the Tubiflorae; these families are particularly closely related and are sometimes considered together as one super-family. The Sympetalae contain a second order, the Ericales, which, like the Tubiflorae, is unusual in being a natural rather than an artificial grouping (cf. Davis and Heywood, 1963).

The flavonoids of the Sympetalae have been reasonably well surveyed (*see* e.g. Bate-Smith, 1962; Lawrence *et al.*, 1939) and a sharp division exists between the families in the first five orders (Diapensiales, Ericales, Primulales, Plumbaginales and Ebenales) and subsequent families and orders. The leaf constituents of these five orders show many similarities to those of the Rosales; thus, flavonols and leucoanthocyanidins are frequent and ellagic acid is occasionally present. The glycosidic patterns of the anthocyanins (galactoside, rhamnoside and arabinoside) also show links with the Leguminosae and Rosaceae. By contrast, plants of other orders in the Sympetalae are characterised by having flavones rather than flavonols and by lacking leucoanthocyanidins and ellagic acid; also the anthocyanin glycosidic patterns are different, being mainly 3-glucoside, 3,5-diglucoside or 3-rutinoside-5-glucoside. These differences in flavonoid pattern fit in with the biological situation. Plants in the orders Ericales to Ebenales are recognised as having affinities with the Archichlamydeae (e.g. Plumbaginaceae with the Caryophyllaceae). They are also often woody in habit and are generally less specialised than the more highly developed Tubiflorae families.

Turning to the individual groups of flavonoids, the anthocyanins of the Sympetalae (Table 6·1) are generally similar to those found in the Archichlamydeae and there is no sharp discontinuity between the two groups. Unusually substituted anthocyanins are, perhaps, more frequent; anthocyanidins with A-ring methylation, for example, do not appear at all in the Archichlamydeae, but are found in three sympetalous families, the Primulaceae, Plumbaginaceae and Apocynaceae. Again, 3-deoxyanthocyanins, which are relatively rare elsewhere, occur with considerable frequency in the Gesneriaceae and are also known in the Bignoniaceae.

3-Glucosides and 3,5-diglucosides are the commonest anthocyanin glycosidic types in the Sympetalae, as in the Archichlamydeae; other types do not often appear. The characteristic occurrence of galactosides and arabinosides in the Ericaceae and Myrsinaceae and of rhamnosides in the Plumbaginaceae should be noted. The only other unusual types are 3-gentiobiosides and 3-gentiotriosides in *Primula sinensis*, but these are, strictly speaking, aberrant types which do not represent the genus or the Primulaceae as a whole. Acylated anthocyanins are restricted to the Tubiflorae, occurring abundantly in the Solanaceae and Labiatae but also being found in the Polemoniaceae and Orobanchaceae.

The flavonol glycosides found in the Sympetalae are listed in Table 6·2. There are few glycosidic types which do not also appear in the Archichlamydeae; the gentiobiosides and gentiotriosides reported in *Primula* and the branched trisaccharides of *Solanum* are the only exceptions. Otherwise, one may note the presence of rare types such as robinin in *Vinca*, also known in *Robinia* in the Leguminosae. 3,7-Diglycosides are not common, having only been noted in the Solanaceae and Compositae. The occurrence of quercimeritrin (quercetin 7-glucoside) in several composites is a point of some systematic interest.

7

TABLE 6·1. Anthocyanins of the Sympetalae

Group, order and family	Pigments present	Organ examined	Reference
ERICALES			
Ericaceae			
Rhododendron spp.	Cy 3-arabinoside Cy 3-galactoside Mv 3,5-diglucoside Cy 3-arabinoside-5-glucoside	Flower	Harborne, 1962a; A and Budin, 196
Vaccinium spp.	Cy, Pn, Dp, Pt and Mv 3-galactosides Cy, Pn, Dp, Pt and Mv 3-arabinosides Cy, Pn, Dp, Pt and Mv 3-glucosides	Fruit	Sakamura and F cis, 1961; Suc lainen and Kera 1961; Francis e 1966.
Empetraceae			
Empetrum nigrum	Dp and Mv 3-galactosides	Fruit	Hayashi *et al.*, 19
PRIMULALES			
Myrsinaceae			
Ardisia crispa	Cy and Pn 3-galactosides	Fruit	Harborne, 1962a.
Bladhia sieboldii	Dp and Mv 3-galactosides	Petal	Yeh and Huang, 1 cf. Harborne, 19
Primulaceae			
Cyclamen spp.	Cy, Pn, Dp, Pt and Mv 3-glucosides Cy, Pn and Mv 3,5-diglucosides	Flower	Karrer and Wid 1927; van Br 1962.
Primula spp.	Pg, Cy, Pn, Dp, Pt and Mv 3-glucosides	Leaf	Harborne and Sherratt, 1961.
	Hs, Mv, Rs, Dp and Pt 3,5-diglucosides	Stem and flower	
	Pg, Cy, Pn, Pt and Mv 3-gentiobiosides		
	Pg, Pn, Pt and Mv 3-gentiotriosides		
PLUMBAGINALES			
Plumbaginaceae			
Armeria spp.	Mv 3,5-diglucoside	Flower	Harborne, 1966a.
Ceratostigma plumbaginoides	Eu 3-glucoside	Flower	Harborne, 1966a.
Limonium spp.	Pt 3-rhamnoside-5-glucoside Cy and Dp 3-glucosides Dp 3,5-diglucoside	Flower	Harborne, 1966a.
Plumbagella micrantha	Cy 3-glucoside	Leaf	Harborne, 1966a.
Plumbago capensis	Cp 3-rhamnoside		
P. europea	Eu 3-glucoside	Petal	Harborne, 1962a,
P. pulchella	Pl 3-glucoside		1966a.
P. rosea	Dp, Cy and Pg 3-rhamnosides		
OLEALES			
Oleaceae			
Ligustrum vulgare	Mv 3-glucoside	Fruit	Hayashi, 1943.

TABLE 6·1—*continued*

ɔup, order and family	Pigments present	Organ examined	References
NTIALES			
ᴦcynaceae			
atharanthus roseus	Hs glycoside	Petal	Forsyth and Simmonds, 1957.
ᵼtiancaeae			
entiana cashmerica	Dp 3-glucoside	Petal	J. B. Harborne, unpublished.
ɔiaceae			
ubia akane	Cy 3-gentiobioside	Fruit	Hayashi, 1944.
ouvardia spp.	Pg 3-sophoroside	Petal	J. B. Harborne, unpublished.
Ioffmannia ghiesbreghtii	Cy 3-rutinoside	Leaf	J. B. Harborne, unpublished.
BIFLORAE			
ᵥvolvulaceae			
pomoea batatis	Cy 3-(caffeoylsophoroside)-5-glucoside	Tuber	C. E. Seaforth, unpublished.
. purpurea	Pg 3,5-diglucoside	Petal	Kataoka, 1936.
ᵊmonaceae			
ilea coronopifolia	Pg 3-(*p*-coumaroylglucoside)-5-glucoside	Petal	Harborne, 1962a.
ʰlox drummondii	Pg 3,5-diglucoside	Petal	Harborne, 1962b.
ᵃginaceae			
nchusa italica	Pt 3,5-diglucoside	Petal	Harborne, 1962b.
ᵇbenaceae			
ʰlerodendron splendens	Pg 3-glucoside	Petal	Harborne, 1962b.
ᵤantana chelsonii	Cy 3-glucoside	Petal	Harborne, 1962b.
ʳerbena spp.	Pg, Cy and Dp 3-glucosides Pg, Cy and Dp 3,5-diglucosides	Petal	Scott-Moncrieff and Sturgess, 1940; Harborne, 1962b.
ᵊiatae			
Ionarda didyma	Pg 3-(*p*-coumaroylglucoside)-5-glucoside	Petal	Harborne, 1958a; Asen, 1961.
ᵊerilla ocimoides var. *crispa*	Cy 3-(*p*-coumaroylglucoside)-5-glucoside	Leaf	Kuroda and Wada, 1935.
ᵃalvia splendens	Pg, Cy and Dp 3,5-diglucosides Pg 3-(*p*-coumaroylglucoside)-5-glucoside Pg, Cy and Dp 3-(caffeoylglucoside)-5-glucosides	Petal	Harborne, 1962; Asen, 1961.
ᴵyssopus officinale	Dp 3-(*p*-coumaroylglucoside)-5-glucoside	Petal	J. B. Harborne, unpublished.
ᵃnaceae			
ᵼtropa belladonna	Pt 3-(*p*-coumaroylglucoside)-5-glucoside	Flower and berry	Harborne, 1960a.
ᵣrowallia speciosa	Dp 3(di-*p*-coumaroylglucoside)-5-glucoside	Flower	Harborne, 1962b.

TABLE 6·1—*continued*

Group, order and family	Pigments present	Organ examined	References
TUBIFLORAE—*continued*			
Solanaceae—*continued*			
Brunfelsia calycina	Mv 3-(*p*-coumaroylrutinoside)-5-glucoside	Flower	Harborne, 1962b.
Cestrum purpureum	Pg and Cy 3-rutinosides	Corolla	Harborne, 1962b.
Iochroma coccinea	Pg 3-glucoside	Corolla	Harborne, 1962b.
I. tubulosa	Mv 3-glucoside	Corolla	Harborne, 1962b.
Lycopersicon esculentum	Pt 3-(*p*-coumaroylrutinoside)-5-glucoside	Leaf and stem	Harborne, 1962a.
Nicotiana tabacum	Cy 3-rutinoside	Corolla	Harborne, 1962b.
Petunia hybrida	Cy, Pn, Dp, Pt and Mv 3-glucosides, Cy 3,7-diglucoside Cy and Pt 3-sophorosides Cy 3-gentiobioside Pn, Dp, Pt and Mv 3-(*p*-coumaroylrutinoside)-5-glucosides	Petal	Harborne, 1960a; Birkofer *et al.*, 1 Meyer, 1964.
Solanum spp.	Pg, Cy, Pn, Dp, Pt and Mv 3-(*p*-coumaroylrutinoside)-5-glucosides Pg, Cy, Dp and Pt 3-rutinosides	Petal, stem and tuber	Harborne, 1960a; Dodds and L 1955.
Scrophulariaceae			
Anarrhinum bellidifolium	Dp 3,5-diglucoside	⎫	
Antirrhinum spp.	Cy 3-rutinoside Dp 3,5-diglucoside		
Asarina procumbens	Cy 3-rutinoside		
Digitalis purpurea	Cy and Pn 3,5-diglucosides		
Digitalis × fulvia	Cy 3-rutinoside		
Gambellia speciosa	Pg 3-rutinoside		
Linaria maroccana	Cy 3,5-diglucoside (in red form) Dp 3,5-diglucoside(in blue form)		
Maurandia speciosa	Dp 3,5-diglucoside	Petal	Harborne, 1962b, 1963c, 1966a.
Mimulus cardinalis	Pg and Cy 3-rutinosides		
M. glutinosus	Pg and Cy 3-glucosides Pg and Cy 3-rutinosides		
M. lewisii	Cy 3-glucoside		
Misopates orontium	Cy 3-rutinoside		
Nemesia strumosa	Cy 3-sambubioside		
Penstemon heterophyllus	Dp 3,5-diglucoside		
Phygelius capensis	Cy 3-rutinoside		
Scrophularia scopoli	Cy 3-rutinoside Cy 3-glucoside	⎭	
Torenia fourneri	Dp, Pt and Mv 3,5-diglucosides	Petal	Endo, 1962.
Verbascum phoenicium	Pg 3-glucoside Pg 3-rutinoside Pn 3-glucoside Pn 3-rutinoside	Pink petal Blue petal	Harborne, 1966a.
Zaluzianskya capensis	Cy 3-glucoside	Petal	
Bignoniaceae			
Arrabidaea chica	Carajurin	Leaf	Chapman *et al.*, 1

TABLE 6·1—*continued*

Group, order and family	Pigments present	Organ examined	References
UBIFLORAE—*continued*			
ignoniaceae—continued			
Campsis radicans	Cy 3-rutinoside	Flower	J. B. Harborne, unpublished.
Jacaranda acutifolia	Dp 3-glucoside	Flower	Billot, 1964.
	Dp 3,5-diglucoside		
esneriaceae			
b-family Gesnerioideae			
Achimenes cvs.	Pg or Mv 3-rutinoside-5-glucoside	Flower	⎫
Alloplectus vittatus	Lt 5-glucoside	Flower	
	Columnin	Leaf	
Columnea × *banksii*	Columnin	Leaf and flower	
Columnea microphylla	Columnin	Leaf and flower	
Columnea cv. "Stavenger"	Columnin	Leaf and flower	
Columnea kucyniakii	Columnin	Leaf	
Episcia reptans	Pg 3-rutinoside	Flower	
	Columnin	Leaf	
Gesneria cuneifolia	Gesnerin	Flower	
	Lt 5-glucoside		
G. ventricosa	Pg 3-rutinoside	Flower	
Koellikeria erinoides	Columnin	Stem	
× *Kohleria*	Mv 3-rutinoside-5-glucoside	Petal	
	Columnin	Leaf	
Kohleria eriantha	Gesnerin, Pg 3-rutinoside	Flower	
	Lt 5-glucoside	Sepal	⎬ Harborne, 1966c.
Nautilocalyx lynchii	Columnin	Leaf	
Rechsteineria cardinalis	Gesnerin	Flower	
	Lt 5-glucoside		
R. macropoda	Gesnerin	Flower	
Sinningia barbata	Columnin	Leaf	
S. speciosa	Mv 3-rutinoside	Flower	
Sinningia cv. (*gloriosa*?)	Pg and Cy 3-rutinosides	Flower	
Trichantha minor	Columnin	Flower	
b-family Cyrtandroideae			
Aeschynanthus obconicus	Pg 3-sambubioside	Flower	
A. parvifolius	Cy 3-sambubioside	Sepal	
	Pg 3-sambubioside	Petal	
Boea hygroscopica	Dp and Cy glucosides	Flower	
Chirita lacunosa	Mv glycoside	Flower	
Dichiloboea speciosa	Dp, Pt and Mv glycosides	Flower	
Ornithoboea wildeania	Pt and Mv glycosides	Flower	
Saintpaulia ionanthe	Mv 3-rutinoside-5-glucoside	Flower	
	Cy 3-sambubioside	Leaf	
Streptocarpus spp.	Mv 3-rutinoside-5-glucoside	Flower	
	Cy 3-sambubioside	Leaf	⎭
banchaceae			
athraea clandestina	Cy and Dp 3-rutinosides	Petal	J. B. Harborne, unpublished.

TABLE 6·1—*continued*

Group, order and family	Pigments present	Organ examined	References
TUBIFLORAE—*continued*			
Orobanchaceae—*continued*			
Orobanche minor	Dp and Pt 3-glucosides Cy and Mv 3-(caffeoylglucosides)	Petal and stem	Barloy, 1963.
O. elatior	Dp and Cy 3-glucosides	Petal	J. B. Harborne, un- published.
DIPSACALES			
Caprifoliaceae			
Sambucus nigra	Cy 3-sambubioside Cy 3-glucoside Cy 3-sambubioside-5-glucoside	Fruit	Harborne, 1962a; Reichel and Reic wald, 1960.
CAMPANULALES			
Campanulaceae			
Campanula sp.	Dp 3,5-diglucoside	Petal	Harborne, 1962b.
Compositae			
Callistephus chinensis	Pg and Cy 3-glucosides	Petal	Willstäter and Bu dick, 1916.
Cosmos sulphureus	Cy 3-rutinoside	Petal	Hayashi, 1941.
Coreopsis tinctoria	Cy 3-glucoside	Petal	Shimokoriyama, 19
Centaurea scabiosa	Cy 3,5-diglucoside	Petal	Bayer, 1958.
Chrysanthemum indicum	Cy 3-glucoside	Flower	Willstäter and Bolt 1916.
Dahlia variabilis	Pg and Cy 3,5-diglucosides	Flower	Hayashi, 1933; H. borne and Sherra 1957.
Happlopappus gracilis	Cy 3-glucoside Cy 3-rutinoside	Tissue culture	Harborne, 1964b; Ardenne, 1965.
Helenium autumnale	Cy 3,5-diglucoside	Flower	Willstäter and Bolt 1916.
Senecio formosus	Cy 3,5-diglucoside	Flower	J. B. Harborne, unpublished.
Solidago vigaurea	Cy 3-glucosylglucoside	Flower	Bjorkman andHol gren, 1958.
Zinnia elegans	Cy 3,5-diglucoside	Flower	Willstäter and Bolt 1916.

Anthocyanidin abbreviations: Pg, pelargonidin; Cy, cyanidin; Pm, peonidin; Dp, delphinid
Pt, petunidin; Mv, malvidin; Hs, hirsutidin; Rs, rosinidin; Eu, europinidin; Cp, capensinidin;
pulchellidin; Lt, luteolinidin.

    A real difference between the flavonols and flavones of the Archichlamydeae and those of the Sympetalae is the variety and abundance of 6- or 8-hydroxylated derivatives in the latter group. Quercetagetin and its methylated derivatives occur in the Ericaceae and Primulaceae, in many families of the Tubiflorae, in the Rubiaceae and with much frequence in the Compositae. Flavones with an extra hydroxyl in the 6- or 8-position are confined in their occurrence almost completely to sympetalous plants. They have been reported in the Labiatae, Bignoniaceae, Scrophulariaceae, Plantaginaceae, Pedaliaceae and the Compositae.

TABLE 6·2. Distribution of Flavonol Glycosides in the Sympetalae

Order, family and species	Flavonols present	Organ examined	References
ERICALES			
Ericaceae			
Arctostaphyllos uva-ursi	Qu 3-galactoside	Leaf	Kawaguchi *et al.*, 1939.
Gaultheria miqueliana	Qu 3-glucuronide	Leaf	Sasaki and Watanabe, 1956.
Ledum palustre	Qu 3-galactoside	Whole plant	Krug and Stepien, 1965.
Rhododendron bauhinnifolium	Qu 3-galactoside	Flower	Harborne, 1962a.
R. campyllocarpum	Qu 3-rhamnoside Qu 3-galactoside Qu 3-arabinoside	Flower	Harborne, 1962a.
R. falconeri	Qu 3-rhamnoside Qu 3-galactoside	Flower	Harborne, 1962a.
R. flavum	Qu 3-arabinoside Qu 3-galactoside	Leaf	Komissarenko *et al.*, 1965.
R. mucronatum and 43 other spp.	Azalein	Flower	Wada, 1956; Harborne, 1962a.
R. niligiricum	Qu 3-rhamnoside	Flower	Rangaswami and Sambamurthy, 1960.
Vaccinium macrocarpon	Qu and My 3-arabinosides Qu 3-galactoside Qu 3-rhamnoside My 3-digalactoside	Fruit	Puski and Francis, 1966.
V. myrtillus	Qu 3-glucosylglucoside ⎤ Qu 3-rhamnoside Qu 3-glucoside ⎬ Qu 3-arabinoside ⎦	Leaf	Ice and Wender, 1953.
V. uliginosum	Qu 3-galactoside Qu 3-glucoside	Leaf	Kawaguchi *et al.*, 1939.
V. vitis-idaea	Qu 3-galactoside Qu 3-glucoside	Leaf	
PRIMULALES			
Primulaceae			
Cyclamen cvs.	Qu 3-rutinoside Km and Qu 3-glucosides	Petal	van Bragt, 1962.
Primula sinensis	My 3-glucoside Qu and Km 3-gentiobiosides Qu and Km 3-gentiotriosides DihydroKm 7-glucoside	Petal	Harborne and Sherratt, 1961.
PLUMBAGINALES			
Plumbaginaceae			
Limonium gmelinii	My and Qu 3-rutinosides My 3-rhamnoside Isorhamnetin	Root	Chumbalov and Kil, 1962.

TABLE 6·2—*continued*

Order, family and species	Flavonols present	Organ examined	References
EBENALES			
Ebenaceae			
Diospyros kaki	Km 3-glucoside	Leaf	} Hattori, 1962.
D. lotus	My 3-rhamnoside	Leaf	
OLEALES			
Oleaceae			
Forsythia suspensa	Qu 3-diglucoside	Flower	Schindler, 1945.
Fraxinus excelsior	Qu 3-rhamnoside		Gintl, 1868.
Nerium oleander	Qu and Km 3-rutinosides	Leaf	Hörhammer, Wa‌ and Luck, 1956
GENTIANALES			
Gentianaceae			
Menyanthes trifoliata	Km 3-rhamnoside		Krebs and Mat‌
	Km 3-galactoside		1957, 1958.
	Qu 3-rutinoside		
Apocynaceae			
Holarrhena floribunda	Qu 3-glucoside	Leaf	Paris and Fouc‌ 1959.
Vinca herbacea	Km 3-rhamnosylgalactoside-7-galactoside	Petal	Rabaté, 1933.
	Qu 3-rutinoside	Leaf	Boichinov *et al.*, 1
V. major and *V. minor*	Km 3-rhamnosylgalactoside-7-galactoside	Petal	Rabaté, 1933.
TUBIFLORAE			
Convolvulaceae			
Calystegia hederacea	Km 3-galactoside	Leaf	} Hattori and Shi‌
C. japonica	Km 3-rutinoside	Leaf	} koriyama, 1956
Cuscuta reflexa	Km 3-glucoside	Stem	Subramanian and Nair, 1963.
Ipomoea sp.	Qu 3-glucoside	Tuber	Sakomoto, 1956.
Solanaceae			
Atropa belladonna	Qu 3-rutinoside	Leaf	Steinegger *et al.*, 1
	Km 3-galactosylrhamnoside		
Hyoscyamus niger	Qu 3-rutinoside	Leaf	Steinegger and S‌ nini, 1960b.
Lycopersicon	Qu 3-rhamnoside	Leaf	Wu, 1957.
esculentum	Qu 3-rutinoside	Leaf	Wu, 1957.
	Qu 3-rhamnoside	Skin	Wu and Burrell, 1
Nicotiana tabacum	Qu 3-OMe, Qu 3,3'-diOMe	Calyx	Yang *et al.*, 1960.
	Qu and Km 3-glucosides	Corolla and	Watanabe and V‌
	Qu and Km 3-rutinosides	leaf	der, 1965.
	Qu and Km 3-rutinoside-7-glucosides		
	Qu 7-glucoside		
Petunia hybrida	Qu and Km 3-sophorosides	Petal	Birkofer and Ka‌
	Qu and Km 3-sophoroside-7-glucosides		1962.
	Km 3-(feruloylsophoroside)		
	DihydroQu 4'-glucoside		

TABLE 6·2—*continued*

er, family and species	Flavonols present	Organ examined	References
BIFLORAE—*continued*			
naceae—*continued*			
olanum spp.	Qu and Km 3-glucosides My, Qu and Km 3-rutinosides Qu and Km 3-sophorosides Qu and Km (2^G-glucosylrutino- side)	Flower and leaf	Baruah and Swain, 1959; Harborne, 1962c.
. tuberosum	Km 3-sophorotrioside-7-rham- noside, Km 3-sophoroside-7- rhamnoside	Seed	Harborne, 1962c.
aginaceae			
accinia glauca	Qu 3-rutinoside	Leaf	Parthasarathy and Seshadri, 1964.
ophulariaceae			
ntirrhinum majus	Km 3,7-diglucoside Qu and Km 3-glucosides	Flower	Harborne, 1964.
PSICALES			
rifoliaceae			
ambucus canadensis	Qu 3-rutinoside	Flower	Sando and Lloyd, 1924.
iburnum opulus	Km 3-glucoside Km 3,7-diglucoside	Flower	Egger, 1961b.
MPANULALES			
npositae			
mbrosia artemisifolia	Isorhamnetin Qu 3-glucoside	Pollen	Hehl, 1919.
rnica montana	Qu and Km 3-glucosides	Flower	Friedrich, 1962a, b.
rtemisia abrotanum	Qu 3-rutinoside	Root	Kranen-Fiedler, 1956.
aeria chrysostoma	Qu 3-glucoside	Petal	Shimokoriyama and Geissman, 1960.
alandula officinalis	Isorhamnetin 3-rutinoside Isorhamnetin 3-glucoside	Petal	Friedrich, 1962a, b.
arthamus tinctorius	Km 3-rutinoside	Petal	Murti et al., 1962.
hrysanthemum coronaria	Qu 7-glucoside	Petal	Anyos and Steelink, 1960.
. segetum	Gossypetin 7-glucoside Qu 7-glucoside	Petal	Geissman and Steel- ink, 1957.
osmos bipinnatus	Qu 3-glucosylglucuronide	Leaf	Nakaoki et al., 1961.
rigeron spp.	Qu 3-rhamnoside	Petal?	Imai and Mayama, 1953.
upatorium cannabinum	Qu 3-rutinoside	Root	Morita, 1957.
elianthus annuus	Qu 7-glucoside	Petal	Sando, 1926.
elichrysum arenarium	Km 3-glucoside Km 3-diglucoside		Jerzmanowska and Grzybowska, 1960; Vrkoc et al., 1959.
atricaria chamomilla	Qu 7-glucoside	Petal	Hörhammer et al., 1963.

7*

TABLE 6·2—continued

Order, family and species	Flavonols present	Organ examined	References
CAMPANULALES—continued			
Compositae—continued			
Nardosmia laevigata	Qu 3-galactoside Qu 3-rutinoside	Leaf	Glyzin and Senov, 1965.
Senecio erraticus	Qu 3-rutinoside	Root	Santavy et al., 1957.
Solidago altissima	Qu 3-xyloside, Qu 3-glucoside, Qu 3-galactoside, Qu 3-rutinoside	Petal	Hirose, 1962.
S. canadensis, S. serotina and S. virgaurea	Qu 3-rhamnoside, Qu 3-rutinoside and Km 3-glucoside	Leaf	Fuchs, 1949; Skrzypczakowa, 1964.
Tagetes erecta	Km 3,7-dirhamnoside	Leaf	Morita, 1957.
Tussilago farfara	Qu 3-rutinoside Qu 3-galactoside	Flower	Olechowska-Baransk and Lamer, 1962.
Viguiera multiflora	Qu 7-glucoside	Flower	Shimokoriyama and Geissman, 1960.

Abbreviations: My, myricetin; Qu, quercetin; Km, kaempferol.

Finally, mention must be made of chalcones and aurones which occur with particular frequency in the tribe Helenieae of the Compositae. They also occur in several families of the Tubiflorae and have been found otherwise only in the Plumbaginaceae (*Limonium*). The distribution in the Sympetalae of all these substances will be discussed in more detail in the following sections.

II. Flavonoids of the Ericaceae

A. GENERAL PATTERN

The Ericaceae, a family of some 50 genera and 1350 species, is usually divided into four sub-families: the Rhododendroideae, the Arbutoideae, the Vaccinioideae and the Ericoideae. The family is mainly of ornamental value (e.g. *Rhododendron* and *Erica*), but *Vaccinium* is grown to a small extent as a fruit crop in North America. Bate-Smith (1962), in his leaf survey, found no obvious differences in flavonoid constituents at the sub-family level, except that in the Arbutoideae, myricetin (4 out of 9 species) and leucodelphinidin (6 out of 9) were particularly abundant. The Arbutoideae, furthermore, was the only sub-family to have ellagic acid (3 out of 9).

From what is known of the individual flavonoids in the Ericaceae, there are again no significant differences at the sub-family level. Thus, anthocyanins with a distinctive glycosidic pattern (3-galactoside) occur in *Empetrum, Erica, Rhododendron* and *Vaccinium*, representing three of the four sub-families.The rare 5-O-methylquercetin, first found in *Rhododendron*, also occurs in *Erica*. Again, quercetagetin occurs in both these genera and there are indications from Bate-Smith's leaf survey that this or related substances also occur in *Cassiope* and *Pieris* in the Arbutoideae. Finally the presence of flavanones in *Rhododendron* can be equated with the dihydrochalcones known in *Pieris*.

B. ANTHOCYANINS

Four anthocyanins have been fully identified in *Rhododendron* petals: cyanidin 3-galactoside (1) in *R. lateritium* (Harada and Saiki, 1956), cyanidin 3-arabinoside in *R. thomsonii* (Harborne, 1962a), malvidin 3,5-diglucoside (2) in *R. reticulatum* (Hayashi, 1944) and cyanidin 3-arabinoside-5-glucoside in the cultivar "Red Wing" (Asen and Budin, 1966). Surveys in the author's laboratory have shown that the first three anthocyanins occur as two mutually

(1)   (2)

exclusive patterns, red-flowered species having the 3-galactoside and 3-arabinoside of cyanidin, blue and mauve forms the 3,5-diglucoside of malvidin. Thus there is a rather remarkable association between diglucosylation and methylation in this genus, although malvidin is accompanied in some blue-flowered species by the less methylated petunidin and the unmethylated delphinidin. Species with red or crimson flowers examined were *arboreum*, *calophytum*, *cerasifolium*, *cinnabarinum*, *euchates*, *fictolacteum*, *glaucophyllum*, *habrotrichum*, *meddianum*, *oldhamii*, *thomsonii* and *venator*. Blue-flowered species having malvin are *augustinii*, *baileyii*, *lateritium*, *ponticum* and *reticulatum*. *R. lateritium* represents a bridge between the two groups as it is exceptional in having malvidin 3,5-diglucoside as well as cyanidin 3-galactoside. The fact that this blue-flowered species has some anthocyanin of the red types would suggest that blue-petalled species were derived from red-petalled species and not vice versa. One of the anthocyanins in red-petalled *Rhododendron* species is also present in the heaths, *Erica*. The bell-shaped flowers of these plants are pigmented by cyanidin 3-galactoside (1).

The only other ericaceous pigments to have been studied are those in the berries of four species of *Vaccinium*. These red and black berries are an extremely rich source of material and provide, incidentally, in most known instances (*see* Table 6·1), a remarkably complex range of anthocyanins. The pigments of the red cranberry, *V. macrocarpon* (formerly *Oxycoccus macrocarpus*), were examined by Grove and Robinson in 1931; these authors, not realising that a mixture of closely similar anthocyanins were present, incorrectly identified, by distribution tests, peonidin 3-glucoside as the colouring matter. The application of chromatographic techniques to these pigments allowed the separation of four substances: the 3-galactosides and 3-arabinosides of cyanidin and peonidin (Sakamura and Francis, 1961; Zapsalis and Francis, 1965).[1] The blueberry, *V. angustifolium*, has recently been examined by modern methods (Francis, Harborne and Barker, 1966) and has yielded the most complex mixture of anthocyanins yet found in a single plant tissue. The 3-glucosides, 3-galactosides and 3-arabinosides of five of the six common anthocyanidins (pelargonidin being the absentee) have been detected and, in addition, traces of 3-diglycosides are also present.

The berries of two other *Vaccinium* species have been examined more cursorily. Arabinose and glucose derivatives of cyanidin, delphinidin, petunidin and malvidin have been noted in *V. myrtillus*, a close relative of *V. angustifolium*, by Suomalainen and Keranen (1961) but none of the pigments was fully characterised. Hayashi (1949) isolated a single pigment (malvidin 3-galactoside) in crystalline form from the berries of *V. uliginosum*, but it seems unlikely that this is the only pigment actually present.

The pattern of the anthocyanins found so far in the ericaceous plants is thus a very consistent one; 3-galactosides and 3-arabinosides are the most widely distributed types and characterise the family.

C. FLAVONOLS

Rhododendron is probably the largest genus in the Ericaceae so it is not surprising that more work has been done on this genus than on any other. The glycosidic patterns of the flavonols isolated from *Rhododendron* show a remarkable similarity to those of the anthocyanins; quercetin 3-galactoside and quercetin 3-arabinoside are regularly present in the flowers. A similar relationship exists in *Vaccinium*, since 3-arabinosides and 3-galactosides of both anthocyanidins and flavonols have been found together in this genus too (Tables 6·1 and 6·2).

A third glycosidic type, showing no relationship to the anthocyanins, is very common in *Rhododendron*: this is the 3-rhamnoside. Besides quercetin and kaempferol 3-rhamnoside, azaleatin (5-*O*-methylquercetin) 3-rhamnoside has been isolated. This latter pigment, named azalein (3), was first identified in the

(3) Azalein   (4) 5-*O*-Methylmyricetin

cream-coloured blooms of *R. mucronatum* (Wada, 1956) but has subsequently been detected in no less than 44 out of 83 *Rhododendron* species examined. Its

[1] An additional five (minor) pigments have recently been detected in cranberries by F. J. Francis (unpublished results): Cyanidin and peonidin 3-glucosides an d3-monosides of delphinidin, petunidin and malvidin.

distribution is broadly related to classification, in that it is more frequent in lepidote species[1] (31/45) than in elepidote (13/38). At the series level, azalein (3) ranges in frequency from high (*Triflorum* 14/16, *Saluense* 5/5, *Lapponicum* 6/7) through medium to nil (*Thomsonii* 0/5, *Falconeri* 0/4). It is interesting that it is replaced by quercetin 3-rhamnoside in those species lacking it, suggesting it is formed from this precursor simply by methylation. Azaleatin has so far been found only in one other Ericaceae, in *Erica vagans* (1 out of 7 *Erica* species surveyed); its occurrence in the Plumbaginaceae will be discussed in a following section (IV.B).

Following the discovery of azaleatin in 1956, Egger in 1962 isolated the related 5-*O*-methylmyricetin (4) in three *Rhododendron* species and two cultivars: *R. catawbiense, R. japonicum, R. obtusum* and cvs. "Grandiflorum"and "Hindegrir". Myricetin also occurs in *Rhododendron* flowers and Egger (1962) theorised, from a rather small sample, that myricetin or its 5-methyl ether would only be found in association with malvidin in blue flowers and would not appear in white-flowered species. In fact a more recent survey (Harborne, 1965b) of seven white-flowered species showed that myricetin occurred in three, *R. aberconwayi, campanulatum* and *irroratum*; it was also detected in one yellow-flowered species, *R. ambiguum*.

D. FLAVANONES

In 1960, Arthur and Tam isolated two *C*-methylflavanones, farrerol (5) and matteucinol (6) from the leaves of *Rhododendron farrerae* and *R. simsii*. Study

(5) Farrerol   (6) Matteucinol

of the leaves of two other species gave negative results; *R. westlandii* had quercetin and myricetin only and *R. simiarum* dihydrokaempferol and quercetin. Leaf surveys in this laboratory would indicate that flavanones are fairly regular constituents in this genus. The interest in these flavanones is that they have only previously been detected in ferns (*see* p. 117) and their occurrence in the Ericaceae would be as the retention of a relatively primitive leaf character.

Other genera of the Ericaceae have not been surveyed as yet for flavanones, but it may be noted that the dihydrochalcones phloridzin (7) and asebotin (8)

---

[1] Those species with scurfy scales underneath the leaf.

have been isolated from *Pieris japonica* and *Kalmia* species in the Arbutoideae (Williams, 1966).

(7) Phloridzin   (8) Asebotin

III. Flavonoids of the Primulaceae

A. ANTHOCYANINS

The Primulaceae are a mainly North temperate family of 25 genera and 550 species, the two best-known genera being *Primula* and *Cyclamen*, plants of which are cultivated for their ornamental flowers.

The anthocyanins of these two genera have been investigated and show several interesting features. In *Primula*, the most distinctive feature is the presence of 7-*O*-methylated anthocyanidins. Hirsutidin (9) was first isolated from petals of *Primula hirsuta* by Karrer and Widmer in 1927. Later surveys

(9) Hirsutidin   (10) Rosinidin

have shown that it occurs fairly widely in the genus, having been recorded in the following species: *alpicola, auricula* (some forms), *cashmiriana, denticulata, hirsuta, japonica* (some forms), *nutans, polyanthus* (blue forms) and *purpurea.* Hirsutidin has been isolated as the 3,5-diglucoside, hirsutin, from *P. hirsuta* and *P. cashmiriana.* The related rosinidin (10) was found in petals of *P. rosea* by Harborne (1958b), but it is relatively rare, only being found otherwise with hirsutidin in blue forms of *P. polyanthus.* It again occurs as the 3,5-diglucoside (Harborne, 1963b).

Nearly all other *Primula* species examined have either malvidin 3-glucoside (11) or malvidin 3,5-diglucoside in their flowers. Thus malvidin 3-glucoside has been isolated from *P. auricula, polyantha* (purple forms), *pubescens* and *sinensis* and malvidin 3,5-diglucoside from *integrifolia, lichiangensis, obconica, poly-neura, secundiflora* and *viscosa.*

Many *Primula* species are very variable in their flower colour and provide excellent material for combined chemical and genetic studies (*see* Chapter 8).

Primula sinensis is a particularly interesting plant from this point of view and a series of 15 anthocyanins has been isolated from colour mutant forms in this species. The most notable feature of these pigments is their glycosidic pattern.

(11) Malvidin 3-glucoside

Besides the 3-glucosides found in "wild type" material, the mutants contain 3-gentiobiosides and 3-gentiotriosides. The glycosidic pattern, in fact, is identical with that of the flavonol glycosides which occur in this and many other *Primula* species. It seems unlikely that these more highly glycosylated anthocyanins are produced *de novo* in mutant *P. sinensis* forms; rather it appears that there has been a change in the specificity of the flavonol glycosylating enzyme.

Another species of genetic interest is *P. obconica*. While blue-petalled types contain malvin, red forms with delphin are known. Hybrids of these two types produce appreciable amounts of petunin. Hybridisation studies have also been carried out on *P. polyanthus* and *P. auricula* (*see* Chapter 8) but little is known of the pigments produced in the hybrids.

The pigments of the cultivated *Cyclamen*, a popular ornamental plant, have been studied, as with *Primula*, because of genetic interest in this plant. Malvidin 3-glucoside (11) was identified in this plant by Karrer and Widmer (1927) but the identification of most of the other pigments present was not completed until recently (van Bragt, 1962). The pigments are 3-glucosides or 3,5-diglucosides; 3-rutinosides reported by Seyffert (1955) are not present. However, there are several peonidin derivatives present, which have so far eluded identification. They all give, on hydrolysis, peonidin and glucose as the only sugar and occur in addition to the 3-glucoside and 3,5-diglucoside (van Bragt, 1962).

Van Bragt (1962) also surveyed 14 wild *Cyclamen* species and found that malvin was uniformly present in 12 of them: *C. africanum, cilicium, coum, cypricum, graecum, libanoticum, neopolitanum, persicum* (from which the cultivated forms are derived), *pseudibericum, purpurascens, repandum* and *rohlfsianum*. The two without malvin, *balearicum* and *creticum*, were white-flowered species. The flavonol glycoside rutin was found in 11 of the species.

B. FLAVONOLS

Flavonols are of frequent occurrence in the Primulaceae. Thus, Bate-Smith (1962), in surveying leaves of 16 species from eight genera (including representatives of all four sub-families), found myricetin in four, quercetin in 14 and kaempferol in 11. Leucocyanidin and leucodelphinidin were also present in 11 species and ellagic acid appeared in *Hottonia palustris*. Similar results have been obtained at the generic level. A survey of over 30 *Primula* species in the author's laboratory has shown that flavonols are present in both leaf and petal. Leucodelphinidin and leucocyanidin have been noted in petals of *P. aucalis* and *P. veris.*

Kaempferol, quercetin and myricetin glycosides have been fully investigated only in *P. sinensis* and *P. obconica*, but the glycosidic patterns present in these species— 3-glucoside, 3-gentiobioside e.g. (12) and 3-gentiotrioside—appear to be widespread in the genus. In a leaf survey of some 30 *Primula* species 3-gentiotrioside or related compounds were regularly present; the only exceptional species was *P. ciocantha*, which unexpectedly has rutin. This is an interesting reversal of the general situation where compounds like rutin are common and 3-gentiotriosides rare. Quercetagetin, which is present as a yellow flower pigment in at least six species, also occurs as the 3-gentiotrioside (13) Quercetagetin and an incompletely identified quercetagetin di(?) methyl ether (Harborne, 1965) have been found in petals of the ordinary primrose, *P. vulgaris*, the oxslip, *P. elatior*, the cowslip, *P. veris*, *P. polyanthus*, *P. sinopurpurea* and *P. silva-taroucana.*

(12) Quercetin 3-gentiobioside   (13) Quercetagetin 3-gentiotrioside

The distribution of quercetagetin derivatives in *Primula* is correlated with yellow flower colour. Thus, in *P. polyanthus*, yellow, brown and purple forms contain quercetagetin whereas white, purple-blue and blue varieties lack it.

The glycosides of kaempferol, quercetin and myricetin found in *P. sinensis* are accompanied by the corresponding dihydro derivatives. Only the dihydrokaempferol derivative has been identified fully, mainly because it accumulates in a particular colour mutant, the orange "Dazzler". It occurs as the 7-glucoside not as a 3-mono, di- or triglucoside as do the flavonols (Harborne and Sherratt, 1961).

C. FLAVONES

Flavone itself (14) occurs as a white, mealy scurf (farina) on the leaves, stems and inflorescences of a number of *Primula* species. It was first isolated by Muller (1915) from plants of *P. pulverulenta* and *P. japonica* and has subsequently been found on at least 25 other species (Blasdale, 1945). It represents a character of taxonomic interest within the genus and has also been recorded in *Dionysia*, which is closely allied to *Primula* (Klein, 1922). Free flavone has never been found elsewhere in the plant kingdom and it must be produced in *Primula* by an aberrant biosynthetic pathway; the simplest flavone produced by the well-recognised acetate–shikimic acid route is 5,7-dihydroxyflavone

(14) Flavone

(chrysin). The very fact that flavone is formed as a scurf on the leaf and stem indicates that it is an abnormal cell metabolite.

Two related compounds have also been reported in *Primula* farinas: 5-hydroxyflavone on *P. imperialis* var. *gracilis* (Karrer and Schwab, 1941) and *P. verticillata* (Blasdale, 1945) and 5,8-dihydroxyflavone on *P. modesta* (Nagai and Hattori, 1930) and *P. denticulata* (Blasdale, 1945). These two hydroxy-flavones are certainly rare and a farina survey in the author's laboratory of species in common cultivation in botanic gardens in this country disclosed only one further occurrence. Primetin or 5,8-dihydroxyflavone was found in the yellow farina of *P. chionantha*.

D. COMPARISON WITH THE MYRSINACEAE

In most taxonomic treatments, the Primulaceae and Myrsinaceae are placed together in the same order (e.g. in the Primulales by Engler and Prantl). The chemical data, as far as it goes, does not indicate a particularly close relationship between the two families. It is true that the general flavonoid pattern in leaves is similar; thus, Bate-Smith (1962), in examining five species of the Myrsinaceae, reports myricetin and quercetin once each, kaempferol in three species and leucodelphinidin in two. However, the anthocyanin glyco-sidic patterns are different in that *Ardisia* and *Bladhia* of the Myrsinaceae have 3-galactosides, instead of 3-glucosides as in Primulaceae and would suggest a link with the Ericaceae rather than with the Primulaceae.

None of the distinctive *Primula* flavonoids, such as quercetagetin, hirsutidin, rosinidin and flavone, appears in the Myrsinaceae. Nor do salicylic acid derivatives (2-hydroxy-4-methoxy- and 2-hydroxy-5-methoxybenzoic acids

(15) Embelin   (16) Rapanone

occur in roots of *P. veris*), although related derivatives are known in the Ericaceae. Finally, benzoquinones, not known in the Primulaceae, occur in the tribe Myrsineae of the Myrsinaceae; they are embelin (15) and rapanone (16).

IV. Flavonoids of the Plumbaginaceae

A. ANTHOCYANINS

The Plumbaginaceae is a small (*ca* 250 species) family which is placed by Engler in an order of its own, the Plumbaginales. The family divides itself naturally into two tribes, the Plumbagineae (*Plumbago, Ceratostigma, Plumbagella*) and the Staticeae (*Aegialitis, Armeria* and *Limonium*). The anthocyanins of species representative of most genera have been examined and the results show several biochemical differences at the tribal level.

Three new anthocyanidins have been isolated from the Plumbagineae, at least two of which are characterised by being methylated on the 5-hydroxyl. Capensinidin (17) was obtained, as the 3-rhamnoside, from the sky-blue petals of *Plumbago capensis* (Harborne, 1962a) and pulchellidin (18), as the 3-glucoside from *P. pulchella*. Europinidin, the anthocyanidin in flowers of *P. europea* and

(17) Capensinidin   (18) Pulchellidin

Ceratostigma plumbaginioides, has not yet been fully identified, but appears to be a dimethyl ether of delphinidin, which is different from malvidin and may be the 5,3'-dimethyl ether. The only species in the Plumbagineae to have common anthocyanidins in the petals is *Plumbago rosea*. This red-flowered plant is very different morphologically from the other *Plumbago* species. The flowers contain delphinidin, cyanidin and pelargonidin 3-rhamnosides; another unusual feature of *P. rosea* is the presence of two galloylglucose derivatives. A common anthocyanin, cyanidin 3-glucoside, has also been found in *Plumbagella micrantha*, but only in the leaves; the petals are too small for chemical study.

No 5-*O*-methylated anthocyanidin has yet been found in the Staticeae; the aglycones are malvidin, petunidin, delphinidin or cyanidin. However, the anthocyanins frequently have a glucose residue attached to the 5-hydroxyl group; 3,5-diglucosides and 3-rhamnoside-5-glucoside are commonly present. Petunidin 3-rhamnoside-5-glucoside (19), only known otherwise in the Vicieae of the Leguminosae, has been found in *Limonium binervosum*, *L. latifolium*, *L. transwallianum* and *L. vulgare*. Delphinidin 3-glucoside and 3,5-diglucoside occur in *L. sinuatum* and cyanidin 3-glucoside in *L. suworowii*.

By contrast, malvidin 3,5-diglucoside (2) is uniformly present in all *Armeria* species studied (*canescens, cantabrica, juncea, labradorica, maritima, montana* and *pubescens*).

(19) Petunidin 3-rhamnoside-5-glucoside

B. FLAVONOLS AND OTHER CONSTITUENTS

Just as 5-*O*-methylated anthocyanins characterise the Plumbagineae and are absent from the Staticeae, so 5-*O*-methylated flavonols are found almost exclusively in petals or leaves of *Plumbago* and *Ceratostigma* species. Azaleatin (5-*O*-methylquercetin), as the 3-rhamnoside azalein, was first noted in *P. capensis* flowers but has subsequently been detected in *P. pulchella, P. zeylanica* and *P. scandens* and in *Ceratostigma plumbaginioides* and *C. willmottianum*. The related 5-*O*-methylmyricetin occurs in traces in the leaves of *P. europea*, but the major flavonol is the 7-*O*-methyl ether (europetin) (20), a new flavonol. By contrast, leaves and petals of *Limonium* and *Armeria* are

(20) Europetin

(21) Myricetin 3-methyl ether

rich in myricetin, quercetin and kaempferol, but methyl ethers have not so far been detected in these organs of these plants. The only members of the Staticeae to have methylated flavonols are *Aegialitis annulata*, which is recognised to be a primitive relic species, and *Limonium gmelinii*. Baker (1948) has shown that *Aegialitis* is the only species in the group with monomorphic pollen; the remainder have dimorphic pollen; the 3-methyl ethers of myricetin (21) and quercetin occur abundantly in the leaves of this plant. A methyl ether of myricetin is also reported in the *tan root* of *L. gmelinii* by Chumbalov and Kil (1962), but the position of substitution is not known. It will be interesting to see if it is the 3-methyl ether as in *Aegialitis*. The leaf of *L. gmelinii* expectedly contains only myricetin.

The flavonol glycosidic pattern, as far as it is known, does not differ in the two tribes. Kaempferol 3-rhamnoside has been found in the flowers of *Plumbago rosea* and myricetin 3-rhamnoside from the leaves of several *Limonium* and *Armeria* species studied. Other glycosides are present but these have not been identified as yet.

The only record of a flavone in the Plumbaginaceae is luteolin in the flowers of *Limonium sinuatum*. There is also a single record of an aurone in the same genus—cernuoside (aureusidin 4-glucoside) (22) in the yellow petals of *L. bonduellii*. The only other flavonoid constituent noted in the family—leuco-delphinidin—occurs regularly in both tribes, it appears regularly in the root, often in the leaf and occasionally in the flowers of *Plumbago*, *Armeria* and *Limonium* species. A substance related to leucodelphinidin, i.e. ellagic acid, is rare, having only so far been noted in the stem of *Aegialitis annulata*.

While the two tribes differ considerably in their flavonoid constituents no single flavonoid can be relied on as a taxonomic marker (i.e., none is present in every taxa of one and absent from all of the other tribe). A related compound,

(22) Cernuoside   (23) Plumbagin

the quinone plumbagin (23) however, appears to fulfil this role. It is primarily a root constituent, though it does occur in leaves (*P. europea, pulchella* and *scandens*) and petals (*P. europea*) on occasion. A careful survey has shown that it occurs in roots of every member of the Plumbagineae obtainable (9 out of 9) but it is uniformly absent from the roots of the Staticeae (0 out of 40). Taking a conservative estimate of the number of species in the family, this represents 40% and 25% coverage, respectively, of the two sub-families. At the generic level, only *Vogelia* of the Plumbagineae and *Limoniastrum* in the Staticeae have yet to be examined.

C. FAMILY AFFINITIES

Before considering the possible significance of chemistry on the taxonomic position of the Plumbaginaceae, it is worth listing some of the chemical characters found in this and other families in this area of the Sympetalae (Table 6·3). The chemical information is still rather fragmentary, in the sense that the particular chemical characters have not been searched for in every family concerned. However, the Ericaceae, Primulaceae and Plumbaginaceae have been fairly well studied and the gaps in these families are probably more meaningful than in the cases of the Myrsinaceae and Ebenaceae. Almost nothing is known of the flavonoid chemistry of the other 10 families in these orders.

An affinity between the Plumbaginaceae and the Primulaceae was recognised by Hutchinson (1959), who placed them in the same order. On the chemical data (Table 6·3), one would not go so far as this but there is a clear link in that these are the only two families to have 7-O-methylated flavonoids. An affinity with the Ericaceae has not usually been recognised by

BLE 6·3. Distribution of Chemical Characters in the First Five Orders of the Sympetalae

Chemical[1] character	Generic distribution in				
	Ericaceae	Primulaceae	Myrsinaceae	Plumbaginaceae	Ebenaceae
thocyanidin galactoside	*Rhododendron* *Vaccinium*	—	*Bladhia* *Ardisia*	—	—
vonoids with 7-methylation	—	*Primula*	—	*Plumbago*	—
vonoids with 6-methylation	*Rhododendron* *Erica*	—	—	*Plumbago* *Ceratostigma*	—
Hydroxylated flavonols	*Rhododendron* *Erica*	*Primula*	—	—	—
icyclic acid derivatives	*Gaultheria* *Monotropa*	*Primula*	—	—	—
agic acid	*Arbutus*	*Hottonia*	—	*Aegialitis*	*Diospyros*
inones	—	—	*Ardisia*	*Plumbago* *Ceratostigma*	*Diospyros* *Drosera*

[1] Characters listed are those occurring in more than one family, but not in all five. Several families have characters not found elsewhere: Ericaceae, C-methylated flavanones (*Rhododendron*); Primulaceae, flavone (*Primula*), 3-glucoside as most common glycosidic pattern; Plumbaginaceae, aurone (*Limonium*), 3-rhamnoside as most common glycosidic pattern.

Characters uniformly present in these orders include flavonols (myricetin being particularly abundant) and leucoanthocyanidins.

taxonomists, but some weight might be given to the occurrence in both families of azaleatin. Indeed, the main conclusion that can be drawn from Table 6·3 is that the Ericaceae and Primulaceae, having three characters in common, are more closely related to each other chemically than any other pair of families. Just as europetin (7-O-methylmyricetin) provides a link between the Plumbaginales and the Primulales, so plumbagin relates the Plumbaginales to the Ebenales. Plumbagin is present in the Ebenaceae in *Diospyros maritima* and *Drosera* spp. and 7-methyljuglone (24) is present in *Diospyros hebecarpa*. Again, the quinone, lapachole (25), occurs in the Sapotaceae (*Illipe longifolia*).

(24) 7-Methyljuglone   (25) Lapachole

The general pattern of flavonoids in the Ebenaceae, for example, does not differ much from the Plumbaginaceae. Thus, Bate-Smith (1962), in surveying four *Diospyros* species, found myricetin in two, quercetin in three, kaempferol in three and leucocyanidin in two. Another link between the two families in these adjacent orders is ellagic acid, occurring as it does in both *Aegialitis* and *Diospyros*.

Chemical affinities between the Plumbaginaceae and the order Centrospermae are not particularly obvious, but on morphological grounds Friedrich (1956) considered placing the family in this archichlamydous order. The most characteristic chemical features of the Centrospermae are the purple betacyanin and the yellow betaxanthin pigments, but these substances certainly do not occur in the Plumbaginaceae (Beck *et al.*, 1962). The only chemical hint of the relationship proposed by Friedrich lies in the presence in the Plumbaginaceae of 3-, 5- and 7-O-methylated flavonols since O-methylated flavonols are also known in the Centrospermae (*see* Chapter 5). Although monomethylated flavonols are relatively rare in the Sympetalae, their occurrence in the Plumbaginaceae only confirms the generally accepted position of the Plumbaginales as one of the least specialised orders in this group.

D. PHYLOGENY OF THE FAMILY

As already mentioned, the chemical data provides striking support for the division of the family into the Plumbagineae and the Staticeae. Five characters divide the two tribes: quinone formation, A-ring methylation of flavonols and anthocyanins, glycosylation pattern of anthocyanins, gallotannin formation and aurone formation. The results are in excellent agreement with Baker's study of pollen morphology: the Plumbagineae are uniformly monomorphic, the Staticeae uniformly dimorphic with the exception of *Aegialitis*. The chemical data supports the inclusion of *A. annulata*, a monotypic relict species found only in New Guinea and North East Australia, in the Staticeae, since it lacks plumbagin or azaleatin. The presence of ellagic acid and 3-O-methylflavonols in this plant, nevertheless confirms that it is more primitive than other members of the Staticeae.

On the basis of incompatibility systems, general morphology and chromosome numbers, Baker (1948) considers the Staticeae to be more highly evolved than the Plumbagineae. If one makes some simple assumptions about the origins of the chemical characters present in the family, the data can be made to fit in very closely with Baker's ideas of the phylogeny (Table 6·4).

The results of the chemical survey of the Plumbaginaceae can also be used to relate plants at the species level. For example, all of the six *Plumbago* species so far studied differ from each other by at least one character. *P. rosea*, the only red-petalled species, lacks azaleatin but contains gallotannins in the flowers. *P. capensis* (blue flowers), *P. zeylanica* and *P. scandens* (both white flowers) are very similar morphologically, but *P. scandens* is unique in having

Table 6·4. Chemistry and Phylogeny in the Plumbaginaceae

Tribe	Pollen form	Genus and chromosome number	Chemical characters[2]
	Dimorphic[1]	*Limonium* ($x = 8, 9$)	Aurone (+) Anthocyanidin 5 – glucosylation (+)
		Armeria ($x = 9$)	Luteolin (+)
Staticeae		*Acantholimon*	
	Monomorphic	*Aegialitis*	3-O-Methylflavonols (−) Ellagic acid (−)
		Ceratostigma ($x = 7$)	Europetin (−)
Plumbaginae	Monomorphic	*Plumbago* ($x = 7$)	Plumbagin (−) Azaleatin (−)
	Common Ancestor	*Plumbagella* ($x = 6$)	Gallotannin (−)

[1] Recent studies (H. G. Baker, unpublished data) have shown that there are a few species in both *Limonium* and *Armeria* which are monomorphic, but, by and large, these genera characteristically have dimorphic pollen.

[2] Chemical characters are considered in this context to be primitive (−) or advanced (+) according to their general distribution in the plant kingdom and to the generally accepted position of the Plumbaginaceae as a member of the Sympetalae.

plumbagin in the leaf, *P. capensis* in having capensinidin in the petals. *P. europea* and *P. pulchella* (both with blue flowers) have the 7-O-methylated flavonol, europetin, in common but *P. europea* has a new dimethyl ether of delphinidin (europinidin) whereas *P. pulchella* has 5-O-methyldelphinidin (pulchellidin).

V. Flavonoids of the Gentianales

A. FLAVONOIDS PRESENT

Relatively little is known about the flavonoids in the families of this order, so that it is convenient to consider all the data that are available together and then to relate them to the taxonomic situation.

Three genera, *Gentiana*, *Catharanthus* and *Vinca*, have yielded anthocyanins of taxonomic interest. The pigment of the deep blue flowers of *Gentiana aucalis* (Gentianaceae) was isolated by Karrer and Widmer in 1927 and reported to be a p-coumaroyl derivative of delphinidin 3-glucoside. In their survey, Lawrence *et al.* (1939) reported an acylated delphinidin in *sino-ornata*, but ordinary delphinidin glycosides in *verna*, *campestris* and *septemfida*. The presence of acylated pigments in the genus, however, has yet to be confirmed by modern methods. A chromatographic study of the blue pigment of *G. cashmeriana* in the author's laboratory showed that the main pigment was simply delphinidin 3-glucoside.

The periwinkle, *Catharanthus roseus* (Apocynaceae),[1] is unusual in having hirsutidin (9) in its flowers (Forsyth and Simmonds, 1954), only known otherwise in the Primulaceae. Its glycosidic combination in this plant is not known but it is reported to be accompanied by traces of petunidin and malvidin. A delphinidin glycoside present in the closely related *Vinca major* has also been studied and it is unusual in apparently having four sugars (rhamnose, xylose, glucose and galactose) attached to the 3-hydroxyl. The original pigment (R_f 0·07 in BAW and 0·71 in 1% HCl) is relatively unstable and is readily converted to a second more stable pigment (R_f 0·20 in BAW and 0·43 in 1% HCl) whilst being purified by paper chromatography. The second pigment has rhamnose, xylose and galactose attached to the 3-position and gives delphinidin 3-galactoside on partial acid hydrolysis. The structure of the tetrasaccharide present in the original glycoside remains to be determined.

The only flavonol of interest in the Gentianales is robinin (26), a kaempferol glycoside first found in *Robinia* (Leguminosae). It has been found in three *Vinca* species, *major*, *minor* and *herbacea*. A study in the author's laboratory

(26) Robinin

(27) Acaciin

of leaf and flower flavonols in these three species has shown that the corresponding quercetin and myricetin glycosides also occur in *Vinca*; the quercetin derivative in leaf of *V. minor* and the myricetin derivative in the flower of *V. minor*. Robinin was also noted in leaf and petal of *Acokanthera spectabilis*, which is in a different tribe from *Vinca*, so that there are indications that it may be of regular occurrence in the Apocynaceae.

Glycoflavones represent a primitive chemical character in plants and it is interesting that one of the few records of glycoflavones in the Sympetalae is their occurrence in *Swertia japonica* (Gentianaceae). The 7-methyl ethers of isovitexin and isoorientin, together with orientin itself, have all been isolated from the leaves of this plant (Komatsu and Tomimori, 1966). Other flavones of

[1] This plant was formerly classified as *Vinca rosea* or *Lochnera rosea* but the correct name is as shown (Lawrence, 1951; Stern, 1966).

interest occur in genera which are now outside the Gentianales as at present constituted: in *Buddleia* (formerly Loganaceae but now in the Tubiflorae as Buddleiaceae, see below) and in *Olea* (Oleaceae formerly classified with the Apocyanaceae etc., but now in its own order, Oleales). The flavone in flowers of *B. variabilis* was first isolated by Yu (1933), who thought it was 3-acetylacacetin. It was re-investigated by Baker *et al.* (1951) who found it was identical with acaciin (27) (acacetin 7-rutinoside). The flavones in the leaves of the olive tree, *Olea europea*, have been examined by Bockova *et al.* (1964), who identified luteolin 7-glucoside and a partly characterised luteolin tetraglucoside. Also occurring in the leaves is the chalcone, olivin, which is reported to have the structure (28).

(28) Olivin

B. FLAVONOID PATTERN AND TAXONOMY

The Gentianales are a much disputed group and most of the families in the order have at one time or another been placed elsewhere in the classification. It is not surprising, therefore, to find that a distinctive flavonoid constituent (Table 6·5) uniting the order is lacking. Of the five families listed, the Loganaceae is particularly anomalous, since it is the only one in which flavonols

TABLE 6·5. Distribution of Flavonoids in Families of the Order Gentianales[3]

Family	Flavonol[1] occurrences	Leucocyanidin	Flower[2] anthocyanins	Flavonol glycosides	Flavone glycosides
Oleaceae	5/8	Absent	Mv (*Ligistrum*) Cy (*Jasminum*)	None studied	Absent
Loganaceae	0/2	Absent	Dp and Cy (*Buddleia*)	Not known	Acaciin (*Buddleia*)
Gentianaceae	2/4	Absent	Dp (*Gentiana*)	Rhamnoside Galactoside Rutinoside	Glycoflavones (*Swertia*)
Apocynaceae	12/15	5/15	Hs (*Catharanthus*) Dp (*Vinca*)	Glucoside Rutinoside Robinobioside	Absent
Asclepidaceae	4/5	1/5	Not studied	Not studied	Absent
Rubiaceae	3/9	5/9	Cy (3/4) Mv (1/4)	Not studied	Absent

[1] Kaempferol and/or quercetin; myricetin is not known in the Gentianales.
[2] Mv = malvidin, Cy = cyanidin, Dp = delphinidin, Hs = hirsutidin.
[3] Data mainly from Bate-Smith (1962) and Lawrence *et al.* (1939).

appear to be replaced by flavones. It is interesting that Engler and Melchior (1964) remove *Buddleia* from the Loganaceae, raise it to family rank to make the Buddleiaceae and place it in the Tubiflorae. The chemical data are very much in favour of this, for besides having acaciin, a flavone otherwise known in *Linaria* (Scrophulariaceae), *Buddleia* contains several other characteristic Tubiflorae constituents, namely the caffeic ester, orobanchin, the water-soluble carotenoid, crocein, and the iridoid, aucubin.

The Oleaceae, like the Loganaceae, contains both flavones and iridoids but is the only family in this group to lack alkaloids. Instead, it contains 6,7,8-trisubstituted coumarins (in *Fraxinus*). Thus, the chemistry justifies its new position in a separate order of its own. The Gentianaceae has no distinctive flavonoid constituents, but is unusual in the order in having xanthones (in the roots). The Apocynaceae and Asclepiadaceae are the only two families to show much similarity in flavonoid constituents (both having leucocyanidin and flavonols in the leaf). They are, in fact, placed together in a separate order by Hutchinson (1959) and are recognised as being closely related from the similarities in their saponin contents (Hegnauer, 1964). Finally, the Rubiaceae, which was formerly placed in the Rubiales with Dipsacaceae and Caprifoliaceae, in its flavonoid pattern fits in reasonably well with its new neighbours. It is interesting that leucodelphinidin, which is relatively rare in the Sympetalae, has been noted in both *Rondeletia cordata* (Rubiaceae) and *Apocynum androsaemifolium* (Apocynaceae).

VI. Flavonoids of the Tubiflorae

A. ANTHOCYANINS

Anthocyanins have been fully identified in 10 families of the Tubiflorae and these are listed in Table 6·1. Additional information about the pigments of these families is available in the surveys of Lawrence *et al.* (1939), Beale *et al.* (1941) and Forsyth and Simmonds (1954). A most striking feature is the difference in anthocyanidin type between families which are mainly temperate in origin (e.g. the Polemonaceae) and those which are largely tropical or subtropical (e.g. the Bignoniaceae). In temperate plants, which are mainly pollinated by bees, there is natural selection for a blue flower colour whereas in tropical plants, which are often bird pollinated, red and scarlet flowers are favoured.

These considerations explain why delphinidin occurs in the flower of almost every species of the Polemoniaceae, Hydrophyllaceae and Boraginaceae that have been surveyed (*ca* 30 species). By contrast, in the Bignoniaceae, of eight species surveyed, only one has delphinidin but five have cyanidin and two pelargonidin. Similarly, in the Acanthaceae, there are seven species recorded with delphinidin, 13 with cyanidin and three with pelargonidin. In the Labiatae, which is partly tropical, partly temperatre, the figures are delphinidin, 12; cyanidin, six; pelargonidin, eight. In New World species of the

Gesneriaceae, selection for red colour has led to the production of 3-deoxy-anthocyanins. Gesnerin, luteolinidin 5-glucoside or columnin have been found in 18 of 21 New World species whereas Old World gesnerads (25 spp. examined) have ordinary anthocyanins, with delphinidin being particularly common (*see* Table 6·1). In *Streptocarpus*, a South African genus, of 18 species studied, 16 are pigmented in their flowers by delphinidin or its methyl ethers.

(29) Columnin

The distribution of methylated anthocyanins is another character which distinguishes some Tubiflorae families from others. Methylated pigments are only frequent in the Gesneriaceae and Solanaceae. While malvidin is common in the first family, petunidin is rather characteristic and unusually abundant in some genera of the second family. It is, perhaps, no coincidence that petunidin was first isolated from the garden *Petunia* (Willstäter and Burdick, 1917), since it also occurs in almost every species of the tuberous *Solanums*; malvidin, by contrast, is rare in this genus. Petunidin is also a major pigment in flowers and fruits of the deadly nightshade, *Atropa belladonna*, and in the stems and leaves of the tomato, *Lycopersicon esculentum*. Petunidin, as the 3,5-diglucoside, also occurs as the major flower pigment of *Anchusa* (Boraginaceae).

Methylated anthocyanins are relatively rare in the Verbenaceae and Scrophulariaceae. In the Verbenaceae, of 25 species surveyed, only one, *Duranta repens*, has a methylated pigment (malvidin)(Forsyth and Simmonds, 1954). Again in the Scrophulariaceae, only three out of 28 species have methylated anthocyanins: *Torenia* (malvidin), *Digitalis* (peonidin) and *Angelonia* (malvidin).

Only six glycosidic patterns have been found so far among the anthocyanins of the Tubiflorae, and two of these are relatively rare. 3,5-Diglucoside is a very common type, being of particularly frequent occurrence in the Convolvulaceae, Polemoniaceae, Labiatae and Scrophulariaceae. 3-Rutinoside-5-glucosides are more or less restricted to the Solanaceae (especially common in *Solanum*) and to the Gesneriaceae (especially common in *Streptocarpus*). 3-Rutinosides are found in these families and also in the Scrophulariaceae and Bignoniaceae. Genera of the latter two families have either 3-rutinoside or 3,5-diglucoside, but rarely both types. *Antirrhinum* is exceptional in this respect but, even here, 3-rutinoside is found in one section of the genus (*Antirrhinum*, Old World) and 3,5-diglucoside in the other (*Saerorhinum*, New World). The 3-sambubiosides of cyanidin and pelargonidin have been found only twice in the Gesneriaceae (in *Aeschynanthus* and *Streptocarpus*) and once in the Scrophulariaceae (in *Nemesia*). 3-Sophoroside is a type which is so far only known in *Petunia*, but it is not typical of this genus. 3-Rutinoside-5-glucoside is a much more characteristic type in *Petunia* and the anthocyanidin 3-sophorosides found in the mutant forms of the cultivated *Petunia* are probably synthesised by the same enzymes which control the production of flavonol 3-sophorosides in this plant.

Acylation of the anthocyanins is a character which distinguishes plants in the Tubiflorae from those in every other sympetalous order. Even in the Tubiflorae, acylation is not common, having been found so far in only four families—the Labiatae, the Solanaceae, the Polemoniaceae and the Orobanchaceae. Of these, only the Labiatae and the Solanaceae regularly have acylated pigments. Acylation is, indeed, probably extremely common in the Labiatae, although such pigments have only been completely characterised in *Monarda, Perilla* and *Salvia*. In the survey of Lawrence *et al.* (1939), acylated pigments were detected in 14 of 23 Labiatae species examined, but this may be an underestimate of the frequency due to the imprecise methods then used. Chromatographic studies in the author's laboratory have failed to reveal a labiate which did not have an acylated anthocyanin.

The best known and characterised labiate pigments are monardein (30), pelargonidin 3-(*p*-coumaroylglucoside)-5-glucoside from blooms of *Monarda didyma* (e.g. cv. "Cambridge Scarlet"), and salvianin (31), pelargonidin 3-(caffeoylglucoside)-5-glucoside from petals of the scarlet sage *Salvia splendens*. Neither pigment contains malonic acid as a second acylating group, as was once thought (cf. Harborne, 1964). Related cyanidin and delphinidin derivatives have also been found: e.g. in *Salvia splendens* cv. "Violet Flame" by Asen (1961).

(30) Monardein

(31) Salvianin

The first acylated pigment to be isolated from the Solanaceae was negretein (32), from the purple-black salad potato "Negresse" by Chmielewska (1936). Complete structures for this and the five related acylated pigments in the potato plant were advanced by Harborne, (1960b, 1964a). The most common pigment present in tuber, haulm, leaf or petal of the cultivated potato, *Solanum tuberosum*, and its related wild species is, however, the petunidin analogue, petanin (33), often accompanied by traces of the delphinidin derivative,

(32) Negretein

(33) Petanin

delphanin. Besides occurring in flowers of some 60 tuber-bearing *Solanum* species, petanin is the principal pigment of flowers and fruits of *Atropa belladonna*, berries of *Solanum quitoense*, *S. guineese*, flowers of *S. seaforthianum* and leaves of tomato, *Lycopersicon esculentum* (Harborne, 1960b). Petanin and delphanin have also been detected in berries or flowers of *S. gsoba*, *S. americanum*, *S. interandinum*, *S. intrusum* and *S. auriculatum* (Briggs *et al.*, 1961). Delphanin alone pigments the skin of most varieties (e.g. "Burma") of the aubergine or egg plant, *S. melongena* (Abe and Gotoh, 1959); it is occasionally replaced by delphinidin 3-rutinoside (e.g. in cv. "Black Beauty"). Negretein (30), petanin and delphanin also occur in abundance in flowers of the garden *Petunia* (Harborne, 1960b), where they are accompanied by several non-acylated pigments which are 3-glucosides or 3-sophorosides (Birkofer *et al.*, 1963; Meyer, 1964).

The cyanidin, peonidin and pelargonidin analogues of negretein, i.e. cyananin, peonanin and pelanin, are restricted in their occurrence to the cultivated potato. Peonanin and pelanin, for example, provide the coloration on the skin of the potato variety "King Edward". Anthocyanins having two *p*-coumaroyl substituents are known in *Browallia speciosa* and in *Solanum guineese* (Francis and Harborne, 1966b). Whereas both *p*-coumaric and caffeic acids appear as acylating groups in the labiate pigments, only *p*-coumaric acid has so far been detected in this role in the Solanaceae. The same acid also appears in the anthocyanin of *Gilia coronopifolia* (Polemoniaceae) whereas caffeic acid is the acyl group of the anthocyanin in *Orobanche minor* (Orobanchaceae).

B. FLAVONOLS

Flavonols are only found regularly in one family in the order, in the Solanaceae. They do appear in the Labiatae, where they are confined to the genera *Lamium* and *Prunella* (Hörhammer and Wagner, 1962), but are otherwise rare. It is interesting that quercetin and kaempferol have been recorded (Bate-Smith, 1962) in the leaf of *Nolana humifusa* (Nolanaceae), because this family is known to be very closely related to the Solanaceae.

Detailed studies have been carried out on the flavonol glycosides of three solanaceous plants, tobacco, potato and petunia. In *Nicotiana tabacum*, their distribution has been studied in different parts of the flower by Watanabe and Wender (1965). Whereas the 3-glucosides, 3-rutinosides and 3-rutinoside-7-glucosides of kaempferol and quercetin all occur in the corolla, only rutin is present in the stamen. Rutin is also present in the calyx, where it is accompanied by the 3-monomethyl and 3,3′-dimethyl ethers of quercetin. No flavonols are present in the pistil. Many of the above flavonols occur in the tobacco leaf and the brown colour produced during the "curing" or processing is due to complex of rutin with chlorogenic acid, scopolin and certain metals (Jacobsen, 1961).

The most interesting flavonol in *Petunia* flowers is petunoside (34), an acylated kaempferol 3-sophoroside isolated by Birkofer and Kaiser in 1962. Isolation of 2-feruloylglucose from its partial acid hydrolysis proves that the ferulic acid is linked $\beta 1 \rightarrow 2$ to the second glucose residue of the glycoside (Birkofer *et al.*, 1965). Petunoside is thus the first acylated flavonoid to be completely characterised. Another unusual constituent in *Petunia* petals is dihydroquercetin, which occurs here as the 4′-glucoside (35).

(34) Petunoside

(35) Dihydroquercetin 4′-glucoside

In the tuber-bearing *Solanums*, the flavonol glycosides of petals and seeds of 60 species and 39 cultivars have been surveyed (Harborne, 1962c). Four glycosidic types are present: 3-glucoside, 3-rutinoside, 3-sophoroside and

3-(2^G-glucosylrutinoside) (Harborne, 1962c and unpublished results). Several points of systematic interest arise from the survey.

(1) Kaempferol 3-glucoside provides a "species marker", since it is confined to flowers of *S. santolallae*.

(2) Species belonging to the series Conicibaccata, Piurana and Demissa are characterised by being rich in kaempferol glycosides. These species are closely related genetically, judging by cytological and "crossability" studies (Marks, 1965). *S. hougasii*, a hexaploid, is exceptional among all species in having higher concentrations of kaempferol than quercetin glucosides in the flower.

(3) The flavone, luteolin 7-glucoside, occurs only in the tetraploid *S. stoloniferum* and in a triploid hybrid species *S. × vallis-mexici* derived from it (Marks, 1958). It is quite characteristic of *stoloniferum*, being present in 34 of 38 clones examined.

(4) Myricetin 3-rutinoside is normally absent from white-flowered species, since it occurs in association with the anthocyanins, delphanin and petanin, in coloured flowers (25% of wild species). It was found exceptionally in a white-flowered form of *S. soukupii*.

(5) Quercetin 3-(2^G-glucosylrutinoside) occurs in flowers of just over one-third of the wild species. It occurs in the flowers of some cultivars (e.g. "Arran Victory" and "Gladstone") and not others (e.g. "Majestic" and "Kerr's Pink") and thus constitutes a useful varietal character.

(6) Rutin, which occurs in 88% of wild species and all cultivars, also occurs regularly in other Solanaceae. It has been recorded in *Solanum aviculare, S. brevidens* and *S. seaforthianum* and in *Datura, Lycopersicon* and *Nicotiana*.

(7) Kaempferol 3-sophoroside-7-rhamnoside (36) and kaempferol 3-sophorotrioside-7-rhamnoside (37) are restricted to the seeds of the potato and

(36)   (37)

its related wild species. These flavonols occur in all tuber-bearing *Solanum* species but have not been detected in any other section of the genus *Solanum* nor in any other member of the Solanaceae.

Few flavonols have been studied in the Tubiflorae outside the Solanaceae. An interesting yellow fluorescent quercetin derivative in the flowers of *Lamium album* is being investigated in this laboratory; it appears to be the 5-glucoside of quercetin. The flavonol glycosides of *Antirrhinum* have been identified in connection with genetic studies (cf. Table 6·2). One may note the occurrence of kaempferol 3,7-diglucoside as providing a link with the Solanaceae, from which 3,7-diglucosides have also been isolated. Quercetagetin derivatives have also been found in the Tubiflorae, but, for convenience, these will be discussed in the following section on flavones.

C. FLAVONES

1. The General Pattern

Flavones replace flavonols as the major leaf and petal flavonoids in most families in the Tubiflorae. Apigenin and luteolin, as the 7-glucosides, occur very frequently in these families. Chrysoeriol (3'-methylluteolin) has been recorded in the Scrophulariaceae and tricin in the Orobanchaceae. Flavones occur as 7-glucuronides, rather than as 7-glucosides, in certain members of the Labiatae and Scrophulariaceae.

Families of the Tubiflorae are particularly rich in flavones with unusual substitution patterns. Flavones with an extra 6-hydroxyl group (baicalein, scutellarein etc.) and 4'-methylated flavones are particularly characteristic, being found in the Labiatae, Bignoniaceae, Gesneriaceae and Scrophulariaceae; 2-hydroxyflavones have recently been isolated from the Labiatae and the Acanthaceae.

2. Flavones of the Verbenaceae

Luteolin and apigenin presumably occur regularly in this family since flavonols are absent from the leaves (Bate-Smith, 1962) but firm records are scanty. Both flavones are definitely present in petals of *Verbena hybrida* (Harborne, 1959). The only other genus to have been investigated is *Vitex* and this because the flavones occur here atypically as *C*-glycosyl derivatives; *Vitex* and *Helichrysum* (Compositae) (Rimpler *et al.*, 1963) are the only genera in the Tubiflorae known to have glycoflavones. Vitexin (38) was first isolated from the heartwood of *Vitex lucens* by Perkin (1900) but its structure was not fully established until 1962 (Rao and Venkateswarlu). Recently, Seikel *et al.* (1966) have thoroughly investigated all the glycoflavones in the species and have isolated, among other compounds, 6-*C*-glucosylapigenin (isovitexin), 6,8-di-*C*-glucosylapigenin (vicenin) (39) and the related luteolin derivatives.

(38) Vitexin   (39) Vicenin

The only other flavone of note found in *Vitex* is casticin, the 3,6,7,4'-tetramethyl ether of quercetagetin (40), which was found in the seeds of *V. agnuscastus* by Belic *et al.* (1961). According to a survey of Hänsel *et al.* (1965), casticin occurs in the leaf or wood of *V. agnus-castus*, *V. negundo* and *V. trifolia* (all of Mediterranean or African origin) but is absent from *V. megapotamica* (a South American plant) and *V. lucens* (from New Zealand), its distribution thus being related apparently to plant geography. Vitexin, by contrast, has been

(40) Casticin

found uniformly in all the above species and also in *V. peduncularis* (Rao and Venkateswarlu, 1962), *V. altissima* and *V. leucoxylon* (Rao, 1965). *V. megapotamica*, besides lacking casticin, also lacks two iridoids, aucubin and agnoside, which occur in the leaves of all other species but it has a triterpene constituent not present elsewhere in the genus.

3. Flavones of the Labiatae

The flavones in leaves and petals of the family were surveyed by Hörhammer and Wagner (1962), who reported that luteolin was the chief flavone of *Lycopus, Lavandula, Teucrium, Mentha, Salvia, Stachys* and *Thymus*. The same authors found that diosmin was confined to *Mentha* and 5,6,7-trihydroxyflavones to *Scutellaria* and *Galeopsis*, these latter substances not occurring in *Teucrium* or *Thymus* as earlier reported. Flavonols kaempferol and quercetin, as already has been noted, are confined to two genera, *Lamium* and *Prunella*. Only four luteolin or apigenin glucosides have so far been isolated from these plants: apigenin and luteolin 7-glucosides from *Lycopus virginicus* (Hörhammer and Wagner, 1962); luteolin 7-glucoside and 7-glucosylglucoside from *Thymus vulgaris* (Awe *et al.*, 1959); and luteolin 3′-glucoside from leaf of *Dracocephalum thymiflorum* (Litvinenko and Sergienko, 1965).

The most interesting labiate flavones, however, are the rare baicalein (41) and scutellarein (42), which occur in leaves and roots of *Scutellaria altissima, S. baicalensis* and *S. galericulata* as the 7-glucuronides (Marsh, 1955). Chrysin

(41) Baicalein   (42) Scutellarein

7-glucuronide accompanies these flavones in the common skullcap, *S. galericulata*. A remarkable diverse group of flavones was found in the leaf of *S. epilobifolia* by Watkin (1960) who was the first to use chromatographic methods

8

of separation with these plant substances. He isolated baicalein, 5,7,8-tri-hydroxyflavone, 5,7-dihydroxy-2′-methoxyflavone (43) and 5,6,7-trihydroxy-flavanone, the first three occurring as the 7-glucuronides.

(43)   (44) Xanthomicrol

That unusual flavones are not confined to *Scutellaria* is shown by the discovery of xanthomicrol (44) in yerba buena, *Satureia douglasii*, by Stout and Stout (1961). Clearly, a range of rare and interesting flavones remains to be discovered in plants of this family. In many species, as in *Scutellaria epilobifolia* (see above), related flavanones will no doubt be found with the flavones. Already the rare 4′-methyl ether of naringenin (isosakuranetin) has been isolated as the 7-rutinoside from leaf of *Monarda didyma* (Brieskorn and Meister, 1965); the related flavone acacetin must surely occur in the same or a related plant.

4. Flavones of the Solanaceae

This family is more noted for its flavonols than for its flavones (see earlier section). Luteolin and apigenin were found in some of 90 *Solanum* and *Nicotiana* species surveyed (Hörhammer and Wagner, 1962). The only glycoside to have been isolated is luteolin 7-glucoside in petals of *Solanum stoloniferum* (Harborne, 1962c).

5. Flavones of the Scrophulariaceae

The flavones of *Digitalis* have been fairly extensively investigated inciden-tally to the general pharmacological interest in these plants as sources of cardio-active steroids, the digitalins. A number of highly substituted 6-hydroxylated flavones and flavonols have been isolated from several species, the most remark-able being digicitrin (45), which occurs in the leaves of the common foxglove, *Digitalis purpurea* (Meier and Furst, 1962). The B-ring substitution pattern (3′-hydroxy-4′,5′-dimethoxy) of digicitrin is a very unusual one, the only

(45) Digicitrin   (46) Calycopterin

other known example being in the isoflavone, irigenin, of *Iris* roots (*see* Chapter 7, p. 244). Digicitrin occurs in the lipid-soluble fraction of the foxglove leaf extract; the flavones in the aqueous phase are luteolin 7-glucuronide and a luteolin 7-glucosylglucuronide; luteolin 7-glucoside, reported in *D. purpurea* by Hukuti (1936), could not be detected in these leaves.

A flavonol related to digicitrin has been isolated from *D. thapsii* and has been named calycopterin (46). Comparison of its proposed structure with that of digicitrin suggests that the methyl group assigned to the 5-position should, perhaps, be placed on the 6-hydroxyl. It is significant that the structure of calycopterin has not yet been confirmed by synthesis and also that a similar rearrangement had had to be made to dinatin, the flavone of *D. lanata* leaves; the original structure (5,6,7-trihydroxy-4'-methoxyflavone) has been shown to be incorrect and dinatin is now formulated as 5,7,4'-trihydroxy-6-methoxy-flavone (47) (Doherty *et al.*, 1963). *D. lanata* leaves also contain the related

(47) Dinatin   (48)

3'-methoxyflavone (48), as well as scutellarein (ApSimon *et al.*, 1963). *D lutea* is the only species to be reported with only luteolin in its leaves.

The only other genus in the Scrophulariaceae to have been surveyed for its flavones is *Antirrhinum*. The flavones in the petals of the garden snapdragon, *A. majus*, have been fully identified because of their genetical interest. Like those of *Digitalis*, they are present as 7-glucuronides and are apigenin, luteolin and chrysoeriol. Apigenin also occurs in this plant as the 7,4'-diglucuronide (Harborne, 1963). The 7-glucuronide of apigenin is widely distributed in the genus having been found in all but one of the eight species surveyed. One may note that the flavanone, naringenin, has been isolated, but as the 7-glucoside and 7-rutinoside, from petals of *A. majus* (Seikel, 1955). It is also interesting that the flavone, acacetin, which occurs in the flowers of the toadflax, *Linaria vulgaris* (a plant in the same tribe as *Antirrhinum*), is present as the 7-rutino-side, not as the 7-glucuronide (Merz and Wu, 1936). A third rutinoside, that of the flavanone hesperetin, is reported in the flowers of *Verbascum phloimoides* (tribe Verbasceae) by Hein (1959).

6. Flavones of the Bignoniaceae

The flavones so far isolated from this family indicate close links with the Labiatae and Scrophulariaceae. Thus, chrysin and baicalein have been found in the rind of *Oroxylum indicum* (Bose and Bhattachanya, 1938) and baicalein 6-glucoside and oroxylin-A (5,7-dihydroxy-6-methoxyflavone) in the seeds of the same plant (Mehta and Mehta, 1954). Again, 6-hydroxyluteolin (49) has recently been isolated from the leaves of *Catalpa bignonioides* and *Tecoma australis* (J. B. Harborne, unpublished results).

(49)   (50)

As in other families of the Tubiflorae, luteolin is probably a common constituent of the Bignoniaceae. Its provisional identification in the leaf of *Campsis radicans* (Bate-Smith, 1962) has been confirmed by the author and it has been found as the 7-glucoside in *Catalpa bignonioides*; an acylated derivative of luteolin 7-glucoside (50) occurs in the latter plant (Birkofer *et al.*, 1965).

7. Flavones of the Orobanchaceae

Luteolin 7-glucoside has been isolated from the whole plant of *Orobanche minor* (Harborne, 1959). Orobanchin, a complex caffeic acid ester, occurs in such quantities in most *Orobanche* plants that detection of flavones in these plants is difficult. A further hazard, in both this family and the Scrophulariaceae, is the presence of iridoids such as aucubin, which tend to interfere with the detection of flavones. Tricin has been found in the seed of *O. ramosa* (formerly *Philypeae ramosa*) and *O. arenaria* but it is of erratic occurrence in the genus since it was absent from eight other species tested (Harborne, 1958c; Harborne and Hall, 1964a).

8. Flavones of the Gesneriaceae

The flavones, apigenin and luteolin, were found in all of 10 species of the Gesneriaceae surveyed, from the genera *Achimenes, Columnea, Kohleria Rechsteineria, Smithiantha, Streptocarpus* and *Tricantha*. The only unusual flavone, so far, found in this family is diosmetin which occurs as the 7-glucoside, in flowers and leaves of *Columnea* × *banksii*. Results of surveys indicate, however, that O-methylated flavones are probably regular constituents of the leaves of gesnerads (Harborne, 1966c).

9. Flavones of the Acanthaceae

Luteolin has been found in the leaves and flowers of 11 species belonging to *Adhatoda, Andrographis, Asteracantha, Barleria, Justicia, Ruellia, Rungia* and *Thunbergia* (Nair *et al.*, 1965). Luteolin 7-glucoside has been identified in *Asystasia gangetica* petals (Harborne, 1966b). Echioidinin (51), present as the

2'-glucoside, and wightin (52) have been isolated from *Andrographis echioides* (Govindachari *et al.*, 1965). Together with 5,7-dihydroxy-2'-methoxyflavone (an isomer of echioidinin) which occur in *Scutellaria*, these two compounds

(51) Echioidinin   (52) Wightin

represent the only known examples of flavonoids with 2'-hydroxy groups in the Sympetalae.

10. Flavones of the Myoporaceae

Some unusual flavonols have recently been isolated from an Australian genus, *Eremophila*, by Jefferies *et al.*, (1962). The 3,6,7,4'-tetramethyl ether (53) of 6-hydroxymyricetin occurs in *E. fraseri*, while pinobanksin (3,5,7-trihydroxyflavanone) and galangin 3-methyl ether (54) are present in *E. ramos-*

(53)   (54)

sissima and *E. alternifolia*. These latter substances are known otherwise as pine constituents. No flavones as such have been reported so far from this family.

11. Flavones of the Pedaliaceae

6-Methoxyluteolin 7-glucoside has been isolated from the leaf of *Sesamum indicum* by Morita (1960).

12. Flavones of the Plantaginales and Rubiales

It is convenient to mention flavones found in these two orders here, since they bear a close relationship to those found in the Tubiflorae. There are, in fact, only five firm records. Scutellarein (as the 7-glucoside) has been isolated from *Plantago asiatica* by T. Nakaoki (unpublished results), indicating a possible relationship with the Labiatae and the Scrophulariaceae. The presence of luteolin 7-rutinoside in leaf of *Lonicera japonica* (Caprifoliaceae) (Nakaoki *et al.*, 1961) is unexceptional but the discovery of the biflavonyl, amentoflavone (55), otherwise only known in the Gymnosperms (*see* Chapter 3), in bark of *Viburnum prunifolium* and *V. opulus* (also Caprifoliaceae) is remarkable (Hörhammer *et al.*, 1965). It appears to represent the retention of what is clearly a primitive flavone character in a relatively highly evolved plant family.

(55) Amentoflavone

Gardenin, one of the few flavones reported from the Rubiaceae, is present in *Gardenia lucida*. Its proposed structure (56) which is reminiscent of digicitrin (45) of *Digitalis*, may not be correct in all details; the lack of a 7-hydroxyl

(56)

group is so unusual that it requires careful checking. Finally, diosmetin is recorded in the shoots of *Galium mollugo* (Kohlmuenzer, 1965).

D. CHALCONES AND AURONES

Chalcones and aurones are of infrequent occurrence in the Tubiflorae, having only been detected in three families, the Gesneriaceae, the Acanthaceae and the Scrophulariaceae.

A simple chalcone glucoside, isosalipurposide (57), has been found in petals of *Asystasia gangetica* (Acanthaceae) and of *Aeschynanthus parvifolius* (Gesneriaceae) during a survey of some 200 species (Harborne, 1966). However,

(57) Isosalipurposide

(58) Pedicellin

the presence of four more complex chalcones in the reddish-brown dust which occurs on the underside of the leaves of *Didymocarpus pedicellata* (Gesneriaceae) has been known for some time and their chemistry has been well studied (for review, *see* Seshadri, 1951). The four substances, pedicellin (58), pedicin (59), methylpedicinin (60) and pedicinin (61) are all closely related structurally and

(59) Pedicin

(60) Methylpedicinin

(61) Pedicinin

are biogenetically interconvertible. The high degree of O-methylation in pedicellin recalls the fully methylated flavones of *Citrus*. Pedicinin and methylpedicinin are unusual in combining in their structures quinonoid and chalcone functions. As taxonomic markers, they are probably of limited value since few plants have coloured deposits on their leaves. However, a similar deposit has been observed on an unnamed *Didymocarpus* species found by Burtt and the pigments present are currently under investigation in this laboratory; they may well bear some relation to the *pedicellata* group.

Aurones are confined to the Scrophulariaceae and the Gesneriaceae. The aurones, aureusin (62) and bracteatin 6-glucoside (63), are relatively rare constituents, having been found in *Antirrhinum* and *Linaria*, both of the tribe Antirrhineae of the Scrophulariaceae, but not being known otherwise in this

(62) Aureusin

(63) Bracteatin 6-glucoside

family. In *Antirrhinum*, the two aurones occur in all five species studied of the section Antirrhinum, including *A. majus*, and in *nuttalianum* (section Saerorhinum); they are not present in *cornutum* and *coulterianum* (also of the Saerorhinum). In *Linaria*, they have been found in flowers of toadflax

L. vulgaris and of *L. maroccana*. A search of 14 other genera in the Scrophulari-aceae revealed aurone-like yellow pigments in *Calceolaria chelidonoides* (Harborne, 1966). The pigments in this plant, although resembling aurones in their spectral characteristics, differed so much in their stability and chroma-tographic properties that they have not yet been identified.

Cernuoside, the 4-glucoside of aureusidin and an isomer of aureusin (62), has been found in two yellow-flowered species of the Gesneriaceae, in *Chirita micromusa* and in *Petrocosmea kerrii*. There are indications that aurones, or chalcones, occur in the flowers of a number of other species in this family. The studies of Bopp and Boll (1960) for example, indicate that the yellow spot on the corolla of certain *Streptocarpus* species is chalcone in nature. Interestingly enough, all the plants mentioned above are in the same sub-family of the Gesneriaceae, the Cyrtandroideae, and may characterise it just as the presence of 3-deoxyanthocyanins (*see* p. 211) characterises the sub-family Gesnerioideae (Harborne, 1966).[1]

E. FLAVONOID PATTERN AND TAXONOMY

In summarising what is known of the flavonoids of the Tubiflorae, it is apparent that there are big differences in flavonoid content between the first five families and most of the remaining ones (Table 6·7). A sharp line can be drawn between families which have flavonols generally present (Group I) and those which have flavones predominantly (Group II). Such a division is supported not only by distribution of other flavonoid characters (e.g. anthocyanins, etc.) but also by the distribution of three other related chemical characters (caffeic esters, iridoids and quinones). A few comments on each of the chemical characters listed in Table 6·7 follow.

While the flavonol records of Bate-Smith, which delineate Group I families, are very reliable, less is known of the distribution of flavones in Group II families. This is mainly because flavones have not been so extensively searched for and also because they are not so readily observed (i.e. do not have the bright colours of flavonols on chromatograms) during leaf surveys. The main point is that Group II families lack flavonols and it is reasonable to assume that their place is taken by flavones in many of these plants.

Differences in the anthocyanin pattern are not quite so clear-cut and are based on rather less data than the flavonol–flavone distinction. The Solanaceae differ from the other Group I families in having anthocyanidin 3-rutinoside-5-glucosides, a glycosidic type common in a Group II family, the Gesneriaceae. Acylation of anthocyanins is a character which cuts across the present classification, since it is present in two Group I families (Polemoniaceae and Solanaceae) and also in two Group II families (Labiatae and Orobanchaceae).

A range of unusual flavonoids clearly occurs in Group II families (*see* previous section), but the absence of such compounds in Group I families is based mainly on negative evidence. Few deliberate searches have been made

---

[1] A recent survey of 10 more Cyrtandroideae in this laboratory has revealed that cernuoside also occurs in flowers of *Cyrtandra oblongifolia* and of *Didymocarpus malayanus* and that isosalipurposide is present in *C. pendula* and *Aeschynanthus tricolor*.

TABLE 6·7. Chemical Characters in Families in the Tubiflorae

Group I families	Group II families
Boraginaceae, Convolvulaceae, Hydrophyllaceae, Polemoniaceae, Solanaceae.	Acanthaceae, Bignoniaceae, Gesneriaceae, Labiatae, Orobanchaceae, Scrophulariaceae, and Verbenaceae.
Flavonols, common (Present in 34 of 46 spp., or 75%)	Flavones, common (Flavonols only in 8 of 97 spp., or 8%)
Anthocyanins: Dp types frequent (all families) 3,5-Diglucoside common (except Solanaceae) Methylation rare	Anthocyanins: Cy and Pg types common (especially Bignoniaceae and Labiatae), 3-deoxyanthocyanins (in Gesn.) 3-Rutinoside or 3-rutinoside-5-glucoside common
Unusual flavonoids: None	Unusual flavonoids: 6-OH flavones (Bign., Lab., Scroph., Verb.) 4′-Me flavones (Lab., Scroph., Gesn.) 2′-OH flavones (Acanth., Lab.) Chalcones and aurones (Acan., Scroph., Gesn.)
Caffeic derivatives: Chlorogenic Caffeoylglucose (Sol.) Isochlorogenic (Conv.)	Caffeic derivatives: Orobanchin (Acan., Bign., Genn., Orob., Scroph., Verb.) Rosmarinic acid (Lab., Acan.)
Iridoids: Absent	Iridoids: In Bign., Orob., Scroph., Verb.
Quinones: Only in Borag.	Quinones: In Bign., Lab., Orob., Scroph., Verb.

for 6-hydroxylated flavones and aurones in the Boraginaceae or Polemoniaceae; nevertheless enough work has been done on the flavonoids of the Solanaceae to make it unlikely that compounds with unusual substitution patterns are regular or even infrequent in this family.

Surveys of the caffeic esters of the Tubiflorae are still in progress, but orobanchin seems to be a particularly good character. This substance, a complex ester of caffeic acid, 3,4-dihydroxyphenylethanol, glucose and rhamnose, was first isolated from *Orobanche minor* in 1925 by Bridel and Charaux. It is now known to occur throughout the Orobanchaceae and the Gesneriaceae, and it has been found fairly frequently in the Acanthaceae (present in 2 out of 5 genera), Bignoniaceae (6 out of 9 genera), Globulariaceae (2 genera), Verbenaceae (3 out of 4 genera) and Scrophulariaceae (2 out of 6 genera). Rosmarinic acid (64), first isolated from *Rosmarinus officinalis* by Scarpati and Oriente (1958), occurs in over half the Labiates so far tested and has been noted in *Thunbergia* (Acanthaceae). It does, however, also occur in the Boraginaceae (*Arnebia* and *Myosotis*) and Hydrophyllaceae (*Eriodictyon* and *Phacelia*) so it is not confined completely to Group II families (Harborne, 1966d).

8*

Iridoids such as aucubin probably represent useful characters in the present context, but have not been widely surveyed, because of difficulties of detection. Aucubin (65) was first isolated from Scrophulariaceae and has been found also in the Orobanchaceae (*Orobanche*) and in the Globulariaceae (*Globularia*). The general distribution of iridoids in families of the Sympetalae is discussed in more detail by Bate-Smith and Swain, (1966).

(64) Rosmarinic acid   (65) Aucubin

As in the case of the iridoids, data on the quinones is scanty and little reliance can be placed on the fact that these pigments have so far been found mainly in Group II families.

Lapachole occurs in both the Verbenaceae (*Tectona*) and the Bignoniaceae (*Bignonia, Tecoma*), digitolutein (66) in the Scrophulariaceae (*Digitalis*) and dunnione and pedicinin in the Gesneriaceae. However, there is one quinone in a Group I family, alkannin (67) in the Boraginaceae.

(66) Digitolutein   (67) Alkannin

VII. Flavonoids of the Compositae

The Compositae are one of the largest of all plant families with 15,000 species and 1000 genera divided into two sub-families and 13 tribes. Phylogenists place it at or near the top of the evolutionary tree and, as a family, it is very rich in chemical constituents. A large number of highly substituted flavones and flavonols have been found in the family. Another distinctive feature is the regular occurrence in a few tribes of chalcone and aurone pigments.

A. ANTHOCYANINS

Anthocyanins have been fully identified in 11 genera (Table 6·1) and from this list it appears that cyanidin is the characteristic anthocyanidin. Pelargonidin only appears in mutant forms of the Chinese aster and the dahlia. The question may be raised as to whether delphinidin occurs in the family, since no glycoside of delphinidin is recorded in Table 6·1. Indeed, delphinidin does not seem to be necessary to give blue flower colour since, in the garden dahlia, the colour of the petals of varieties such as "Bonny Blue" is due to cyanidin 3,5-diglucoside, perhaps present as a metal complex and/or with a co-pigment. However, delphinidin glycosides were reported in no less than seven genera by Lawrence *et al.* (1939), so it seems unlikely that it is completely absent. In fact, the pigment in the flowers of one of the plants mentioned by Lawrence *et al.*, namely the chicory, *Cichorium intybus*, has been examined by the author and the presence of delphinidin confirmed. R_f data of the delphinidin glycoside in chicory petals indicate that it may be the rare 3-rhamnoside-5-glucoside but the sugars have yet to be identified.

Glycosidic patterns found in the Compositae are otherwise simple ones, being usually 3-glucoside or 3,5-diglucoside. The 3-arabinoside-5-glucoside of cyanidin reported in *Dahlia variabilis* (Nordström, 1954) is, in fact, the 3,5-diglucoside, arabinose being incorrectly identified as a sugar due to its production as an artifact of chromatography (*see* Chapter I). The report of cyanidin 3-glucoside and 3-rutinoside in *Happlopappus gracilis* is one of the few instances where pigments of plants grown in tissue culture have been fully identified. In confirming the presence of two anthocyanins in the tissue cultures, Ardenne (1965) incorrectly assumed the second pigment to be the 3,5-diglucoside; However, the spectral data reported by this author for the pigment shows that it must be a 3-, not a 3,5-diglucoside.

B. Flavonols

According to Bate-Smith's leaf survey, kaempferol and quercetin are quite common constituents, occurring as they do in 10 and 21 species respectively of 39 surveyed. Myricetin is, however, absent (compare the relative rarity of delphinidin as a flower pigment, *see* above). One may note too that ellagic acid has been reported once, in *Tagetes patula*, as has leucocyanidin (*Cosmos bipinnatus*) but that leucodelphinidin is not known. Although methylated anthocyanidins have not been reported in the Compositae, isorhamnetin, the flavonol analogue of peonidin, occurs in *Ambrosia* and *Calendula*. The glycosidic patterns of the flavonols fully identified in plants of the Compositae (Table 6·2) are simple ones, 3-glucoside being usual. The 7-glucoside of quercetin, a relatively rare glucoside otherwise, occurs with some frequence (in *Chrysanthemum*, *Helianthus*, *Matricaria* and *Viguiera*).

(68) Quercetagetin

(69) Atanisin

No less than nine different 6-hydroxylated flavonols have been found in leaf, flower or root of Composites. They range (Table 6·8) from quercetagetin itself (68) to atanisin, which is a flavonol with a furan ring fused on the A-ring (69).

TABLE 6·8. Quercetagetin and Derivatives in the Compositae

Flavonol	Source	Reference
QUERCETAGETIN and methyl ethers		
Parent compound	*Tagetes erecta* (as 3- and 7- glucoside) *T. patula* petals *Chrysanthemum coronarium*	Morita, 1957.
6-Methyl ether (patuletin)	*Tagetes patula* petals *Matricaria chamomilla*	Rao and Seshadri, 1941.
3,6-Dimethyl ether	*Xanthium pennsylvanicum* leaf	Taylor and Wong, 1965.
3,6,3'-Trimethyl ether (centaureidin)	*Centaurea jacea*	Farkas *et al.*, 1963a.
3,6,7,3'(?)- Tetramethyl ether (polycladin)	*Lepidophyllum quadrangulare*	Marini-Bettolo *et al.*, 1957.
3,6,7,3',4'-Pentamethyl ether (artimetin)	*Kuhnia eupatoroides* *Artimesia absinthium* root *A. arborescens*	Herz, 1961. Mazur and Meisels, 1955.
GOSSYPETIN Parent Compound	*Chrysanthemum segetum* (as 7-glucoside) in petals	Geissman and Steelink, 1957.
3,6,7,8,3',4'-Hexa- methoxy-5-hydroxy flavone (?) (erianthin)	*Blumea eriantha*	Seshadri and Venka- teswarlu, 1946.
6-HYDROXYKAEMPFEROL methyl ethers		
3,6,7-Trimethyl ether (penduletin)	*Brickellia pendula* (as 4'-glu- coside) in root *B. squarrosa*	Flores and Herran, 1958. Flores and Herran, 1960.
Atanasin (69)	*Brickellia squarrosa*	Flores and Herran, 1960.

The structures of several of these compounds are not completely proven. In the case of polycladin, reported to be the 3,6,7,3'-tetramethyl ether of querce-tagetin, comparison with synthetic material (Hörhammer *et al.*, 1965) showed that the original formulation must be incorrect. The structure of erianthin from *Blumea* is also still suspect.

From the systematic point of view, it is interesting that these compounds occur mainly in three tribes, the Helenieae (*Tagetes*), the Anthemideae (*Chrys-*

anthemum, Artemisia, Matricaria) and the Cynareae (*Centaurea*). A proper phytochemical survey of plants in these tribes would, no doubt, reveal a number of new structures of the quercetagetin type.

The synthetic versatility of the Compositae is well illustrated by the discovery of a most unusual flavonol derivative in fruits of *Carduus marianus*. This is silybin, a derivative of 3′-methyl ether of dihydroquercetin, which has recently been shown to have structure (70) (Wagner *et al.*, 1965, and unpublished results):

(70) Silybin

C. FLAVONES

Apigenin and luteolin 7-glucosides are widely distributed in the Compositae, having been reported in 26 species representing 19 genera (Table 6·9). Acacetin, as the 7-rutinoside, occurs in *Cirsium* and *Chrysanthemum* and genkwanin in *Artemisia*. Besides these simple flavones, five 6-hydroxylated flavones have been found in the family. While scutellarein has only been reported once in *Centaurea scabiosa* as the 7-glucuronide in leaf and flower (Charaux and Rabaté, 1940), its 6,4′-dimethyl ether, pectolinaringenin (71), has been recorded in the leaf of eight *Cirsium* species (Nakaoki and Morita, 1960; Wagner *et al.*, 1960). It occurs as the 7-rutinoside in *Cirsium microspicatum, otayae, yoshizawae, japonicum, kagamontanum, hipponicum, inun-*

(71) Pectolinaringenin   (72)

datum and *oleraceum*. The isomeric 6,7-dimethyl ether (72) has been isolated, interestingly enough, from *Cirsium maritimum* where it occurs as the 4′-glucoside (Morita and Shimizu, 1963). Two other methylated derivatives of scutellarein have also been found in the Compositae: the 6-methyl ether, hispidulin, in *Ambrosia maritima, A. hispida* and *Parthenium incanum* (Herz and Sumi, 1964) and the 6,7,4′-trimethyl ether, mikanin, in *Mikania cordata* (Kiang *et al.*, 1965).

TABLE 6·9. Distribution of Apigenin and Luteolin Glycosides in the Compositae

Species	Flavone, present in leaf or ray floret	References
Achillea millefolium	Apigenin 7-glucoside	
Anacyclus officinarum	Apigenin 7-glucoside	Hörhammer and Wag 1962.
Anthemis nobilis	Apigenin 7-glucoside	
Artemisia sacrorum	Genkwanin	Chandrashakar, *et al.*, 1¶
Bellis perennis	Apigenin 7-glucoside	Hörhammer and Wag 1962.
Carthamus tinctorius	Luteolin 7-glucoside	Nakaoki and Morita, unpublished.
Chrysanthemum morifolium	Acacetin 7-rutinoside	Hsu and Wong, 1965.
C. maximum	Apigenin 7-glucoside	
C. leucanthemum	Apigenin 7-glucoside	
C. parthenium	Apigenin 7-glucoside	Hörhammer and Wagne 1962.
C. uliginosum	Apigenin 7-glucoside	
Centaurea cyanus	Apigenin 7-glucoside	
Cosmos bipinnatus	Apigenin 7-glucoside	Nakaoki, 1936.
Cirsium purpureum	Acacetin 7-rutinoside	Nakaoki and Morita, 19⅃
Cynara scolymus	Luteolin 7-glucoside	Masquelier and Micha
Coreopsis gigantea	Luteolin 7-glucoside	1958.
Dahlia variabilis	Apigenin 7- and 4'-glucosides and 7-rutinoside	Nordström and Swain, 1⅃
	Luteolin 7-mono and diglucosides	
Elephanthopus scaber	Luteolin 7-glucoside	Ghanim *et al.*, 1963b.
Echinops gmelinii	Apigenin 7-glucoside	
Erigeron annuus	Apigenin 7-glucuronide	
Gnaphalium affine	Luteolin 4'-glucoside	Aritomi *et al.*, 1964.
Hieracium pilosella	} Luteolin 7-glucoside	Hörhammer and Wag¶ 1962.
H. staticifolium		
Lactuca repens	Luteolin 7-glucoside	Nakaoki and Morita, 19€
Matricaria discoidea	} Apigenin and luteolin 7-glucosides	
M. chamomilla		Hörhammer and Wag¶ 1962.
M. inodorata	Apigenin 7-glucoside	
M. moderna	Apigenin 7-glucoside	
Onopordon acanthium	Luteolin 7-glucoside	Bogs and Bogs, 1965.
Taraxacum officinale	Apigenin and luteolin 7-glucosides	Hörhammer and Wag¶ 1962.
Zinnia elegans	Apigenin 7-glucoside	Nakaoki, 1941.

D. CHALCONES AND AURONES

That chalcones and aurones are important yellow flower pigments in certain members of the Compositae was recognised by Gertz (1938) before the chemistry of these substances was fully understood. However, the botanical surveys of Gertz inspired Geissman in California to isolate and identify these rare flavonoid constituents and he and his co-workers have been responsible for most of our present knowledge of these pigments. Detailed studies have been carried out on *Baeria, Bidens, Carthamus, Coreopsis, Cosmos, Dahlia, Helichrysum* and *Vigueira* (cf. Shimokoriyama, 1962).

Chalcone glycosides often occur with the related aurone glycosides in the flower petals. *Baeria chrysostoma*, for example, has two related pairs of pigments, coreopsin (73)–sulphurein (74) and marein (75)–maritimein (76) (Shimokoriyama and Geissman, 1960). *Cosmos sulphureus* and *Viguiera*

(73) Coreopsin

(74) Sulphurein

(75) Marein

(76) Maritimein

multiflora have coreopsin and sulphurein (Shimokoriyama and Geissman, 1960). A rich mixture of these and other chalcone–aurone pairs have been isolated from the petals of various *Coreopsis* species (Geissman *et al.*, 1956; Shimokoriyama, 1957). Of the eight species examined, three have only chalcones: *douglasii* with coreopsin, *drummondii* with coreopsin and an unidentified pentahydroxychalcone, and *stillmanii* with stillopsidin (77). Of the other five,

(77) Stillopsidin

(78) Lanceolin

three, *gigantea, maritima* and *tinctoria*, have coreopsin–sulphurein and marein–maritimein. *Maritima* and *gigantea* are, in fact, very similar morphologically and the species names refer more to the fact that one grows on the coastal plains of California and the other further inland rather than to any real difference in shape and habit between the plants. The remaining two species, *lanceolata* and *saxicola*, have a different pair of pigments, the chalcone, lanceolin (78), and the related aurone, leptosin.

Dahlia variabilis is another plant with coreopsin and sulphurein in its flower petals. Also present are the corresponding diglucosides and also 2',4',4-trihydroxychalcone (Nordström and Swain, 1956). *Helichrysum bracteatum* is the only composite known to have an aurone with the 3',4',5'-trihydroxylation pattern, bracteatin (79) having been isolated from the ray florets as the 4-glucoside by Hänsel *et al.* (1962); related chalcone glucosides also occur in the flowers. Finally, mention must be made of the two chalcone pigments that

(79) Bracteatin   (80) Carthamin

have been isolated from *Carthamus tinctorius*. These are carthamin, a yellow quinol (80), and carthamon, the orange quinone related to it (Seshadri, 1956).

Chalcones and aurones are certainly of systematic value in the Compositae, since they have been found mainly in the sub-tribe Coreopsidinae of the tribe Heliantheae. Other records, based on simple colour tests, of anthochlor pigments in the Coreopsidinae include species from the genera *Chrysanthellum*, *Glossogyne*, *Guizotia*, *Heterospermum*, *Hidalgo*, *Isostigma*, *Microlecane* and *Thelesperma*. However, these pigments are not confined to the Coreopsidinae, since they also occur in *Viguiera* (sub-tribe Verbesininae), nor to this tribe, since they occur also in *Baeria* (Helenieae), *Carthamus* (Cynareae) and *Helichrysum* (Inuleae).

CHAPTER 7

FLAVONOIDS OF THE MONOCOTYLEDONEAE

I. Introduction

The monocotyledoneae comprise some 53 families of plants, divided into 14 orders, and the flavonoids of plants of only about 10 families representing 6 orders have been investigated in any detail. Much more survey work is clearly required; the Orchidaceae is an example of a family of considerable potential interest but of which only the anthocyanins of a few species have been examined. Attention has naturally focused mainly on plants of economic importance (the grasses and cereals, the banana and the onion) or on plants of ornamental beauty (crocuses, irises, tulips, etc.). All the anthocyanins that have been completely identified in monocotyledonous plants are listed in Table 7·1. Fewer flavones and flavonols have been studied, however, and these are mentioned in the text below.

The general pattern of flavonoids in the monocotyledons does not differ significantly from that in the dicotyledons. With regard to the anthocyanins, all the six common anthocyanidins are regularly present, but other types are rare. The only unusual anthocyanidins so far detected are the 3-deoxyanthocyanidins, apigeninidin and luteolinidin, found by Stafford (1965) in *Sorghum vulgare*. There are indications of other unusual anthocyanidins: the author has noted a yellow 3-deoxyanthocyanidin (R_f 0·51 in Forestal) in the sedge *Carex riparia* (Cyperaceae), and E. C. Bate-Smith (private communication) has evidence of a new highly hydroxylated anthocyanidin ($R_f < 0$·30 in Forestal) in *Dactylorchis coccinea* (Orchidaceae).

TABLE 7·1. Anthocyanins of the Monocotyledons

Order, family and genus	Pigments present[1]	Organ examined	References
LILIIFLORAE			
Liliaceae			
Allium cepa	Pn 3-arabinoside	Bulb scale	Brandwein, 1965.
Colchicum autumnale	Cy 3-rutinoside and Cy 3-sophoroside-5-glucoside	Petal	J. B. Harborne, published.
Fritillaria spp.	Cy 3-rutinoside and Cy 3,5-diglucoside	Petal	Harborne, 1963a.
Hyacinthus orientalis	Pg, Cy and Dp 3-(p-couma-roylglucoside)-5-glucosides	Petal	Harborne, 1964a.
	Cy 3-(p-coumaroylglucoside)	Bulb scale	Harborne, 1964a.
Lilium regale	Cy 3-rutinoside	Petal, pollen	J. B. Harborne, published.
Scilla non scripta	Cy and Dp 3-(p-coumarsyl-glucoside)-5-glucosides	Petal and bulb	R. W. Riding, published.
Urginea maritima	Cy 3-glucoside and Cy 3-(caffeoylglucoside)		Vega and Martin, 19
Tulipa spp. and cvs.	Pg, Cy and Dp 3-glucosides, Pg, Cy and Dp 3-rutinosides	Petal	Shibata and Ishiku 1960; Halevy, 196 Halevy and Asc 1959.
Amaryllidaceae			
Cyrtanthus angustifolia	Pg 3-sophoroside	Petal	J. B. Harborne, published.
Clivia miniata	Pg 3-rutinoside and Pg 3(2G-glucosylrutinoside)	Petal	Harborne and Ha 1964b.
	Cy 3-rutinoside	Red berry	
Hippeastrum cvs.	Pg and Cy 3-rutinosides	Petal	J. B. Harborne, published.
Lycoris radiata	Cy 3-glucoside and Cy 3-sambubioside	Petal	Hayashi, 1942.
Nerine bowdenii	Pg 3-glucoside and Pg 3-sambubioside	Red berry	J. B. Harborne, published.
Iridaceae			
Babiana stricta	Mv 3,5-diglucoside	Petal	Harborne, 1963a.
Chasmanthe ethiopica	Cy and Pn 3-rutinosides	Petal	Harborne, 1963a.
Crocosma masonii	Cy 3,5-diglucoside and Cy 3-rutinoside-5-glucoside	Petal	Harborne, 1963a.
Crocus spp.	Dp 3,5-diglucoside	Petal	Hayashi, 1960. J. Harborne, unpub lished.
Freesia cvs.	Mv 3-glucoside	Petal	Harborne, 1963a.
Gladiolus gandavensis	Pg 3-rutinoside, Pg 3-sophoroside-5-glucoside, Pn 3-rutinoside-5-glucoside Cy, Dp and Mv 3,5-diglucosides	Petal	Shibata and Nozak 1963.
Iris spp. and cvs.	Dp and Mv 3-(p-coumaroyl-rutinoside)-5-glucosides	Petal	Harborne, 1964a.
Lapeyrousia cruenta	Cy 3-rutinoside	Petal	Harborne, 1963a.
Tritonia cv. "Prince of Orange"	Pg 3-gentiobioside	Petal	Harborne, 1963b.
Watsonia meriana *W. rosea* *W. tabularis*	Pg 3-glucoside, Pg 3-sophoroside and Pg 3-sophoroside-7-glucoside	Petal	Harborne, 1963b.

TABLE 7·1—*continued*

der, family and genus	Pigments present[1]	Organ examined	References
MMELINALES			
nmelinaceae			
Commelina communis	Dp 3-(*p*-coumaroylgluco-side)-5-glucoside	Petal	Mitsui *et al.*, 1959.
RAMINALES			
aceae			
Hordeum vulgare	Cy 3-arabinoside	Husks	Metche and Urion, 1961.
	Cy 3-glucoside	Pericarp and aleurone	Mullick *et al.*, 1958.
Nardus stricta	Dp 3,5-diglucoside	Leaf	Clifford and Harborne, 1966.
Oryza sativa	Cy 3-glucoside, Cy 3,5-diglu-coside, Cy 3-rutinoside, Mv 3-galactoside	Leaf	Nagai *et al.*, 1960.
Pennisetum japonica	Cy 3-glucoside	Spike hair	Shibata and Sakai, 1958.
Phalaris arundinacea var. *picta*	Cy and Pn 3-arabinosides	Root	Clifford and Harborne, 1966.
Poa annua	Cy 3-arabinoside and Cy 3-glucoside	Leaf and inflorescence	Clifford and Harborne, 1966.
Sorghum vulgare	Cy 3-glucoside, Ad 5-gluco-side, Lt 5-glucoside	Leaf	Stafford, 1965.
Zea mays	Pg 3-glucoside, Cy 3-gluco-side	Endosperm	Straus, 1959.
PATHIFLORAE			
aceae			
Anthurium scherzerianum	Pg 3-rutinoside	Spathe	J. B. Harborne, un-published.
mnaceae			
Spirodela minor	Cy 3-glucoside	Leaf	Harborne, 1963a.
S. oligorrhiza	Pt 3-glucoside	Leaf	Ng and Thimann, 1962.
CITAMINEAE			
nnaceae			
Canna indica	Cy 3-rutinoside	Petal	Hayashi *et al.*, 1954.
usaceae			
Musa velutina	Cy and Dp 3-glucosides, Cy and Dp 3-rutinosides	Bract	J. B. Harborne, un-published.
Strelitzia regina	Dp 3-rutinoside	Petal	J. B. Harborne, un-published.
CROSPERMAE			
chidaceae			
Anacamptis pyramidalis	Cy 3,5-diglucoside	Petal	Harborne, 1963a.
Orchis mascula	Cy 3,5-diglucoside	Petal	J. B. Harborne, un-published.

[1] Abbreviations: Pg, pelargonidin; Cy, cyanidin; Pn, peonidin; Dp, delphinidin; Pt, petunidin; v, malvidin; Ad, apigeninidin; Lt, luteolinidin.

Most of the anthocyanidin glycosidic patterns found in the dicots are also known in the monocots. The most notable absentees are 3-rhamnosides and 3-rhamnoside-5-glucosides, which occur in the Leguminosae and Plumbaginaceae (*see* Chapters 5 and 6). Some parallel occurrences of rare glycosidic types in mono- and dicots are shown in Table 7·2.

TABLE 7·2. Co-occurrence in the Monocotyledons and Dicotyledons of Rare Flavonoids or Rare Flavonoid Types

Chemical character	Typical monocotyledon source	Typical dicotyledon source
Anthocyanins		
3-Deoxyanthocyanidins (luteolinidin, apigeninidin)	*Sorghum vulgare* (Gramineae)	*Gesneria cuneifolia* (Gesneriaceae)
Pelargonidin 3-sophoroside-7-glucoside	*Watsonia tabularis* (Iridaceae)	*Papaver orientalis* (Papaveraceae)
Pelargonidin 3-gentiobioside	*Tritonia* cv. (Iridaceae)	*Primula sinensis* (Primulaceae)
Cyanidin 3-α-arabinoside	*Hordeum vulgare* (Gramineae)	*Rhododendron thomsonii* (Ericaceae)
Pelargonidin-3-(2^G-glucosylrutinoside)	*Clivia miniata* (Amaryllidaceae)	*Rubus strigosus* (Rosaceae)
Flavonols		
Flavonol 3-sophoroside-7-glucoside	*Galanthus nivalis* (Amaryllidaceae)	*Helleborus foetidus* (Ranunculaceae)
Flavonol 3-rutinoside-7-glucoside	*Crocus fleischeri* (Iridaceae)	*Baptisia sphaerocarpa* (Leguminosae)
Flavonol 4'-methyl ether	*Alpinia officinarum* (Zingiberaceae)	*Tamarix africa* (Tamaricaceae)
Flavones		
Acacetin	*Crocus laevigatus* (Iridaceae)	*Linaria vulgaris* (Scrophulariaceae)
Tricin	*Triticum dicoccum* (Gramineae)	*Orobanche ramosa* (Orobanchaceae)
Other Flavonoids		
Isoflavones	*Iris germanica* (Iridaceae)	*Genista tinctoria* (Leguminosae)
Chalcones	*Xanthorrhoea australis* (Liliaceae)	*Paeonia trollioides* (Ranunculaceae)

Table 7·2 also lists other rare flavonoids or types which occur in both groups. These all represent examples of the parallel evolution of chemical characters. Isoflavones, for example, occur in both the Iridaceae and the Leguminosae. It is interesting that these families are also rich in glycoflavones and both synthesise the glycoxanthone, mangiferin (*see* Section IV.B).

The pattern of flavonols and flavones found in the monocots corresponds fairly closely to that of the dicots. Highly hydroxylated flavones such as quercetagetin of *Tagetes* (Compositae) and scutellarein of *Scutellaria* (Labiatae)

have not so far been discovered in monocots; nor are aurone pigments known here. Otherwise, the synthetic capacities of the two major groups of angiosperm are very similar.

II. Flavonoids of the Liliaceae

A. ANTHOCYANINS

The Liliaceae is a vast heterogeneous family (250 genera, 3700 species), best known chemically by its alkaloids (e.g. colchicine) and saponins. The anthocyanin pigments of eight genera have been examined (Table 7·1). The glycosidic patterns present are simple, being 3-glucoside, 3,5-diglucoside or 3-rutinoside. Another simple or "primitive" aspect of the anthocyanins is the almost complete absence of methylated pigments, the main anthocyanidins detected so far being pelargonidin, cyanidin and delphinidin. Lawrence *et al.* (1939) did report, it is true, a malvidin 3-bioside in the Autumn crocus, *Colchicum autumnale*, but re-investigation showed the blue petals to contain two cyanidin glycosides instead. Peonidin has, in fact, been recorded more recently in *Allium*, but this and related genera in the Allioideae are known to differ to some extent from the rest of the family in their chemistry (cf. Hegnauer, 1963). Acylated pigments have been isolated from *Hyacinthus, Scilla* and *Urginea*, three genera which are in the same tribe, the Scilloideae. The distribution of these pigments in the bluebell, *Scilla non scripta*, has been the special study of Riding (1961), who found the delphinidin pigment in blue flowers and the cyanidin derivative in bulb and leaf sheath of blue forms and in pink flowers. The delphinidin pigment, the 3-(*p*-coumaroylglucoside)-5-glucoside, also occurs in *Hyacinthus* and has been isolated as a blue magnesium complex from blue flowers of *Commelina communis* (Commelinaceae) (Mitsui *et al.*, 1959). Otherwise acylated anthocyanins are known in the monocots only in the grasses and in *Iris* (Iridaceae).

The anthocyanins of two lilaceous genera, *Fritillaria* and *Tulipa*, have been extensively surveyed. *Fritillaria* pigments were studied by the author, following the report by Shibata (1958) of a cyanidin 3-xylosylrhamnoside (or 3-rhamnosylxyloside) in the dark purple petals of the Japanese fritillary *F. kamchatchensis*. It was immediately of interest to see if an anthocyanin with such an unusual sugar combination was present elsewhere in the genus. Examination of the English wild fritillary, *F. meleagris*, showed only the presence of the common 3-rutinoside of cyanidin (1) with traces of the 3,5-diglucoside. A survey of all the cyanic species that could be obtained (23) showed that the same pigment, the 3-rutinoside was always the main pigment. Finally an authenticated specimen[1] of *F. kamchatkensis* became available and this

---

[1] The author is grateful to L. F. La Cour, the cytologist, for providing authentic material of this and the other *Fritillaria* species examined. They were as follows: *attica, assyriaca, californica, elwesii, falcata, graeca, gracilis, latifolia, libanotica, liliflora, lanceolata, messanensis, mutica, obliqua, parviflora, recurva, rhodaxanthus, tenella, tuntasia, thunbergia, tubiformis* and two unnamed spp.

only had the 3-rutinoside like all the others. The sugars produced on hydrolysis of the anthocyanin were carefully checked as glucose and rhamnose. Thus, unless *F. kamchatkensis* is a very variable species chemically, it can only be concluded that Shibata's pigment contained a xylose derivative as an impurity. It is significant that no R_f data were published to support the structure of the xylosylrhamnoside; a dipentoside would certainly have a different mobility from the 3-rutinoside. It seems much more probable that the anthocyanidin glycosidic pattern is uniform throughout the genus.

(1) Cyanidin 3-rutinoside

Tulipa is probably similar to *Fritillaria* in having anthocyanidin 3-rutinosides as the main pigments of the flowers. This is not completely certain because species were surveyed during the period when the sugar residues could not be fully identified. Beale *et al.* (1941), working with Hall (1940), found that a pelargonidin 3-pentoseglycoside was restricted to species in the sub-genus *Leiostemones* (26 species examined) and a delphinidin 3-pentoseglycoside was restricted to species in the sub-genus *Eriostemones* (6 species examined). Cyanidin derivatives were uniformly present throughout the genus, as was delphinidin as the pigment responsible for colour in the basal blotch of the petal. All three anthocyanidins occur in flowers of the garden tulip, which is derived from the *Leiostemones* sub-genus, so that the restriction of delphinidin to the basal blotch has apparently been lost as a result of artificial selection for colour.

More recently, attention has been given entirely to the anthocyanins of the cultivated tulip. Six pigments have been fully identified: the 3-glucosides and 3-rutinosides of pelargonidin, cyanidin and delphinidin. In addition Shibata and Ishikura (1960) report a third delphinidin glycoside (the 3,5-diglucoside) and Halevy (1962) a third pelargonidin glycoside which is isomeric with the 3-rutinoside. Halevy's pigment differs in R_f value in all solvent systems from the 3-rutinoside and may be the related 3-neohesperidoside, in which the rhamnose and glucose are linked $\alpha,1\rightarrow2$ instead of $\alpha,1\rightarrow6$. Alternatively, this pigment could be an artifact of chromatography—some of the 3-rutinoside held back on the chromatogram by a persistent impurity. In this connection, it should be mentioned that "isomeric" 3-rutinosides were detected in this laboratory in petals of *Chasmanthe* and *Lapeyrousia*, but repeated purification showed that they were identical with the 3-rutinosides, which were also present in these flowers.

Pigment inheritance in the cultivated tulip is unusual in that most varieties contain mixtures of pelargonidin, cyanidin and delphinidin glycosides. Pure anthocyanidin strains, which are usual in most other garden plants, are completely absent. Compare the garden hyacinth, for example, in which the variety "Scarlet O'Hara" is pigmented almost entirely by pelargonidin, the variety "Mauve Queen" by pure cyanidin and the pale blue variety "Spring-time" by pure delphinidin. In *Tulipa*, Shibata and Ishikura (1960) analysed no less than 107 varieties but failed to find a pure strain. In a survey in this laboratory of 19 varieties commonly grown in this country, no less than 11 contained glycosides derived from all three anthocyanidins. Even the purple-black variety "Queen of the Night" contains some pelargonidin and cyanidin derivatives. Typical are the varieties analysed by Halevy (1962): "President Eisenhower" with pelargonidin and delphinidin, "Smiling Queen" with pelargonidin and cyanidin, and "Pride of Haarlem" with a mixture of all three. In spite of the presence of such mixtures, there is reasonably good correlation between flower colour and anthocyanidin type in the garden tulip (*see* Chapter 9).

B. Flavonols

No flavones have been fully identified in the Liliaceae, but flavonol glycosides have been isolated from four genera, *Allium, Convallaria, Lilium* and *Tulipa*. Thus, the bulb of the onion, *Allium cepa*, contains the 4'-glucoside, 3,4'-diglucoside and 7,4'-diglucoside of quercetin and the leaf an unidentified 3,7-glycoside (Herrmann, 1958; Harborne, 1965a). Petals of *Lilium regale* and *L. leucanthum* have the following kaempferol glycosides: a 3-diglucoside, the 7-rhamnoside, the 3-glucoside and the 3-glucoside-7-rhamnoside (Asen and Emsweller, 1962). These flavonols are not, however, responsible for yellow flower colour in *Lilium*; carotenoids pigment both yellow and orange forms. Dark red varieties of *L. regale* do, however, have cyanidin 3-rutinoside and the same pigment is present in dark-coloured stamens. The flavonol isorhamnetin 3-rutinoside, has been isolated from pollen of *L. auratum*, and the related isorhamnetin 3-rhamnosylgalactoside from *Convallaria keiski* (Komissarenko *et al.*, 1964).

Petals and leaves of the garden tulip contain quercetin 3-rutinoside (Kawase and Shibata, 1963), quercetin 3-rutinoside-7-glucuronide (2) and kaempferol 3-rutinoside-7-glucuronide (Harborne, 1965a). The latter two pigments are

(2) Quercetin 3-rutinoside-7-glucuronide

probably widely distributed in the genus, since E. C. Bate-Smith (unpublished data) has found the kaempferol and quercetin 7-glucuronides in acid-hydrolysed extracts of 23 species. The reason why the free aglycones were not found is that the glucuronic acid attached in the 7-position of the original flavonols is very resistant to acid hydrolysis, whereas the rutinose in the 3-position is readily hydrolysed.

The few fragmentary results regarding the flavonols of the Liliaceae support the conclusion from a consideration of the anthocyanins that the family has a rather primitive flavonoid pattern. The apparent absence of flavones is noteworthy (compare Poaceae and Iridaceae). A characteristic feature of the flavonols in the Liliaceae is their occurrence as 3,7-diglycosides, but this type is also known in the Amaryllidaceae (*see* below).

C. OTHER FLAVONOIDS

Other primitive characters found only in the Liliaceae among the monocotyledons, are the chalcones (3) and (4) which occur in the resins of *Xanthorrhoea australis* (Duewell, 1954) and *X. preissii* (Birch and Hextall, 1955) respectively. Since the resins, which are used commercially as a varnish,

(3)   (4)

trickle out of the bases of old leaves, it is not clear whether these chalcones occur in the living plant or whether they are artifacts, formed perhaps from flavanones present in the fresh leaf. Chalcones are not uniformly present in the genus, since they were absent from a third resin studied, that of *X. resinosa*.

Another interesting resin pigment in this family is dracorhodin (5) the red anhydrobase of a 5-methoxy-7-hydroxyanthocyanidin, which occurs in dragon's blood resin, the exudate of trees of *Dracaena* (Liliaceae) and *Daemonorops* (Palmae) (Frankel and David, 1927). More complex red pigments (e.g. dracorubin) have also been isolated from the resin.

(5) Dracorhodin

Yet another interesting chemical character found in the Liliaceae is the glycoxanthone, mangiferin, which occurs in *Anemarrhena* roots (Morita *et al.*,

1965). This substance, although not a flavonoid itself, is closely related to the flavonoids and has been found also in the Iridaceae (see Section IV.B).

III. Flavonoids of the Amaryllidaceae

The Amaryllidaceae is closely allied to the Liliaceae and does not differ markedly from it in flavonoid composition. Only a handful of species have been studied in detail so that it is not possible to make many generalisations. However, the following three points are worth noting:

1. Pelargonidin is an unusually frequent pigment since five of the eight anthocyanins identified are pelargonidin glycosides.

2. The only rare pigment so far found in the family is pelargonidin 3-(2^G-glucosylrutinoside) in *Clivia* blooms. This pigment is only known elsewhere in *Rubus* (Rosaceae).

3. The flavonol glycosides are similar to those in the Liliaceae; only two have been examined. One is kaempferol 3-sophoroside-7-glucoside in petals of the snowdrop, *Galanthus nivalis* (Harborne, 1965a), and the other is isorhamnetin 3-rutinoside in daffodil corolla sepals (Kubota and Hase, 1956), also known as a constituent of lily pollen.

IV. Flavonoids of the Iridaceae

A. ANTHOCYANINS

The Iridaceae is a family of 57 genera and 800 species and contains many plants of ornamental value. Anthocyanins have been identified in 10 genera; all six anthocyanidins are represented and a considerable range of glycosidic types have been found. Two particularly rare types are present in *Tritonia* (3-gentiobioside) and *Watsonia* (3-sophoroside-7-glucoside). The pigments of *Crocus*, *Gladiolus* and *Iris* have been studied in some detail and warrant further discussion.

Cyanic colours (mauve and blue) in *Crocus* flowers are due mainly to delphinidin 3,5-diglucoside. This glycoside has been positively identified in *Crocus sativus* (Hayashi, 1960) and *C. laevigatus* (J. B. Harborne, unpublished). A survey showed it to be present in five other spp. although it is accompanied by traces of the corresponding petunidin and malvidin glycosides. Lawrence *et al.* (1939) also record a delphinidin 3,5-diglycoside in six of eight species surveyed. It is interesting that a predominantly malvidin type is missing and further that no mutations to cyanidin or pelargonidin forms have been observed considering that *Crocus* has been in cultivation for a considerable period and that hybridisation is common in the genus.

A very similar situation exists in *Iris*, where the main pigment is again a delphinidin 3,5-diglycoside, the 3-(*p*-coumaroylrutinoside)-5-glucoside (6) (Harborne, 1964). No cyanidin or pelargonidin types have been detected in species or hybrids (Werckmeister, 1960). This delphinidin glycoside occurs throughout the cyanic forms of the garden *Iris*, its mauve colour being modified by metal complexing, by co-pigmentation with the yellow xanthone mangiferin, by

(6) Delphanin

admixture with yellow carotenoid or by dilution. Unlike the situation in *Crocus*, a malvidin-containing *Iris* is known; *Iris ensata* is reported to have a partly characterised malvidin derivative (Hayashi, 1940), as is *I. chrysographes* (Werckmeister, 1960) and *I. delavayi*. Seven other *Iris* species examined (Harborne, 1964a) had the same pigment as the cultivars. This pigment delphanin occurs in some white-petalled varieties of the garden *Iris* in pseudo-base form, a rather unusual phenomenon; treatment of petal extract with dilute acid at room temperature causes complete conversion to the normal anthocyanin.

By contrast with the situation in *Iris* and *Crocus*, the garden *Gladiolus* contains the whole range of common anthocyanidin types (Shibata and Nozaka, 1963). Of 10 varieties examined by these workers, five were red and thus pelargonidin types; one, the claret "Red Signal" also contained malvidin. The remainder (purple, plum and violet varieties) contained mixtures of cyanidin, peonidin, delphinidin and malvidin (the nature of the glycosides present is indicated in Table 7·1). One variety, the hollyhock-coloured "Hawaii" contained pelargonidin, cyanidin, peonidin and malvidin. Such complex mixtures are reminiscent of the situation in *Tulipa* (*see* p. 238).

B. FLAVONES

Extensive surveys of the flavonoids in *Crocus* and *Iris* have been carried out by Dr. Bate-Smith, partly in collaboration with the author, during the last 10 years; the results are still largely unpublished (but *see* Bate-Smith and Harborne, 1963). The data indicate a number of similarities between the genera, although they are usually placed in different tribes of the Iridaceae. For example, glycoflavones occur widely in both taxa. In *Crocus* they have been found in leaves of 35 out of 53 species surveyed and they are probably as frequent in the *Iris*. Furthermore, the glycoxanthone mangiferin (7) has been found in both these genera but, with one exception, nowhere else in the Iridaceae. The exception is *Belamcanda chinensis*, which is known to be closely related to *Iris*. Within the *Iris*, mangiferin has been found in most species of the sub-section *Pogoniris* (which includes the garden form *I. germanica*), and in subsections *Apogon* (*I. pseudacorus, I. kaempferi, I. unguicularis*), *Pardanthopsis* (*I. dichotoma*) and *Oncocyclus* (*I. sari*). These occurrences are in the leaf or petal, but it has been found curiously in high concentration in the rhizome of *I. unguicularis*. In *Crocus*, mangiferin occurs in only two species, *Crocus stellaris*[1] and *C. aureus*, as a leaf constituent.

(7) Mangiferin

The survey of *Crocus* leaves and petals has shown that a rich diversity of flavonoids exist in the genus. The rare tricin of the grasses, for example, has been detected in *C. cambessedesii* and may be present in two other species. Acacetin (4'-O-methylapigenin) has been isolated from leaf of *C. laevigatus* and an unidentified highly hydroxylated flavone observed in leaf of *C. corsicus* and *C. chrysanthus*. The flavonols kaempferol and quercetin occur regularly in small amounts in the leaves of about half of the genus, but myricetin is restricted to a related group, *C. aureus, C. candidus, C. olivieri* and *C. stellaris*,[1] all of which are in the section *Nudiflori*. Bate-Smith argues from the presence of these primitive flavonoids—mangiferin and especially myricetin—that this group of species, which are native to the Balkan and Aegean, represent the centre of origin of the *Crocus* genus. He suggests that as the genus spread westwards to Spain and the Balearic Islands, these flavonoids were replaced by glycoflavones and the cinnamic acid sinapic acid. The distinct differences in the flavonoid pattern in leaves between Eastern and Western Mediterranean species is paralleled by differences in petal, or rather throat, colour. Throat colour is mainly orange or yellow (carotenoid) in Eastern species, mainly violet (anthocyanin) or white in Western species.

Crocus petals, besides having the anthocyanin delphin and the rare water-soluble carotenoid, crocein, are rich in flavonol glycosides. Kaempferol 3-sophoroside was obtained in high yield from petals of *C. laevigatus* and was detected in six other species. This glycoside is very acid-labile, being converted to the aglucone and glucose on standing in aqueous acid at 0°. This is the probable explanation why Price *et al.* (1938) were able to isolate free kaempferol from the petals of several yellow-flowered species (e.g. *C. asturicus, C. speciosus*). A re-investigation by paper chromatography showed that no species contained free kaempferol. A more complex kaempferol glycoside, the 3-rutinoside-7-glucoside, also occurs in *Crocus* flowers and has been isolated

---

[1] The origin of this "species" is not certain and it has been suggested that this plant is a hybrid involving *C. aureus*.

from *C. fleisheri* and *C. etruscus*. The only other flavonol known in *Crocus* is isorhamnetin 3,4'-diglucoside, isolated from pollen of the variety "Sir John Bright" by Kuhn and Low (1944).

Isoflavones have not so far been detected in *Crocus* although they are well known in the *Iris*. Irigenin (8) was first isolated from the rhizome of the cultivated Iris in the nineteenth century, and the related tectorigenin (9) from *Iris tectorum* in 1927 by Shibata. Two other isoflavones, irisolone (10) and irisolidone (11), have been detected in *I. nepalensis* (Gopinath *et al.*,

(8) Irigenin

(9) Tectorigenin

1961; Prakash *et al.*, 1965) where they occur with irigenin. Irigenin has also been isolated, as the 7-glucoside iridin, from *I. pallida* (Carles, 1935) and *I. kumaonensis* (Ghanim *et al.*, 1963a). Thus isoflavones seem to be of erratic dis-

(10) Irisolone

(11) Irisolidone

tribution in the genus, although surveys of Carles (1935) and Bate-Smith (1956) suggest that they are only *regular* constituents in two subsections, Evansia and Pogoniris.

V. Flavonoids of the Poaceae

A. ANTHOCYANINS

The Poaceae (or Gramineae) is one of the largest of all plant families and is certainly the most important from the economic point of view. From their superficial appearance, the grasses would not seem to be good sources of anthocyanin pigments but, in fact, anthocyanin coloration is frequently present in inflorescence, glume, root or seed. It has been detected in one organ or another of all the cultivated cereals and in common weed grasses such as *Poa annua* and *Festuca pratensis*. The colour may be disguised in some cases by admixture with plastid or other pigments and appear as brown, grey or black. The anthocyanins of only eight grass species have been fully identified (Table

7·1) but a systematic survey (Clifford and Harborne, 1966) showed that cyanidin glycosides were present in 21 other species drawn from some 19 genera.

The characteristic anthocyanins of grasses seem to be three in number and are all cyanidin derivatives. Cyanidin 3-glucoside has been found in nearly every grass examined (Clifford and Harborne, 1966). It may be replaced by the 3-galactoside in some species; care has not always been taken to distinguish between these closely related glycosides. For example, Straus (1959) reports cyanidin 3-glucoside in *Zea mays*, whereas Baraud *et al.* (1964) describe cyanidin 3-galactoside as being present. The second characteristic anthocyanin is the 3-arabinoside of cyanidin, first isolated from barley, *Hordeum vulgare*, by Metche and Urion (1961). It has since been detected in *Poa annua*, *Phalaris arundinacea* and six other species. A third pigment is an acylated cyanidin glycoside present in some 10 species but it has not been fully characterised. Acylated pigments have already been reported in *Sorghum vulgare* and *Zea mays*. The pigment is *Sorghum* is apparently a cyanidin 3-glucoside, acylated with an unidentified aliphatic acid (Stafford, 1965), whereas that in *Zea* is reported to be cyanidin 3-galactoside acylated with *p*-coumaric acid (Baraud *et al.*, 1964).

Anthocyanins based on aglycones other than cyanidin appear to be uncommon in grasses. Most interesting taxonomically is the occurrence of delphinidin 3,5-diglucoside in mat-grass, *Nardus stricta*, an anthocyanin not present elsewhere in the family as so far sampled. This species is also unusual in possessing forked silica bodies in the leaves (Parry and Smithson, 1964) and having a grain rich in mannose (McLeod and MacCorquodale, 1958). In his recent classification of the family, Prat (1962) placed *Nardus* in a group by itself and the chemical data are in accord with this viewpoint.

Peonidin has so far only been found in the reed-grass *Phalaris arundinacea*. It is interesting, again, that this species and *P. tuberosa* are unusual among grasses in containing the tryptamine derivative, bufotenine (Culvenor *et al.*, 1964). A delphinidin monomethyl ether is reported in sugar cane, *Saccharum officinale* (Dasa Rao *et al.*, 1938), but has not been fully characterised. The only record of malvidin is in the grain pericarp of cultivated rice, *Oryza sativa*. *Oryza* is the only genus to have been systematically surveyed but the results are uninteresting; all six species examined contained the 3-glucoside and 3-rutinoside of cyanidin (Nagai *et al.*, 1960) in leaf-sheath or stigma.

B. FLAVONES

The most characteristic flavones of grass leaves are *C*-glycosyl compounds. Derivatives of vitexin and orientin have been found to be widely distributed, being present in 29 out of 34 species representing 32 genera which were surveyed recently (J. B. Harborne, unpublished results). Glycoflavones have been isolated and fully characterised in the leaves of three common cereals: barley, for example, contains saponarin, lutonarin and lutonarin 3′-methyl ether (12) (Seikel and Bushnell, 1959; Seikel *et al.*, 1962). Oats, *Avena sativa*, contains an 8-*C*(arabinosylglucosyl)-apigenin and wheat, *Triticum dicoccum*, a 6-*C*-(rhamnosylglucosyl)-luteolin (Harborne and Hall, 1964a). The germ of the wheat contains two related apigenin *C*-glycosides (King, 1962).

(12) Lutonarin 3′-methyl ether   (13) Tricin

Glycoflavones are otherwise uncommon in the monocots. They have been reported in the duckweeds, *Spirodela* (Lemnaceae) (Geissman and Jurd, 1955; Alston, 1966); in the Iridaceae (*see* below) and possibly in Araceae (*Arum*) (cf. Hegnauer, 1963). In addition, a *C*-glycosylchromone has been reported in *Aloe* (Liliaceae) (Haynes, 1965). Glycoflavones occur in such abundance in *Spirodela oligorrhiza* that they make the anthocyanin present difficult to isolate and identify. Such is the contamination of the anthocyanin by glycoflavone that Thimann and Edmondson (1949) originally thought it was a cyanidin glycoside and Geissman and Jurd (1955) a pelargonidin derivative. After repeated purification, the pigment was eventually identified as petunidin 3-glucoside by Ng and Thimann (1962), a result that was confirmed in this laboratory.

The only notable simple flavone in grasses is tricin (13). This substance was first isolated from the rust-resistant wheat variety "Khapli" by Anderson (1934), who named it tricin after the genus. It has since been found to occur frequently in the family, i.e. in 16 of 20 grass genera surveyed (Harborne and Hall, 1964a). It is otherwise rare, having been reported recently in a *Crocus* species (*see* above); it has been found in two dicotyledonous families, the Orobanchaceae, where it occurs free in the seeds, and the Leguminosae, where it occurs as the 7-glucuronide in lucerne leaf, *Medicago sativa*. Difficulty was experienced by E. C. Bate-Smith (private communication) in detecting tricin in grasses because it was obscured by the much larger amounts of glycoflavone present, but a technique using paper chromatography and water as solvent has been developed for its identification. Tricin occurs in six *Triticum* species as the 5-glucoside and 5-glucosylglucoside. The 5-glucoside was also found in leaf of *Oryza sativa* (Harborne and Hall, 1964a), but a 7-glucoside and 7-rutinoside have also been described in this species (Kuwatsuka and Oshima, 1964). It is curious that no *C*-glycosyltricin has yet been found, particularly since apigenin and luteolin occur in grasses almost exclusively as *C*-glycosyl derivatives. Exceptionally, *O*-glycosides of luteolin are reported to occur in *Paspalum conjugatum* (Paris, 1953) and in *Arthraxon hispidus* (Sannié and Sauvain, 1952).

By contrast with the frequent occurrences of flavones, flavonols are rare in grasses as leaf constituents. Only four species, of some 50 surveyed, yielded flavonols. Rutin was discovered in *Festuca pratensis*, quercetin 3-glucoside in *Poa pratensis* and kaempferol and quercetin glucosides in *Panicum bulbosum* and *Lolium perrene* (J. B. Harborne, unpublished results). Flavonols, however, do occur, perhaps more frequently, in other parts of the grass plant. Quercetin 3-glucoside is present, for example in *Zea mays* seed, isorhamnetin 3,4'-diglucoside (dactylin) (14) occurs in *Dactylis glomerata* and *Phleum pratense* pollens (Inglett, 1956, 1957) and quercetin 7-glucoside pigments the fruit of *Andropogon sorghum* (Hirao, 1935). Leucoanthocyanidins, like the flavonols, are rare in the leaves of grasses.

(14) Dactylin

The unusual leucoluteolinidin has been detected in the leaf of *Hyparrhenia filipendula*, *H. hirta* and *Andropogon gerardii* by E. C. Bate-Smith (unpublished results). Leucopelargonidin and leucocyanidin have been found not in the leaf but in seed of *Zea mays*, particularly in colour mutants when anthocyanin synthesis is blocked (Reddy, 1963). Leucocyanidin is also present in barley seed (Metche and Urion, 1961).

VI. Flavonoids of the Musaceae

The Musaceae are a small tropical family of six genera, and contain the economically important genus, *Musa*. The anthocyanins of *Musa* were studied by Simmonds (1954) in connection with work on the evolution and breeding of the cultivated banana. He examined the anthocyanidins of the bracts and noted four patterns of distribution: (i) pelargonidin–cyanidin mixtures, in *Musa coccinea*; (ii) cyanidin–delphinidin mixtures, in *M. laterita*, *M. balbisiana* and *M. velutina*; (iii) partly methylated cyanidin and delphinidin mixtures in *M. acuminata*; and (iv) peonidin and malvidin mixtures in *M. ornata* and *M. violascens*. Bracts of the edible banana, which is a cultivar derived from *M. acuminata*, contained pigments similar to this species.

Simmonds did not study the glycosidic pattern of the pigments, apart from noting that glucose was one of the sugars present. Bracts of *M. velutina* have, however, been recently examined in this laboratory and the 3-glucosides and 3-rutinosides of cyanidin and delphinidin identified. It remains to be seen whether this glycosidic pattern occurs throughout the genus; the distribution tests of Simmonds would indicate that this is so. The anthocyanins present in other parts of the banana have not been studied in detail but it is known that the red colour in the skin of certain banana fruit is due to peonidin and malvidin.

Flavones, the distributions of which are related to taxonomy, have been noted in the banana but not further identified. The only other *Musa* flavonoids to have been examined in detail are leucoanthocyanidins. The unripe green fruit is a very abundant source of leucocyanidin and leucodelphinidin, but these substances are universally present, albeit in lesser amounts, in bracts, flowers, sheaths and leaves. Curiously, these leucoanthocyanidins are replaced in the seed of the banana by leucopelargonidin (15). This substance has been isolated in quantity from seed of *M. balbisiana* and also detected in *M. acuminata* in this laboratory. The "methylated leucopelargonidin" reported by Chadha and Seshadri (1962) in seed of *M. acuminata* is almost certainly not, in fact, methylated. There was no evidence of any unusual properties in the material isolated by the author and there are no other reports in the literature of methylated leucoanthocyanidins.

(15) Leucopelargonidin

VII. Flavonoids of the Zingiberaceae

The Zingiberaceae are a tropical family of some 45 genera and 500 species and include several economic plants such as the ginger, *Zingiber*. No anthocyanin has been fully identified in the family but cyanidin derivatives were noted in three *Costus* species and in *Paeomeria speciosa* by Forsyth and Simmonds (1954) during their survey of the Trinidad flora. The most remarkable feature of the flavonoids found in the family is the absence from their structures of free hydroxyl groups on the B-ring. In fact, all but one have no B-ring substituents. The exception is kaempferide (16), the 4'-methyl ether of kaempferol, which occurs in the rhizome of *Alpinia officinarum* (Heap and Robinson, 1926).

(16) Kaempferide

(17) Galangin

Kaempferide is accompanied in *A. officinarum* by galangin (17) and its 3-methyl ether. The 7-methyl ether of galangin has been found in the seeds of

A. chinensis and *A. japonica*, accompanied by flavones. Seed of *A. chinensis* contains alpinetin, the flavanone (18), whereas seed of *A. japonica* has alpinone, the flavonol corresponding to galangin 7-methyl ether (Gripenberg *et al.*, 1956). Alpinetin has been found in the rhizomes of another plant of the family, in *Kaempferia pandurata*, where it is accompanied by its isomer, pinostrobin (19)

HO.

MeO

(18) Alpinetin

MeO.

HO

(19) Pinostrobin

(Mongkolsuk and Dean, 1964). It remains to be seen whether production of flavonoids lacking B-ring hydroxyls is a special feature of the Zingiberaceae or whether compounds such as galangin and alpinetin reflect only their occurrence in such specialised organs as rhizomes and seeds. It is of biogenetic interest here that *p*-methoxycinnamic acid, which is a likely precursor of the B-ring of kaempferide, has been found in this family, in *Kaempferia* and *Hedychium*.

9

CHAPTER 8

INHERITANCE AND BIOSYNTHESIS OF
FLAVONOIDS IN PLANTS

I. Introduction

From the preceding four chapters on flavonoid distribution, it is clear that a basic pattern of flavonoid synthesis is common to all higher plants. Superimposed on this basic pattern is the production of variants, which differ only slightly in structure from common types (i.e. cyanidin and quercetin) and which are usually restricted in distribution to a few species. Examples are azaleatin, formed by methylation of quercetin in the 5-position, and quercetagetin, the 6-hydroxy derivative of quercetin. These substances often replace the common flavonoid from which they are derived and presumably require one or more extra enzymes than the usual complement for their synthesis.

The question now arises whether these substances represent satisfactory chemical characters to be used in systematic studies. Are they stable? What is the effect of environment on them? Is their production strictly controlled and if they are genetically variable compounds, how important is this to taxonomic work? Even more interestingly, what do flavonoid characters represent in terms of evolutionary advancement? Can the various flavonoid classes be placed in order in a biosynthetic sequence?

Partial answers to most of these questions are available from studies of the inheritance, biosynthesis and physiology of flavonoids in plants. It is clear, for example, from genetic studies that many steps in synthesis are closely controlled by single independent genes, which in turn produce highly substrate-specific enzymes. Again, biosynthetic studies have shown that the flavonoids, which are linked to primary plant metabolism through the aromatic amino acids, are formed by sequential branching from a common C_{15} precursor. Finally, physiologists have shown that, while the environment has no qualitative effect on flavonoids, light is very important for their production in quantity by plants.

Much work has been done on the biological aspects of flavonoid formation in plants and it is the purpose of this chapter to provide a brief and up-to-date summary of this field. Unfortunately, the early work on flavonoid inheritance led to bitter controversy and the first physiological experiments were largely contradictory in their results. For the sake of clarity, historical aspects are largely omitted. Excellent accounts of the earlier theories on flavonoid biogenesis can be found in the reviews of Geissman and Hinreiner (1952) and Blank (1958) and in the book of Onslow (1925).

II. Inheritance

A. General Summary

More is known about the inheritance of flavonoids than that of any other plant constituents. This is due to the importance of flavonoids as flower pigments, and the fact that flower colour variation is such a common phenomenon and constitutes one of the easiest of plant characters to score. Flower pigment genetics, first studied by Bateson and others at the turn of the century, flourished during the 1930's and it was natural that biochemists should apply themselves to the chemistry of the plants concerned. Many examples were found (Scott-Moncrieff, 1936) of single gene differences being related to simple modifications in anthocyanin synthesis and one of the fruits of these early studies was the enunciation by Beadle (1945) of the one gene–one enzyme hypothesis, a cornerstone of present-day molecular biology. More recent studies of flavonoid inheritance have provided information on biosynthesis, indicating what are likely to be discrete steps in biosynthesis and what kind of specificities the enzymes required are likely to possess.

The experimental approach to flavonoid inheritance is basically simple. It consists of choosing a plant which has a range of flower colour forms, carrying out breeding experiments to establish single gene differences between pairs of mutants and then analysing the various genotypes for their pigments. With the aid of paper chromatography and absorption spectroscopy, the separation and identification of the complex mixtures of flavonoids usually present in such flowers is now readily accomplished.

TABLE 8·1. List of Genes Controlling Flavonoid Production in Higher Plants

Plant species	Genes controlling synthesis or structural modification of flavonoids	Leading references
Antirrhinum majus Snapdragon (diploid)	P, anthocyanidin and flavonol Y, suppression of aurone N, general flavonoid production M, 3'-hydroxylation of anthocyanidin, flavonol and flavone	Geissman *et al.*, 1954; Sherratt, 1958.
Cyclamen Cultivars (diploid)	W, anthocyanin F, increase in flavonol production M, 5-glucosylation of anthocyanin	Van Bragt, 1962.
Dahlia variabilis (octaploid)	A, B, anthocyanin Y, chalcone and aurone I, flavone	Lawrence and Scott-Moncrieff, 1935; Bate-Smith *et al.*, 1955.
Dianthus caryophyllus Carnation (diploid)	A, S, M, anthocyanin R, 3'-hydroxylation of anthocyanidin and flavonol	Geissman *et al.*, 1956.
Impatiens balsamina Garden balsam (diploid)	H, anthocyanidin and leuco-anthocyanidin L, 5'-hydroxylation of anthocyanidin and flavonol	Alston and Hagen, 1957; Clevenger, 1959.
Lathyrus odoratus Sweet pea (diploid)	C, R, general anthocyanin K, M, flavone D, acidity of cell sap E, 5'-hydroxylation of anthocyanidin and flavonol Sm, 3'-hydroxylation of anthocyanidin and flavonol	Beale *et al.*, 1939; Harborne, 1960a, 1963b.
Matthiola incana Stock (diploid)	B, 3'-hydroxylation of anthocyanidin L, 5-glucosylation of anthocyanin V, acylation and glucosylation of anthocyanin	Seyffert, 1960.
Phaseolus vulgaris French bean (diploid) seed coat	C, (allelic series) anthocyanin and flavonol V, 5'-hydroxylation of anthocyanidin and flavonol Sh, leucoanthocyanidin	Prakken, 1940; Feenstra, 1960.
Petunia hybrida (diploid)	M, 5'-hydroxylation of anthocyanidin F, acylation, glucosylation and 3'-methylation of anthocyanin K, 5'-methylation of anthocyanidin	Meyer, 1964; Birkofer *et al.*, 1962.

TABLE 8·1—*continued*

Plant species	Genes controlling synthesis or structural modification of flavonoids	Leading references
Pisum sativum Garden pea (diploid)	**B**, 5'-hydroxylation of anthocyanidin **Cr**, methylation of anthocyanidin and glycosylation	Lamprecht, 1957; Dodds and Harborne, 1964.
Primula sinensis Chinese primrose (diploid)	**K**, 5'-hydroxylation of anthocyanidin and flavonol **Dz**, increases anthocyanin **B**, increases flavone (i.e. co-pigments) **R**, changes pH of cell sap **D**, inhibits anthocyanin	Scott-Moncrieff, 1936; Harborne and Sherratt, 1961.
Raphanus sativus Radish root (diploid)	**H**, 3'-hydroxylation of anthocyanidin **I**, distribution of anthocyanidin	Uphof, 1924; Harborne and Paxman, 1964; Hoshi *et al.*, 1963.
Solanum tuberosum Cultivated potato (diploid)	**I**, inhibits anthocyanin (in tuber) **P**, 5'-hydroxylation of anthocyanidin and flavonol **R**, (allelic series), cyanidin in flower, pelargonidin in tuber **Ac**, acylation, 5-glucosylation and 3'-methylation of anthocyanidin **Ac'**, 5'-methylation of anthocyanidin **Gl**, flavonol glucosylation	Dodds and Long, 1955; Harborne, 1960b, 1964a, Simmonds and Harborne, 1965.
Streptocarpus hybrida Cape primrose (diploid)	**V, F**, general or localised anthocyanin **O**, 5'-hydroxylation of anthocyanidin and flavone **D**, 3-rhamnosylation of anthocyanidin 3,5-diglucoside **X, Z**, 5-glucosylation of anthocyanin **P**, 3-xylosylation of anthocyanin **Q**, basic anthocyanidin glucosylation **M**, methylation of anthocyanidin	Lawrence and Sturgess, 1957; Harborne, 1963b.

Key: 3'-Hydroxylation indicates production of cyanidin (and quercetin) instead of pelargonidin (and kaempferol); 5'-hydroxylation indicates further oxidation to delphinidin (and myricetin); 3'- and 5'-methylation indicate the synthesis of petunidin and malvidin, respectively, from delphinidin.

Almost every plant that has been brought into cultivation shows some variation in colour, but the most suitable for genetic studies are those which are diploid, have a reasonably rapid life cycle and produce large flowers. In fact, flower colour genetics has been studied in about 100 plants (Paris *et al.*, 1960), but the number that have been chemically studied is much less. Satisfactory chemical and genetical information is available for only some 14 plants (Table 8·1). A number of other plants have also been investigated in less detail and, where the results have provided interesting new data, they will be mentioned in subsequent discussion. Anthocyanin colour variation is not restricted to floral organs and of the 14 plants listed, most show some variation in other parts; in the case of the French bean, potato and radish, colour variation in seed, tuber and root respectively is of primary interest.

The origin of colour variation in many of these plants is still partly obscure, but is presumably due in most cases to mutations within the species being preserved in cultivation. This is certainly true of the Chinese primrose, *Primula sinensis*, which was introduced into this country in 1819 as a mauve-flowering form. The 30 or so mutants available today, of which a third differ in colour from the wild type, must all have arisen in cultivation, since the species will not hybridise with other *Primulas*. The flower colour now ranges from blue and crimson through orange and pink to white. Colour variations in both the Dahlia and in the Cape primrose, *Streptocarpus*, is the result of hybridisation of closely related species and the origins of the garden forms in both plants have been successfully traced (Lawrence and Sturgess, 1957; Lawrence and Scott-Moncrieff, 1934; *see* also Stearn, 1965).

The data presented in Table 8·1 on flavonoid inheritance in the 14 most thoroughly investigated species represent the results of continued investigations, often by several different groups of workers, over the course of a number of years. Details of the flavonoid aglycones detected in different genotypes of *Antirrhinum majus*, the most thoroughly investigated of all plant species, are given in Table 8·2. The results in other plants are discussed in more detail by Scott-Moncrieff (1936), Harborne (1962f) and Alston (1964). Here, a few general comments on the data are needed before discussing (in subsequent sections) particular aspects of flavonoid inheritance.

It is immediately clear (Table 8·1) that different gene symbols are used in different plants for the same biochemical effect. For example, 3′-hydroxylation of anthocyanidins, i.e. the oxidation of pelargonidin to cyanidin, is controlled by gene **M** in *Antirrhinum* and gene **Sm** in *Lathyrus*. Plainly, the situation would be much clearer if the one symbol could be used for the same effect in both plants. However, the situation is not as simple as it appears since, in *Antirrhinum*, **M** also controls the oxidation of kaempferol to quercetin and of apigenin to luteolin, whereas in *Lathyrus*, **Sm** only carries out the former oxidation, flavones being absent. Furthermore, **M** in *Antirrhinum* is independent in its action of the other pigment gene, but **Sm** in *Lathyrus* is hypostatic to the gene for 5′-hydroxylation, **E**, and its effect is repressed in **EE** genotypes.

TABLE 8·2. Flavonoid Aglycones of *Antirrhinum majus* Colour Types[1]

Genotype	Flower colour	ANTHOCYANIDINS Cyanidin	Pelargonidin	FLAVONOLS Quercetin	Kaempferol	FLAVONES Luteolin	Chrysoeriol	Apigenin	FLAVANONE Naringenin	AURONES Aureusidin	Bracteatin
PPMMYY	Magenta	+	−	+	−	+	+	+	−	tr	tr
PPMMyy	Orange-red	+	−	+	−	+	+	+	−	+	+
PPmmYY	Pink	−	+	−	+	−	−	+	+	tr	tr
PPmmyy	Orange-yellow	−	+	−	+	−	−	+	+	+	+
ppMMYY	Ivory	−	−	−	−	+	+	+	×	tr	tr
ppMMyy	Yellow	−	−	−	−	+	+	+	×	tr	tr
ppmmyy	Ivory	−	−	−	−	−	−	+	+	tr	tr
ppmmyy	Yellow	−	−	−	−	−	−	+	+	+	+
nn	Dead white	−	−	−	−	−	−	−	−	−	−

[1] Key: +present, −absent, tr trace. All genotypes are **NN**, except dead-white albino mutant.

Thus, it is necessary that genes apparently controlling the same step in flavonoid biosynthesis should be given different symbols in different plants. In fact, every plant that has been extensively studied has a unique genetic system for controlling flavonoid synthesis; unique in the sense that the extent of gene interaction present varies from one plant to the next. What is reprehensible is the fact that different symbols have been employed by different workers for pigment genes in one and the same plant. This is most noticeable in *Antirrhinum* which has been studied in Germany, England, America and Canada; the symbolism used here for this plant follows that of the American workers (*see* Geissman *et al.*, 1954).

Table 8·1 also shows that genes controlling flavonoid synthesis fall into two classes, those which modify chemical structure and those controlling general production. Yet other genes are known for controlling distribution, either within the flower or within the whole plant. The genetics of anthocyanin distribution has been studied particularly extensively in *Streptocarpus* (Lawrence and Sturgess, 1957) and in the potato (Dodds and Long, 1956). What has emerged in each case is that patterning is controlled by a series of closely linked genes, which contrasts with the independent genetic control of flavonoid production and modification.

It is doubtful whether further studies of flavonoid inheritance will add much to the body of information available today on biosynthesis of these pigments. What are most needed at the present time are investigations at the enzymic level, and the genetic material available would appear to be ideal for such purposes. Clearly, the mechanisms of pigment formation in higher plants will continue to be of purely genetic interest. Pigment formation in snapdragon flowers, for example, is being used at the present time for measuring the effects of environment on frequencies of somatic and germinal mutation in higher plants (Harrison and Fincham, 1964).

B. Genetic Control of Flavonoid Class

Most work on flavonoid inheritance has been devoted to anthocyanins. The mutation from a coloured cyanic variety to a white acyanic form has been detected in most garden plants and so genes controlling general anthocyanin synthesis have been described very frequently. There may be a single independent gene as in the potato, two or more complementary genes, as in dahlia petals, or an allelic series of genes at one or more loci, as in corn endosperm (Reddy and Coe, 1962). White is not necessarily recessive to coloured, since a dominant gene **D** exists in *Primula sinensis* for suppressing anthocyanin synthesis in the flower. The phenotypic expression of a block in anthocyanin synthesis is not always clear-cut. Some white mutants may contain traces of pigment and a "blush" of flower colour can sometimes be induced in such plants by putting them into an environment strongly favouring flavonoid synthesis. This suggests that the blockage in synthesis sometimes involves an inhibitor rather than the loss of an essential enzyme.

Nearly all white-flower mutants contain the flavones or flavonols present in the coloured forms, and the inheritance of flavones and flavonols is, thus, less easy to study. Flavonols appear to be more closely related biosynthetically to the anthocyanins than the flavones are, since the gene **P** in *Antirrhinum* controls anthocyanidin and flavonol (but not flavone) synthesis. Flower mutants completely lacking flavonoids are very rare; the only fully substantiated example is the dead-white albino mutant "nivea" in *Antirrhinum*. Other examples may occur in *Pisum sativum* and *Petunia*. In *Dahlia*, a number of bottom recessive white varieties are known, which fall into one of three classes: (1) those with the flavones of cyanic types, (2) those with flavanones (e.g. the cultivar "Clare White") and (3) those with kaempferol or quercetin (Bate-Smith *et al.*, 1955). The production of flavonols here is unusual because this flavonoid class is otherwise absent from flowers of other *Dahlia* material, both varieties and species. A very similar situation occurs in *Streptocarpus*, the flowers of which contain the common anthocyanidins and the flavones luteolin and apigenin. The flavonol kaempferol makes its only appearance in the bottom recessive pink form (genotype **oorrdd**) of the garden hybrids and is not known otherwise in the garden forms and does not occur elsewhere in the genus or even in the same family (Gesneriaceae) (Harborne, 1966c).

A general, as yet unexplained, characteristic of white flower mutants of coloured forms is their unthriftiness as plants. Most would certainly be lost under natural conditions and some are even difficult to keep alive in cultivation.

The inheritance of the minor flavonoids has scarcely been studied, but some information is available on aurones, leucoanthocyanidins, isoflavones and flavanones (dihydroflavonols). Thus, aurone production in *Antirrhinum majus* is known to be suppressed by the dominant gene **Y** and yellow pigments accumulates, by default, in **yy** types. Aurone occurs on the lip of the flower in **YY** types and is only completely absent from the true albino form, in which synthesis of all C_{15}-molecules is blocked. Again leucoanthocyanidin synthesis in the French bean seedcoat is controlled in a straightforward manner by the **Sh** gene, which is independent of anthocyanin. Isoflavone synthesis in clover leaf also appears to be monofactorial (Francis and Millington, 1965). Finally, although no gene for flavanone synthesis is known, dihydrokaempferol, which occurs in traces in all *Primula sinensis* genotypes, accumulates in one mutant, the Dazzler **kkDzDz** form, perhaps because enzymes for its conversion to the corresponding flavonol or anthocyanidin are in short supply.

Although the genetic regulation of individual flavones, flavonols or chalcones is poorly understood compared to that of the anthocyanins, many examples are known of genes affecting the amounts of these substances produced in flower petals. Usually a limited amounts of precursor is available for flavonoid synthesis in the petal and different pathways compete for this common precursor. The classic example here is the dahlia, the flowers of which contain a nicely held balance of ivory flavone, yellow chalcone (and aurone), pink pelargonidin and magenta cyanidin (Lawrence and Scott-Moncrieff, 1935). Although the accumulation of dominant pigment genes in a particular colour variety increases the overall amount of flavonoid present, an analysis of the individual substances in dahlia has indicated that one pigment is formed in bulk at the expense of another. Thus, deeply cyanic varieties contain little flavone or chalcone which accumulate, instead, in pale-coloured varieties. Very similar results were obtained by Jorgensen and Geissman (1955) in *Antirrhinum*; the genes **P** and **Y** compete in such a manner that plants with high anthocyanin content are low in yellow aurone pigment and vice versa.

By contrast, some plants have a mechanism of increasing the concentration of one flavonoid class without affecting that of another. Thus, the **Dz** gene in *Primula sinensis* increases the concentration of pelargonidin glycoside threefold, without affecting the amounts of flavonol and dihydroflavonol present in the petal. The reverse of this also occurs in the same plant. The gene **B** increases flavonol three- or fivefold (leading to a blueing effect on flower colour) with little or no effect on anthocyanin concentration.

9*

C. Genetic Control of Hydroxylation

Most flower colour variation in cultivated ornamental plants is due to mutations at the genetic loci controlling anthocyanidin hydroxylation. In the general case, a wild type plant species has mauve or blue flowers and gives rise to magenta, scarlet, pink and orange forms in cultivation. Good examples are the Chinese primrose (already mentioned, p. 254) and the sweet pea. The reputedly wild form of the latter plant, *Lathyrus odoratus*, collected in Palermo, Sicily, has mauve (delphinidin-containing) petals, whereas modern sweet pea varieties fall into three classes: those with delphinidin (e.g. the deep purple-flowered "Jupiter"), those with cyanidin (e.g. the carmine "Harrow") and those with pelargonidin (e.g. the scarlet "Air Warden").[1]

In practically all plants studied, mutations occur in the direction delphinidin→cyanidin→pelargonidin; or cyanidin→pelargonidin (in plants where delphinidin types are absent, e.g. *Rosa*). Furthermore, dominant-recessive relationships operate in the same direction, delphinidin being dominant to cyanidin and cyanidin being dominant to pelargonidin production (Beale, 1941). The dominance of delphinidin can clearly be traced to natural selection among temperate plants, for a blue flower colour favoured by bee pollinators. Indeed, practically the only exception to the above rule occurs in the one tropical plant that has been studied, *Salvia splendens*. Here, pelargonidin is dominant to delphinidin but this is hardly surprising since this *Salvia* species, which in the wild has bright scarlet corollas, is known to be bird pollinated (van der Pijl, 1961). The unusual mutation from pelargonidin to cyanidin which has been noted in *Rosa polyantha* (Scott-Moncrieff, 1936) is due to a back-mutation and is probably cytoplasmic in origin.

Genes controlling the hydroxylation of pelargonidin to cyanidin and of cyanidin to delphinidin which have been uncovered in most plants studied (Table 8·1) often also control flavonol hydroxylation. This is certainly true in *Antirrhinum* (*see* Table 8·2), in sweet pea, carnation, *Impatiens*, French bean and in the cultivated potato (Table 8·1). In sweet pea the genes **E** and **Sm** act in the following manner:

In the potato, it is clear from surveys (and breeding experiments) that the gene **P** controlling 5'-hydroxylation is widespread in related wild species as well as the cultivars. Mauve-flowered species consistently have delphinidin and

---

[1] Throughout this section, the factor of methylation is ignored for the sake of clarity. Delphinidin is used here to include pigments based on petunidin and malvidin, and cyanidin includes pigments based on peonidin.

myricetin in the petals, whereas white-flowered species have only quercetin and/or kaempferol. The exceptional presence of myricetin in a white mutant of the mauve-flowered *Solanum soukupii* indicates that anthocyanin is suppressed at a locus other than **P**, a result which has been confirmed by hybridisation studies (G. J. Paxman, private communication).

Genetic control of hydroxylation is usually tissue specific, being limited in most cases to floral organs. For example, in *Impatiens*, the flowers may have delphinidin, cyanidin or pelargonidin but the sepals of most genotypes have the simplest anthocyanidin, cyanidin. This tissue specificity may be lost in organ culture and flowers of *Impatiens balsamina*, which are genotypically pelargonidin or delphinidin producers, can be induced to produce cyanidin under artificial conditions (Klein and Hagen, 1961). Even when genetic control of hydroxylation affects pigment synthesis throughout the plant, the expression of the genes may differ from one tissue to another. Thus, in *Primula sinensis*, while **K** types have delphinidin in both leaf and petal, **kk** types have mainly cyanidin in leaf but mainly pelargonidin in petal. Again, in the potato, gene **R** controls production of cyanidin in the flower but pelargonidin in the tuber. Pelargonidin has never been found as a flower pigment in the potato (though it does occasionally occur in the stamens); this is probably a reflection of the fact that the selection for colour in this plant has centred on the tuber.

In spite of the amount of information available on genetic control of hydroxylation, it is not known with certainty at what stage hydroxylation occurs. Most (but not all) biosynthetic studies indicate that hydroxylation occurs after elaboration of the C_{15}-flavonoid skeleton. If this is so, then hydroxylation of anthocyanidins and flavonols almost certainly occurs at the dihydroflavonol stage, i.e. before the pathways to the two types of pigment diverge. The most striking result to come out of the genetic work is that while the hydroxylations of anthocyanidins and flavonols are intimately connected, those of other flavonoids are unrelated. In *Antirrhinum*, for example, aurones with the 3',4'- and 3',4',5'-hydroxylation patterns occur uniformly in all genotypes (Table 8·2). As regards the flavones (as distinct from flavonols) it is true that in the snapdragon luteolin is only found in conjunction with quercetin and cyanidin. However, in *Streptocarpus*, luteolin is found with both cyanidin and delphinidin, so that the gene controlling 5'-hydroxylation does not seem to affect flavone. Presumably enzymes similar to the well-known phenolases are responsible for catalysing the oxidation of 4'-hydroxy- to 3',4'-dihydroxyflavonoids and of 3'4'-dihydroxy- to 3',4',5'-trihydroxyflavonoids, but attempts to isolate them have so far failed.

D. Genetic Control of Methylation

Methylation is usually restricted to the anthocyanins in the flowers of garden plants, since the flavones and flavonols occurring with peonidin, petunidin and malvidin have never been found methylated. The only system that has been fully studied from the genetic point of view is the methylation of delphinidin via petunidin to malvidin:

Delphinidin   Petunidin   Malvidin

Recent studies of pigment inheritance in both the cultivated potato and the garden petunia indicate that there are two distinct processes involved, controlled by different alleles. In *Petunia*, the pleiotropic gene **F**, which also determines glycosylation (see section E), controls petunidin synthesis (i.e. 3'-O-methylation) and gene **K** the further methylation of petunidin to malvidin (Hess, 1964). Again, in the potato, **Ac**, a complex locus like **F**, controls 3'-methylation and **Ac'** 5'-methylation.

The inheritance of anthocyanin methylation has been studied in a number of other plants, but nearly always results have been inconclusive because of practical difficulties. Even in the case of the potato, it was necessary to reduce the chromosome number of the tetraploid salad potato variety "Congo", the only known source of malvidin, from tetraploid to diploid level before it could be crossed with a petunidin-containing diploid cultivar of known genotype (Simmonds and Harborne, 1965). The ploidy barrier has, in fact, so far prevented the study of methylation in *Primula sinensis*, the wild type "Standard Stellata" malvidin form being diploid and the only known delphinidin-containing form, "Duchess Fern Leaf", being tetraploid (Harborne and Sherratt, 1961). Studies in *P. sinensis* have, however, shown that methylating gene(s) are particularly susceptible to gene interaction. Thus, the co-pigment (flavonol-producing) gene **B** interacts with the methylating system in such a way that **BB** mauve-flowered types are more highly methylated (75% malvidin, 25% petunidin) than **bb** maroon-flowered types (63% malvidin, 28% petunidin and 9% delphinidin).

Studies of methylation in other *Primulas* have likewise been only partially successful. Working with the tetraploid *P. malacoides*, Seyffert (1959) found a mendelian factor for controlling methylation of petunidin to malvidin. In this laboratory, indications have been obtained (but final proof is lacking) that two genes are responsible for methylation of delphinidin to malvidin in *P. obconica*. Finally, indications have been obtained from hybridisation in *P. auricula* (Gairdner, 1936) that a third gene is present for controlling the further methylation of malvidin in the 7-position to yield hirsutidin (1).

7-O-methylation of anthocyanins is more or less confined to the Primulaceae and the comparable 5-O-methylation which occurs in the Plumbaginaceae is likewise a rare process. In this latter instance, it is possible that this methylation represents the loss of substrate-specificity in a flavonol 5-O-methylating system rather than the establishment of a new methylating locus. Thus, 5-O-methylated flavonols are more common than related anthocyanins and flowers of *Plumbago capensis*, the source of 5-O-methylmalvidin (capensinidin), contain large amounts of 5-O-methylquercetin (azaleatin). Unfortunately, the genetic control of flavonol methylation has not been studied at all but a consideration of azaleatin distribution, particularly in *Rhododendron*, where it is widespread, suggests that it is formed by a single-step process from quercetin. Although methylation of flavonols has not been examined, that of isoflavones has and 4'-O-methylation of daidzein and genistein (to give formononetin and biochanin-A) is known to be controlled by a single gene **M** in subterranean clover (Francis and Millington, 1965).

(1) Hirsutidin   (2) Digicitrin

The whole question of how plants regulate O-methylation of flavonols and other flavonoids poses many intriguing problems. It hardly seems likely, for example, that the biosynthesis of digicitrin (2), the pigment in *Digitalis purpurea* leaf, involves six separate methylating steps; rather the 5- and 3'-positions must be protected in some way from a powerful general O-methylating system. Some of the many methylating patterns in the flavonol series may be produced by direct methoxylation, i.e. introduction of a methoxyl into the flavone nucleus. Such a step would account, for example, for the synthesis of patuletin (6-methoxyquercetin) in *Tagetes patula*. Methoxylation was at one time suggested as a process in the anthocyanin series (i.e. peonidin from pelargonidin, etc.) but the genetic data outlined above overwhelmingly indicate that hydroxylation and methylation are separate processes.

The genetic results have not so far indicated at what stage methylation takes place in synthesis. Hess (1965) argues, from his failure to detect specific anthocyanin methylating enzymes in *Petunia* flowers, that methylation must occur at the cinnamic acid stage and the genes uncovered simply regulate the incorporation of methylated cinnamic acids into anthocyanin synthesis. However, negative findings are always open to question and the experimental difficulties in the field of plant enzymes are well known to be formidable. The fact remains that enzymes which will O-methylate other phenolic substrates have been isolated from plants and the presence of a positive charge on the anthocyanin molecule provides a methylating enzyme with a simple means of finding the right substrate. It is also difficult to accept Hess's theory because it does not allow for methylation of the A-ring of anthocyanins, and because it indicates a closer relationship between cinnamic acid and anthocyanin synthesis than is known to exist at the present time.

E. Genetic Control of Glycosylation and Acylation

Only in a limited number of plants has evidence been obtained that the attachment of sugars to flavonoid pigments is under genic control. As usual, more is known about the anthocyanins than other classes and it is clear that 3,5-diglucoside production is normally dominant to 3-glucoside and 3,5-tri-glucosides to 3,5-diglucosides. A total of five genes have been uncovered in *Streptocarpus hybrida* which control anthocyanidin glycoside synthesis (Lawrence and Sturgess, 1957; Harborne, 1963b). Apart from the fact that there is a pair of complementary genes controlling 5-glucosylation, each of the other genes can be allotted a separate step in the biosynthesis of the 3-rutinoside-5-glucoside, the glycosidic type present in flowers of the wild type plant:

$$\text{Anthocyanidin} \xrightarrow{Q} \text{3-Glucoside} \xrightarrow{X,Z} \text{3,5-Diglucoside} \xrightarrow{D} \text{3-Rutinoside} \\ \text{precursor} \qquad\qquad \downarrow P \qquad\qquad\qquad\qquad\qquad\qquad\qquad \text{-5-glucoside}$$

3-Sambubioside

Glycoside synthesis is under particularly strict genic control in *Streptocarpus* since the genes described above form an epistatic series: $D > X$, $Z > P > Q$. This explains why "mixed" glycosidic types (i.e. 3-sambubioside-5-glucosides, 3-rutinosides) have not been detected in these plants.

Less is known about the inheritance of glycosylation in other plants. A gene for the 5-glucosylation of anthocyanidin 3-rutinosides has been described in potato, *Petunia* and egg plant; in each case the gene is tightly linked to one controlling acylation and forms part of a compound locus. By contrast, the gene **M** controls the 5-glucosylation of malvidin 3-glucoside in *Cyclamen* in a straightforward manner and is a simple locus (van Bragt, 1962).

Information about the genetic control of flavonol glycosylation has been obtained by studying the variation in glycoside content which occurs between different clones of two wild potato species, *Solanum chacoense* and *S. stoloniferum* (Harborne, 1962). In each species, a single gene, **Gl**, controls the transfer of glucose to quercetin 3-rutinoside to give a branched triglycoside, with a $\beta 1 \rightarrow 2$ (sophorose) glucose–glucose linkage:

$$\text{3-O—Glc—O—Rha} \xrightarrow{Gl} \text{3-O—Glc} \begin{array}{l} {}^{\diagup}\text{O—Rha} \\ {}_{\diagdown}\text{O—Glc} \end{array}$$

The same gene also controls the transfer of glucose to quercetin 3-glucoside:

$$\text{3-O—Glc} \xrightarrow{Gl} \text{3-O—Glc—O—Glc}$$

since small amounts of 3-sophoroside occur in dominant types, replacing 3-glucoside present in the recessives. The enzyme controlling flavonol glycosylation in this plant is thus not completely substrate specific and is able to use either the 3-glucoside or the 3-rutinoside for its action.

All the above genetic results indicate that glycosylation is the last step in flavonoid synthesis, a conclusion which has been confirmed by enzymic studies in mung bean, *Phaseolus aureus*. Barber (1962a, b) has isolated from seedlings of this plant, two separate enzymes which catalyse in turn the transfer of glucose (from uridinediphosphateglucose) to quercetin to give the 3-glucoside and the transfer of rhamnose (from thymidinediphosphaterhamnose) to quercetin 3-glucoside to give the 3-rutinoside.

Data on the inheritance of glycosylation of other flavonoid classes is lacking but there is circumstantial evidence to indicate that enzymes controlling glycosylation are usually specific as to flavonoid class. The best example here is *Antirrhinum*, in which plant all five classes of pigment have different patterns (Table 8·3). Anthocyanidins occur typically in snapdragon flowers as 3-rutinosides, flavonols as 3-glucosides, flavanones as 7-rutinosides, flavones as 7-glucuronides and aurones as 6-glucosides.

TABLE 8·3. Glycosidic Patterns of the Flavonoids
of the Snapdragon, *Antirrhinum majus*

Flavonoid Type	Aglycones	Glycosidic types
Anthocyanidin	Pelargonidin Cyanidin	3-Rutinoside
Flavonol	Kaempferol	3-Glucoside and 3,7-diglucoside
	Quercetin	3-Glucoside and 3-rutinoside
Flavone	Apigenin	7-Glucuronide and 7,4'-diglucuronide
	Luteolin Chrysoeriol	7-Glucuronide
Aurone	Aureusidin Bracteatin	6-Glucoside
Flavanone	Naringenin	7-Glucoside and 7-rutinoside

The flavonols and anthocyanidins are more nearly related in structure and biosynthesis than other flavonoids and their glycosidic patterns do occasionally correspond rather closely. A striking example is the co-occurrence of flavonol *and* anthocyanidin 3-arabinosides and 3-galactosides in *Rhododendron* flowers (*see* Chapter 6, p. 196). It is difficult to believe that in such cases the glycosyltransferases do not operate on both types of substrate. The ability of glycosyltransferases of one class to catalyse the synthesis, on occasion, of glycosides of the other class is the only rational explanation for the unexpected production in *Lathyrus* mutants of anthocyanidin glycosidic types not present in the wild type form. These crimson and scarlet mutants (Table 8·4) have the 3-rhamnosides and 3-rhamnoside-5-glucosides of purple forms and, in addition, 3-lathyrosides (lathyrose is xylosylgalactose), 3-galactoside-5-glucosides and 3-galactosides. *L. odoratus* is a strong inbreeder and contamination of stocks by hybridisation can be ruled out; in any case, all other wild *Lathyrus* species examined have only 3 rhamnosides or 3-rhamnoside-5-glucosides. The only possible source of enzymes for the synthesis of these anthocyanins is from the flavonol side and indeed examination of the major glycoside in wild type sweet pea (kaempferol 3-lathyroside-7-rhamnoside) showed that the necessary enzymes were present.

TABLE 8·4. Examples of Changes in Specificity of Glycosyltransferases

Plant	Change in pattern		Occurrence
	Usual types	New types	
	Flavonol to anthocyanidin		
Lathyrus odoratus	3-Rhamnoside	3-Lathyroside	Genotypes
	3-Rhamnoside-5-glucoside	3-Galactoside	recessive for **E** (5'-hydroxylation)
		3-Galactoside-5-glucoside	
Primula	3-Glucoside	3-Gentiobioside	*P. sinensis* (all genotypes)
	3,5-Diglucoside	3-Gentiotrioside	
	Anthocyanidin to flavonol		
Antirrhinum majus	3-Glucoside	3-Rutinoside	Genotypes recessive for **pal** (conc. gene)

A similar explanation probably holds true for the production of anthocyanidin 3-gentiobiosides and 3-gentiotriosides in *Primula sinensis*; all other *Primula* species examined have only 3-glucosides or 3,5-diglucosides (*see* Chapter 6, p. 198). The reverse situation, loss of specificity of anthocyanidin glycosyltransferase to admit synthesis of a related flavonol glycoside, would explain the anomalous synthesis of quercetin 3-rutinoside in certain genotypes of *Antirrhinum*, those which are recessive at the pallida locus (a gene controlling anthocyanin concentration) (Fincham, 1963).

That acylation (attachment of a cinnamic acid residue) of anthocyanins is related to glycosylation is borne out by the remarkable finding that in all four plant species in which a gene for acylation has been uncovered, the acylating function is closely linked with a glycosylating function. This is true of gene **F** in *Petunia*, **Ac** in *Solanum tuberosum*, an unnamed gene in *S. melongena* (Abe and Gotoh, 1959) and **V** in *Matthiola incana*. The acylating group in all cases is p-coumaric acid and the glycosylating function is 5-O-glucosylation. The fact that the locus is compound is shown by the fact that it does not involve transfer of an acylating sugar but that the acyl group is attached to the rhamnose in the 3-position and the glucose to the 5-hydroxyl. Furthermore, the same locus has a methylating function in both potato and petunia:

Unfortunately, genetic evidence proving that the **Ac** locus in the potato consists of three closely linked genes is not available and is not likely to be obtained as it would involve raising some 10,000 plants. Nevertheless, the close association of these three biochemical effects would suggest (but does not prove) that they occur sequentially as one of the last stages in anthocyanin synthesis.

F. Hybrid Studies

Little information on the effect of plant hybridisation on flavonoid synthesis has been obtained from chemogenetical studies. Most work has been carried out on plant material arising from spontaneous gene mutation within a species (e.g. sweet pea, Chinese primrose, snapdragon, etc.) and new substances produced in mutants not explainable on genetic grounds are the results of simple alterations in enzyme specificities (see section E). Relatively few garden varieties have been produced by deliberate hybridisation; examples are *Viola* × *wittrockiana* (cf. Stearn, 1965), *Delphinium* (Legro, 1964) and *Streptocarpus hybrida* (Lawrence and Sturgess, 1957). In these plants, dominant-recessive relationships control flavonoid production in the flowers and F_1 plants have the same pigments as the dominant parent; only in the F_2 are new substances likely to be discovered.

The general situation may be illustrated by reference to *Streptocarpus*, the garden cultivars being derived essentially from the cross of the delphinidin-containing *S. rexii* (genotype **Or**) and the cyanidin-containing *S. dunnii* (genotype **oR**). Only in the F_2 will pink-flowered pelargonidin-containing forms (genotype **or**) segregate out, the production of a new anthocyanidin being simply the result of the accumulation of two recessive genes. The production of kaempferol in **or** forms is more difficult to explain, since neither parent contains any flavonols, and this substance may be considered a "true" hybrid constituent.

The only deliberate attempt to study the fate of flower flavonoids in hybrids is that of Parks (1965), who worked with the genus *Gossypium*. He first studied colour mutant forms of *G. arboreum*, *G. hirsutum* and *G. barbadense* and noted that the species could be identified, even in their mutant forms, by the residual array of flavonoids not affected by mutation. Parks then examined hybrids derived from crossing pairs of nine diploid species in different combinations and as a result of examination of the flower flavonoids in a few crosses, was then able to predict successfully the flavonoid complement that would be produced in other hybrid combinations.

Hybrids are occasionally obtainable from crossing plants of different ploidy, but the resulting chromosomal abnormalities may cause a particular flavonoid constituent to disappear in the F_1. For example, the tetraploid *Solanum stoloniferum* when crossed with the hexaploid *S. demissum* gave rise to a tetraploid hybrid by double chromosome reduction in the *stoloniferum* parent (Marks *et al.*, 1965). Not unexpectedly, the hybrid had the flavonol complement of *demissum* in its flowers but lacked luteolin 7-glucoside, a characteristic petal constituent of *stoloniferum*. Its absence from this particular hybrid must be due to loss of chromosomal material since luteolin 7-glucoside was found in hybrids of *stoloniferum* with other tetraploid species and in the triploid hybrid *S.* × *vallis-mexici*. The hybrid nature of the abnormal tetraploid hybrid involving *demissum* was established by studying the alkaloids in the leaves, which showed characters from both parents.

Inheritance of flavonoids in the leaves of higher plants is usually very straightforward and hybrids usually contain all (or most) of the constituents of the two parents. However, there may be considerable quantitative alterations as a result of hybridisation and this may give what appears to be new compounds in the hybrid. For example, leaves of the cross of *Phaseolus vulgaris* × *P. coccineus* contain 12 flavonoid constituents, four from each parent and four "new" hybrid constituents (Schwarze, 1959). In fact, the four "new" constituents occur as trace components in *coccineus* and are simply produced as major constituents in the hybrid.

Production of completely new flavonoids in leaves of hybrid plants is very rare. Extensive studies in the genus *Baptisia* (Alston and Turner, 1962) and in *Vitis* (Yap and Reichardt, 1964) have shown that, in most cases, hybrids contain flavonoid patterns obtained by superimposing those of one species on the other. In fact, in *Baptisia*, hybrids are so difficult to distinguish morphologically that chromatography of the flavonoid patterns has been used extensively in this group for recognising hybrid plants. The only new flavonoid— quercetin 3-rutinoside-7-glucoside—detected by Alston *et al.* (1965) arose in leaves of a hybrid of *leucantha* and *sphaerocarpa*, quercetin 3-rutinoside being present in the former and quercetin 7-glucoside in the latter parent.

III. Biosynthetic Pathways to Flavonoids

A. Tracer Experiments

Largely as a result of tracer experiments with plants, it is possible to derive with some confidence a biosynthetic pathway for the flavonoid constituents (Fig. 8·1). This scheme in no way conflicts with genetic data reported in the last section. In fact, it is partly based on information from other sources, including studies of inheritance, physiology, chemical "Comparative anatomy" (Geissman and Hinreiner, 1952) and so on. Many of the finer details of flavonoid biosynthesis still remain to be elucidated and, in particular, the enzymology of biosynthesis requires much attention in the future.

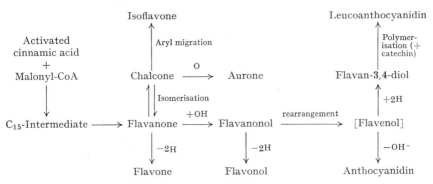

Fig. 8·1. Biosynthetic pathways to the flavonoids.

Since excellent detailed reviews of flavonoid biosynthesis have appeared elsewhere (Swain, 1962; Grisebach, 1965; Neish, 1964), it is the present purpose to provide a brief summary and consider the several problems in this field which are still outstanding.

Tracer studies involve feeding plants with ^{14}C-labelled precursors and measuring the incorporation of radioactivity in the flavonoids produced after various time intervals. Degradation of the isolated flavonoid will further indicate the position(s) in the molecule where radioactivity is specifically located. Clearly, such studies have yielded much valuable information, but, as in the biosynthesis of alkaloids (Battersby, 1961) and terpenoids (Goodwin, 1965), the results obtained require rather careful interpretation. Compounds fed artificially to plants through cut surfaces may be detoxified and metabolised by other than the normal synthetic pathways. Again, the efficiency with which a fed precursor is taken up may depend not on its importance as a natural intermediate but rather on its ability to penetrate to the site of synthesis. Results from ^{14}C experiments have to be accepted with a certain amount of caution (cf. Swain, 1965) and require support based on other approaches.

B. Synthesis of the C_{15} Skeleton

By feeding labelled acetate, phenylalanine, cinnamic acid or shikimic acid to tobacco, buckwheat and red cabbage, Neish and others (cf. Neish, 1964) have demonstrated quite conclusively that the C_{15} skeleton of flavonoids is derived from two separate pathways—from acetate and from shikimic acid (Fig. 8·2). The A-ring arises by head-to-tail condensation of two malonyl CoA units and acetyl CoA as in the synthesis of aromatic rings in fungi. Thus, feeding labelled acetate to red cabbage seedlings leads to incorporation of activity in only the A-ring of quercetin and cyanidin; furthermore, $^{14}CH_3CO_2H$ gives a flavonoid with activity at C1, C3 and C5 whereas $CH_3{}^{14}CO_2H$ gives a flavonoid with activity at C2, C4 and C6 (Grisebach, 1957).

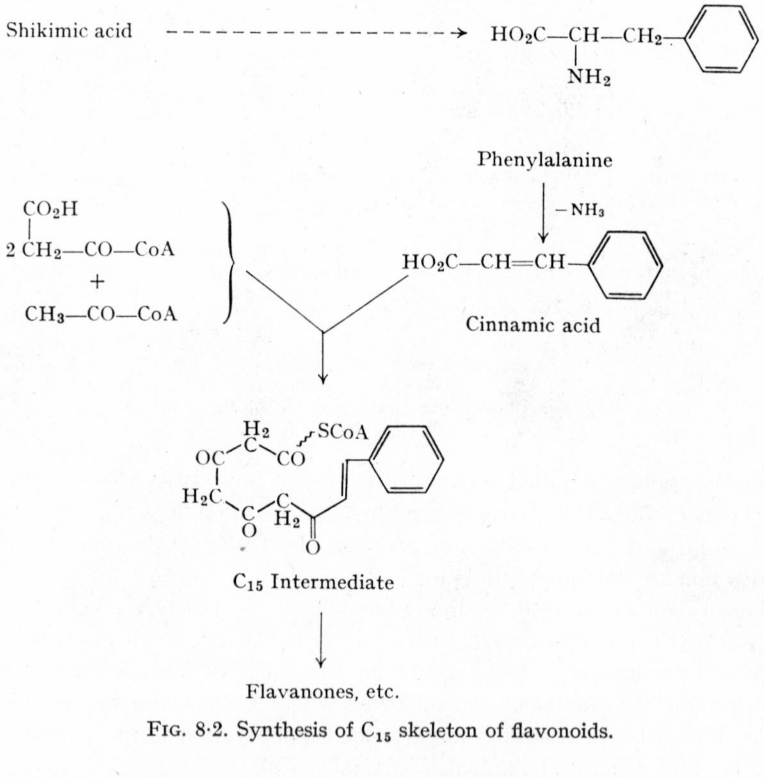

Fig. 8·2. Synthesis of C_{15} skeleton of flavonoids.

The B-ring and the central C_3 unit of flavonoids comes from a C_6–C_3 precursor derived ultimately from sedoheptulose via the shikimic acid pathway and which may be cinnamic acid itself. This precursor presumably requires activation, but it is not known how the plant does this. Both cinnamic acid and phenylalanine are excellent flavonoid precursors and the enzyme for the one-step deamination of phenylalanine to cinnamic acid (Koukol and Conn, 1961) has been found to occur widely in plants (Young et al., 1966). Whether this enzyme is closely involved in flavonoid synthesis is, however, a matter for further study. The comparable enzyme tyrase (Neish, 1961), which deaminates tyrosine to p-hydroxycinnamic acid, is also fairly common in plant leaves and p-hydroxycinnamic acid is, in fact, nearly as good a precursor of quercetin in buckwheat as cinnamic acid. However, neither tyrosine nor caffeic acid is an efficient precursor of quercetin and so it appears that hydroxylation of the B-ring occurs after the formation of the C_{15}-intermediate.

More recent experiments by Meier and Zenk (1965), however, indicate that the possibility that hydroxylation occurs at the C_9 rather than the C_{15} stage must still be borne in mind. Thus hydroxycinnamic acids prove to be good precursors of delphinidin in *Campanula medium* flowers and furthermore 3,4,5-trihydroxycinnamic shows better incorporation ($1\cdot4\%$) into delphinidin than either caffeic (3,4-dihydroxycinnamic) ($0\cdot7\%$) or p-hydroxycinnamic ($0\cdot35\%$). Unfortunately, comparable data for cinnamic acid itself was not provided by these authors. The 3,4,5-trihydroxycinnamic acid, which is still unknown as a natural constituent of plants, was also incorporated (to the extent of $0\cdot71\%$) into myricetin when it was fed into seedlings of *Myrica gale*.

The genetic data on hydroxylation of flavonoids would allow monohydroxylation of the B-ring at the C_9 stage but other hydroxyls would appear to be introduced at the C_{15} level. In support of this view, Patschke and Grisebach (1965) have shown by comparing incorporation rates of 4-hydroxy- and 3,4-dihydroxychalcones in three unrelated plants (aster, buckwheat and cabbage) that the 3'-hydroxyl group of quercetin (and cyanidin) is introduced after chalcone formation. In addition, Patschke *et al.* (1966) have shown that dihydrokaempferol is an efficient precursor of quercetin and cyanidin in buckwheat. It is interesting that 3'-hydroxylation must occur at or before the dihydroflavonol stage, since kaempferol, in contrast with its dihydro derivative, is not converted by this plant to quercetin or cyanidin.

C. Modification of the Central C_3 Unit

Once the C_{15}-intermediate is formed, it is modified in a variety of ways, but in a limited number of steps, to yield the range of different flavonoid classes encountered in nature. No precise data are available about these interrelationships, but the steps outlined in Fig. 8·1 are all mechanistically plausible. Crude enzyme preparations have been isolated which will oxidise chalcones to aurones (Shimokoriyama and Hattori, 1953) or which will convert chalcones to isoflavones (Grisebach and Brandner, 1962); otherwise, little is known of the enzymes required for these various oxidations and reductions.

It remains to be seen whether or not the central C_{15}-intermediate shown in Fig. 8·1 is an as yet unidentified substance which is so rapidly turned over that it has escaped detection, or alternatively, is an activated, perhaps phosphorylated, flavanone or chalcone. Grisebach and Patschke (1961) support the latter view with experiments which show that when ^{14}C-labelled chalcones are fed to buckwheat, they are incorporated into quercetin and cyanidin. The same authors (Patschke et al., 1964), however, have also demonstrated that chalcones, when fed to plants, undergo degradation to C_6 and C_6–C_3 fragments so the above results are by no means unequivocal. The isomeric flavanones are perhaps more acceptable intermediates in flavonoid synthesis than chalcones and recent feeding experiments by Grisebach and his co-workers show that flavanones are specifically oxidised in the 3-position to give flavanonols, which are then efficiently converted by plants to flavonols or anthocyanidins (Grisebach and Kellner, 1965; see also Wong et al., 1965). Further, they have shown (Patschke et al., 1966) that flavanone incorporation is stereospecific and that (−)-5-glucosyloxy-7,4'-dihydroxyflavanone-2-^{14}C is incorporated, with 16 times the efficiency of the (+)-form, into quercetin in buckwheat.

Many feeding experiments have been carried out to establish that isoflavones are formed in plants from chalcones by aryl migration, the phenyl migration probably occurring after closure of the heterocyclic ring but before introduction of a 3-hydroxyl group (Grisebach, 1965). The facts that phenylalanines, separately labelled at each position in the C_3 side-chain, are correctly incorporated into formononetin by clover seedlings according to the scheme:

and that the label from chalcones fed to legumes can be found in the subsequently isolated isoflavones (see above), would seem to prove the case for aryl migration. However, the efficiency of incorporation of chalcones is very low and there is still a lingering doubt (see Swain, 1965) about what appears to be a proven biosynthetic route. Plausible alternative pathways involving modification of precursors before formation of the C_{15}-skeleton can be suggested and can, by no means, be completely eliminated at the present time.

The intervention of chalcones in aurone synthesis has not yet been proved by feeding experiments but both in vitro and enzymic studies favour such an oxidation step. Indeed, Wong (1966b), on adding chalcone to a cell-free extract of soybean seedlings (which make the aurone hispidol), was able to isolate two unstable hydrated aurones (optical isomers?) which were rapidly converted to hispidol on keeping:

Isoliquiritigenin

.6,4'-Dihydroxy aurone
(hispidol)

D. METHYLATION AND GLYCOSYLATION

Methylation and glycosylation presumably occur towards the end of synthesis because these processes affect some flavonoids and not others. Enzyme specificity must depend, in some cases, on the difference between the positively charged anthocyanidin substrate and the neutral flavonol molecule. It is difficult to eliminate the possibility that some methylation or glycosylation occurs at the C_{15}-intermediate stage or even before this, bearing in mind that glycosylation may be necessary for improving the sap solubility of these intermediates. However, most experiments that have been carried out so far indicate that these processes terminate synthesis.

That methylation of anthocyanidins proceeds as expected from S-adenosylmethionine has been demonstrated in petunias by Hess (1964a). Petunidin, peonidin and malvidin all labelled at the methyl groups were isolated from petunia flower buds which had been earlier supplied with ^{14}C-labelled methionine. Enzymes which catalyse the O-methylation of caffeic acid have been found in the same plant (Hess, 1964b) but these are not apparently involved in anthocyanidin methylation; the isolation of specific anthocyanidin O-methylating enzymes has still to be carried out.

The glycosylations of flavonols and anthocyanidins presumably require two different groups of enzymes, since the glycosylation patterns of these pigments, when they co-occur, are usually different. The two types of pigment certainly require different hydrolytic enzymes for the breakdown of the respective glycosides: the common β-glucosidases will hydrolyse flavonol glycosides but a specific anthocyanase is required for the anthocyanins. Enzymic studies of synthesis have so far been confined to rutin production in mung bean (see p. 263) but it is hoped that the presence of anthocyanidin glycoside synthetases in plants will soon be demonstrated.

IV. Physiology of Biosynthesis

A. GENERAL

It has long been recognised that environmental factors can have profound effects on anthocyanin synthesis in plant tissues. In fact, innumerable papers have been published on the effect of light or added sugar on anthocyanin production in leaf discs or in young seedlings (see Blank, 1958). Only recently have such studies been extended to cover other types of flavonoids, such as the flavonols, and to look for differential effects in flavonoid production. In many instances, stimulatory or inhibitory substances have been added to the system under study but rarely is it possible to distinguish between their effect on general growth or metabolism of the tissue and their specific intervention in flavonoid synthesis. In the past, it has not usually been possible to separate the effect of one environmental factor (e.g. temperature) from another (e.g. light) but today, with the modern growth-room facilities that are available, this is no longer a problem.

Environmental effects are most marked in leaf tissue and favoured material for physiological experiments are young seedlings known to accumulate pigment in some quantity, e.g. red cabbage, cress, etc. However, leaves of most plants which are normally acyanic will accumulate anthocyanin under environmental stress and mature leaves, in which growth is more or less stationary, may be used quite effectively. Discs of (green) strawberry leaves floating on nutrient solution will, for example, synthesise anthocyanin under suitable light conditions (Creasy et al., 1965). Flavonoid precursors or inhibitors may be added to the solution on which the leaf discs float but a more "natural" system for such experiments is one in which the plant material is accustomed to growing on the surface of water. It is not surprising to find, therefore, that the most extensive series of physiological experiments on anthocyanin biosynthesis have, in fact, been carried out with the water plant, *Spirodela oligorrhiza* (Thimann and Edmondson, 1949, and later papers).

A recent trend in plant physiology is the use of organ and tissue culture for experimental purposes. In the case of flavonoids, petals of *Impatiens balsamina* (Hagen, 1966) and endosperm of *Zea mays* (Straus, 1959) have been used for organ culture. Many plants in tissue culture give rise to anthocyanin-synthesing strains and these may be used for studying flavonoid biosynthesis. Anthocyanins have been positively identified in tissue cultures of potato, *Happlopappus gracilis* and *Parthenocissus* and have been reported in those of rose, carrot and grape. Cyanidin 3-glucoside and 3-rutinoside, for example, occur to the extent of 3% of the dry weight in cultures of *H. gracilis* and traces of quercetin derivatives have been detected in the same material (Harborne, 1964b). Although plant tissue growing as calluses on agar slopes or in liquid "shake" culture appears to be the ideal system for biosynthetic studies, disconcerting features are the lack of uniformity in pigmentation at the cellular level and the fact that anthocyanin-producing strains may inexplicably lose their pigment after several months in culture.

The main conclusions that can be drawn from physiological studies are that light is important for initiating or stimulating flavonoid synthesis and that both nitrogen and carbohydrate metabolism are related to pigment formation. Such studies have not so far added a great deal to our knowledge of biosynthetic pathways, but obviously physiological factors have to be borne in mind when planning tracer experiments. These factors also cannot be ignored when carrying out phytochemical surveys since it is essential to know at what stage in the life cycle a plant is likely to produce flavonoids in most abundance.

B. Effect of Light

Light is the most important external factor controlling flavonoid synthesis in plants. It is necessary, for example, for the full development of anthocyanin colour in fruits such as the apple and strawberry and in most garden flowers. Even flavonol synthesis in foliage is light-dependent; winter-grown pea plants, for example, have only a small fraction in amount of the flavonol glycosides produced by spring-grown plants.

In the case of the apple, anthocyanin synthesis in the skin is completely light-dependent and normal red-fruited varieties left to mature in total darkness remain green (Siegelman, 1964). Light-catalysed reactions are known to be required for anthocyanin synthesis in at least a dozen other tissues and they may also be needed for production of other flavonoids. Kaempferol and quercetin glycosides, for example, are only produced in freshly cut potato discs following exposure to a high light source. In *Pisum sativum* seedlings, light has a differential effect on flavonol synthesis. Kaempferol derivatives are formed in dark-grown tissues whereas both kaempferol and quercetin glycosides are found in plants grown under lights; thus, it is the 3'-hydroxylation of kaempferol (or precursor) to quercetin that is light-catalysed (Bottomley et al., 1965). Oddly enough, exactly the reverse holds in the case of anthocyanin synthesis in rye seedlings (Metche and Gay, 1964); cyanidin and pelargonidin are formed in the light whereas pelargonidin is absent from dark-grown seedlings.

That light is necessary for the full development of flower colour can be demonstrated by examining pigment concentrations in *Antirrhinum* flowers which have been kept in darkness from the primary bud stage (Schmidt, 1962). This treatment considerably reduces the production of all flavonoids; again there is a differential effect, yellow aurone pigment being least and magenta anthocyanin most affected. Studies of flowers which change colour with age (e.g. the rose "Masquerade", *Cheiranthus mutabilis*, etc.) also indicate that a light reaction is involved in pigment formation.

Many experiments have been carried out on the light dependency of anthocyanin synthesis in plants. For example, Creasy et al. (1965) investigated cyanidin 3-glucoside formation in strawberry leaf discs floated on a 0·1 molar sucrose solution at 25–30°. The discs contained no initial pigment and synthesis on exposure to light was subject to a lag period of 15 hr (the inductive photochemical period). Once started, pigment contained to be formed at a linear rate for several days.

The wavelengths and types of light source required for anthocyanin have been examined in detail by the Beltsville group (Siegelman, 1964; Hendricks and Borthwick, 1965) in America and by Mohr (1962) in Germany. Both groups of workers have shown that the photochemical reaction involves phytochrome, the partly characterised light receptor which is apparently universal to all plants (Butler *et al.*, 1965). This pigment system exists in two forms, one absorbing maximally near 660 mμ (red light), P_r, and the other near 730 mμ (far-red light), P_{fr}, which are converted one to the other by light of the appropriate wavelength:

$$P_r \underset{730}{\overset{660}{\rightleftarrows}} P_{fr}$$

Studies of the action spectra of anthocyanin synthesis in several plants have indicated that the P_{fr} form is the one usually involved in pigment synthesis.

In most plants studied, there are at least two light reactions; one or more associated with the initial short period of rapid synthesis (photoresponse I) and one with the longer slower linear phase (photoresponse II). These two responses have different energy requirements and it is photoresponse II which is usually (but not always) associated with phytochrome. The controlling influence of phytochrome is demonstrated by determining effective wavelength maxima (660 or 730 mμ) and the photoreversibility of the response. Both photoresponses have been observed in red cabbage, mustard, milo, balsam and buckwheat plants; on the other hand, only photoresponse I could be detected in experiments with apple skin or turnip leaf. In these atypical tissues, phytochrome is associated with photoresponse I.

Attempts have been made to indicate at what stages flavonoid synthesis in apple and turnip are light-dependent, but the schemes proposed must certainly be considered tentative ones. In apple, anthocyanin synthesis has been suggested (Siegelman and Hendricks, 1957) to take the following course (times given in hours):

Glucose $\xrightarrow{0\cdot5}$ Pyruvate $\xrightarrow{1}$ Active acetate $\xrightarrow{\text{Photoresponse I}}$

First radiation product $\xrightarrow{4}$ Polyacetate intermediates $\xrightarrow{P_{fr}}$

Product of phytochrome action $\xrightarrow{8}$ Dark reactions $\longrightarrow$ Cyanidin

Evidence supporting the positioning of the first light reaction is the discovery that apple skins in darkness accumulate C_2 compounds, notably ethanol and acetaldehyde.

Grill and Vince (1964), working with seedlings of turnip *Brassica rapa*, have obtained evidence that pigment synthesis begins in the cotyledons, in which tissue a light reaction is required for synthesis of a precursor. This precursor is then translocated to the hypocotyl where a second photochemical reaction is required for pigment synthesis. If translocation to the hypocotyl is prevented, anthocyanin only forms in the cotyledons. The nature of this hypothetical precursor is not known but it may be a C_{15}-molecule, since phenylalanine, acetate, shikimic acid and sugars were ineffective substitutes. The idea of a translocated intermediate, the synthesis of which is light-catalysed, is an attractive one and would explain pigment formation in subterranean plant organs, e.g. in radish roots and potato tubers. Flavonoid synthesis has been shown to be light-dependent in every plant tissue that has been so far studied and this would seem to be the general rule.

C. NUTRITIONAL EFFECTS

A large number of different substances, both inorganic and organic, have been added (usually in low concentration) to anthocyanin-synthesising systems to see if they have any effect on pigment formation. Not unexpectedly, the majority have either an inhibitory or stimulatory effect. It is, however, often difficult to interpret the results, since the effects recorded in many cases are undoubtedly indirect, the added substance having a primary effect on growth or nitrogen metabolism. Furthermore, mutually contradictory results have been obtained from studies in different tissues; whether these really represent differences in tissue response is, of course, another matter. Clearly, the concentration of added substance, its ability to penetrate to the site of synthesis and the presence of the right environment are all factors that have to be taken into account in such feeding experiments.

In spite of a rather confusing situation, two main points are clear from the data available. First, anthocyanin synthesis requires the presence of free sugar and addition of carbohydrate (up to a certain concentration) to the system always stimulates production. Second, anthocyanin synthesis is inversely related to protein synthesis, conditions favouring rapid protein build-up inhibit anthocyanin and vice versa. Much less is known about the relationship between flavonoid synthesis and the electron transport system (Krogmann and Stiller, 1962) or RNA synthesis (Radner and Thimann, 1963); these are clearly matters for future study.

In summarising the effects of added metabolites on flavonoid synthesis, it is convenient to consider the separate effects of sugars, protein inhibitors and precursors.

1. Sugars

The general enhancement of anthocyanin formation by added sucrose has long been recognised. The "spring flush" of anthocyanin colour in young leaves of many plants and likewise the production of autumnal reds in dying leaves have been ascribed to the fact that on both occasions, the leaves are particularly rich in free sugar. Under artificial conditions, glucose at an optimum concentration of 4% can increase anthocyanin four and a half fold over controls.

The relationship of sugar to anthocyanin synthesis has been studied in most detail by Eddy and Mapson (1951) for cress seedlings. They found that glucose, fructose, sucrose, galactose, sorbose and arabinose were equally effective (at concentration of 1%) in stimulating pigment formation in the dark. The effect of sugar is thus clearly an indirect one, and there can be no suggestion of the direct incorporation of carbohydrate into the flavonoid pathway. Rather, feeding excess sugar upsets the intermediary metabolism in a direction favouring anthocyanin synthesis. Pigment formation may in fact, be a built-in "safety" mechanism in plants to prevent excessive accumulation of free sugar. Certainly, treatment of plants with solutions containing over 2% glucose interfere with cell elongation, perhaps because of their osmotic effects.

2. Protein Inhibitors

Conditions favouring rapid protein synthesis in plants are unfavourable to anthocyanin production so that addition of urea or inorganic nitrogen to an anthocyanin-synthesising system generally reduces pigment formation. Conversely, when protein synthesis is low, anthocyanin production is high and similar additions of inhibitors of protein synthesis (ethionine, chloramphenicol, etc.) enhance pigment production. Many experiments have been carried out which confirm these generalisations. For example, Eberhardt (1959) has shown that urea added to leaf discs of *Saxifraga crassifolia* depresses anthocyanin and increases protein.

Many workers have shown that chloramphenicol and related antibiotics (at concentrations of 10^{-4} to 10^{-5} M) increase anthocyanin in radish, rape, turnip and mustard seedlings (*see* Siegelman, 1964). In two cases, those of white mustard (Pramer and Wright, 1955) and *Saxifraga* (Eberhardt, 1959), treatment with protein inhibitors also inhibit anthocyanin. This was probably due to the rather high concentrations (1.5 to 3×10^{-3} M) of chloramphenicol used. Faust (1965) working with apples, found that chloramphenicol when applied as a dip to whole fruits enhanced anthocyanin but when applied to skin discs inhibited it. He assumed that in the latter case sufficient chloramphenicol penetrated the cells to disturb both protein synthesis and RNA metabolism, the latter disturbance not unnaturally interfering with pigment production as well.

The effects of added amino acids are generally very difficult to interpret because of their effect on plant metabolism generally. The results are very contradictory: thus ethionine, a methionine antagonist, inhibits anthocyanin instead of enhancing its production but so does methionine (*see* Thimann and Radner, 1955). Feeding experiments with purines and pyrimidines, carried out with *Spirodela* by Thimann and Radner (1958) and with corn endosperm (Straus, 1960), are similarly contradictory. The bases obviously first exert an effect on nucleic acid synthesis but eventually they are also bound to interfere with protein turnover; long-term feeding with these substances are bound to give misleading results.

3. Precursors

It is a curious fact that there are few reports of flavonoid precursors stimulating pigment synthesis in plant tissues. Phenylalanine and cinnamic acid, for example, which seem to be on the direct biosynthetic pathway, do not markedly increase anthocyanin formation when added to suitable systems. This may be because they are rapidly metabolised in other directions or else they are rarely limiting factors in pigment production.

On the assumption that cinnamic acid is on the main pathway to flavonoids, Grisebach and Kellner (1964) fed p-fluorocinnamic acid to red cabbage seedlings and found that, at a concentration of 100 γ per 25 seedlings, it inhibited both anthocyanin and flavonol synthesis. The fact that radioactivity from ^{14}C-labelled p-fluorocinnamic acid was incorporated into cyanidin and quercetin in these experiments was assumed to be due to degradation of the fluorinated acid in the plant before incorporation; no trace of fluorine could be detected in the cyanidin formed. In similar experiments in which p-fluorophenylalanine was fed to buckwheat or *Scutellaria*, Watkin *et al.* (1965) isolated 4'-fluorochrysin. The substance however was not a precursor of quercetin; the fluorine atom again effectively prevents p-hydroxylation taking place.

4. Miscellaneous Factors

The effects on anthocyanin synthesis of growth substances, herbicides, metal ions and respiratory inhibitors have all been studied; the results vary somewhat from tissue to tissue and are not interpretable in general terms. Gibberellic acid, for example, is inhibitory in *Spirodela* (Furuya and Thimann, 1964) whereas kinetin is stimulatory in *Impatiens* tissue cultures (Klein and Hagen, 1961). Herbicides (e.g. 2,2-dichloropropionic acid), when applied to foliage, generally appear to inhibit anthocyanin synthesis specifically in the petal; this applies to *Salvia splendens* and at least six other plants (Gowing and Lange, 1962; Asen *et al.*, 1963).

Copper may be needed for anthocyanin synthesis, since phenylthiocarbamide, a copper chelate, inhibits anthocyanin production in *Spirodela* (Edmondson and Thimann, 1950). Faust (1965) found that nickel and cobalt ions have a slight stimulatory effect on colour development in the apple skin. As expected, respiratory inhibitors inhibit anthocyanin; this is certainly true of dinitrophenol (7.5×10^{-5} M) and sodium fluoride (0.02 M) when they are added to the strawberry leaf disc system (Creasy *et al.*, 1965).

Injury and virus infection both disturb the nutrition of plant cells so it is not surprising that both also affect anthocyanin synthesis. Injury quite frequently stimulates anthocyanin synthesis or even initiates pigment formation in tissues otherwise devoid of anthocyanin colour (Bopp, 1959). Virus infection usually affects flavonoid production, presumably by disturbing RNA metabolism, but it can both initiate or stimulate synthesis. In the case of the well-known virus breaks in tulip, the infected areas lack pigment. By contrast, *Ustilago* infection in corn and leaf roll infection in potato stimulate anthocyanin, and virus infections in cherry and peach increase flavonol production threefold (Geissman, 1956). Finally, different viruses are known in garden stock *Matthiola incana* for initiating anthocyanin in white flowers and inhibiting colour development in red flowers. Studies of the pigments formed in the virus-infected red petals, however, indicate that flavonol synthesis is not seriously disturbed by the virus (Feenstra *et al.*, 1963).

D. RATES OF SYNTHESIS

Feeding experiments with radioactive precursors indicate that flavonoids are not stationary end-products of metabolism but are constantly "turned over" in the plant. For example, Grisebach and Bopp (1959) studied the incorporation and subsequent disappearance of ^{14}C-labelling in quercetin and cyanidin in buckwheat to see if there was any evidence of interconversion.

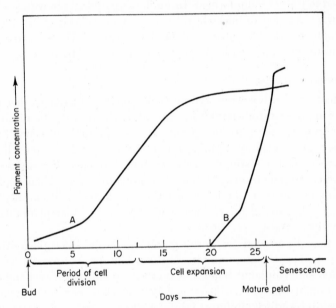

FIG. 8·3. Relative rates of flavonol and anthocyanidin synthesis in petals of *Primula sinensis* (after Reznik, 1961). Curve A, flavonol; curve B, anthocyanidin.

They found that the label was incorporated in both substances at the same time and disappeared also at the same rate. Labelled rutin was not converted to cyanidin and there was no evidence of interconversion. It was clear, however, from these experiments that both flavonoids were in a state of active metabolism.

In buckwheat seedlings, flavonols and anthocyanidin seem to be synthesised at approximately the same rate but there is much indirect evidence indicating that flavonol and anthocyanidin synthesis are independent of each other in other tissues. This has been confirmed for petal tissue by Reznik (1961), who examined the development of pigment in flowers of *Primula obconica*. His results (Fig. 8·3) show that the time of initiation and rate of synthesis of kaempferol and malvidin glycosides in these flowers are quite unrelated. Little is known about the rates of synthesis of other flower flavonoids but the competition observed in flowers of *Dahlia* and *Antirrhinum* (*see* p. 257) of different pathways for a common C_{15} precursor could certainly be explained on the basis of rate differentials.

Examination of the flavonoids formed at different stages of flower development have indicated fairly clearly what are the last stages in pigment synthesis. Thus, in *P. obconica*, Reznik (1961) noted that kaempferol 3-monoglucoside was formed in the early bud stage and that kaempferol 3-gentiotrioside could only be detected in petals at 14 days or older. Similarly, he found the first anthocyanin to be formed was a delphinidin glucoside and that methylation and 5-glucosylation of this to give the malvidin 3,5-diglucoside of the mature petal (at 26 days) occurred during the last few days of synthesis.

From studies in *Petunia* petals, Hess (1963) noted that anthocyanidins were formed (as glucosides) during petal development in the sequence: cyanidin→ delphinidin→peonidin→petunidin→malvidin. Again, Hagen (1966) working with *Impatiens* petal, observed that pelargonidin 3-monoglucoside was the only pigment present in buds four days before anthesis, whereas an acylated pelargonidin 3,5-diglucoside occurred in buds at anthesis and in mature petals. Studies of the relationship between flavonoids and ontogeny in other plants would, no doubt, yield similar data.

E. THE BEARING OF PHYSIOLOGICAL STUDIES ON PHYTOCHEMICAL SURVEYS

Bate-Smith, in choosing material for his surveys of the flavonoids in dicotyledons (1962), collected leaves from mature plants growing in a well-lighted position. Physiological studies indicate that these are minimal requirements for obtaining representative material for flavonoid investigation. There is much to be said for sampling, if possible, leaves at different ages from plants growing in more than one environment. Certainly, plants which have been treated with herbicides or which are diseased should not be used in such investigations. Although petal flavonoids are less affected by environmental factors, it is important to use mature petals to avoid the developmental effects already touched upon (*see* above).

That flavonoids are very stable plant constituents has been demonstrated by McLure and Alston (1964) who sampled *Spirodela oligorrhiza* grown in axenic culture under 25 different regimes. Culture conditions were varied by altering the light, the nature of the medium, the pH, the temperature and by adding various inhibitors and stimulators. The majority of flavonoids present in the leaf tissue were unaltered by these various treatments. The few qualitative differences that occurred only involved minor or trace constituents and probably only represented quantitative changes.

CHAPTER 9

FUNCTION OF FLAVONOIDS

I. Pigment Function

A. BIOLOGICAL SIGNIFICANCE OF FLOWER AND FRUIT COLOURS

Flavonoids are the most important group of colouring matters in plants, being responsible for most orange, scarlet, crimson, mauve and blue colours as well as contributing much to yellow, ivory and cream flowers. However, to state that all flavonoid colours in nature are of biological importance is not possible on present evidence. Flavonoids occur in all parts of the plant and pigmentation must in some cases be fortuitous and without immediate function. Nevertheless, the importance of flavonoid colours in flower petals as a means of attracting bees, butterflies and other animals to ensure fertilisation is now well established. Indeed, the association between bee activity and flower colour has long been studied (Sprengel, 1793; Darwin, 1876; Manning, 1956). The fact that flower colour in temperate plants has evolved towards production of blue hues, colours most favoured by bees, is clear evidence of this association.

Petal colour, of course, is only one factor which, together with scent, shape and presence of guide-lines, ensures the fertilisation of the plant by animal vectors and hence the preservation of the species. Animals involved include birds, butterflies, moths, bees and flies; colour preferences have only been studied in the case of birds and bees. Birds are sensitive to red and are important pollinating agents for many tropical plants, particularly of members of the Bignoniaceae, Gesneriaceae and Labiatae (van der Pijl, 1961). Humming birds and the bananaquit (*Coreta flaveola*), for example, are known to visit the bright orange and scarlet-red flowers of *Gesneria* and *Kohleria* species (Gesneriaceae) in Jamaica (R. Sutton, 1965), the pigments attracting them being pelargonidin and apigeninidin derivatives. Bees prefer blue colours and are major pollinators of temperate plants, most of which contain delphinidin or its derivatives, e.g. *Primula, Campanula, Anchusa*, etc. Bees can also distinguish yellow pigments and those which absorb in the ultra-violet (von Fritsch, 1950; Carthy, 1964), and accordingly visit yellow or ivory flowers as well as blue. Their ability to "see" in the ultra-violet explains why bees also pollinate certain red poppies, although they are red-blind; the poppy petals have a high reflectance for ultra-violet.

Flavonoid petal colour is most important in plants which are outbreeders and those which have incompatibility systems which prevent selfing. The fact that many inbreeding plants, such as the sweet pea, have brilliant anthocyanin colours does not contradict this, since as Stebbins (1959) has pointed out, self-pollinating plants are almost certainly derived from cross-pollinating ancestors. The mechanism for cross-pollination has not had time to be lost and the continued presence of floral colour may in fact be advantageous to the species. Some self-fertile plants are visited by insects with a resultant increase in seed production. The corollary that flavonoids are minor constituents of or absent from petals of plants having other pollinating mechanisms is well illustrated by the wind-pollinated grasses. The flowers of grasses are small and inconspicuous and generally lacking in anthocyanin. A survey of those grasses with pigment has indicated that, as expected, most have cyanidin and there is no selection towards other colours (Clifford and Harborne, 1966).

The importance of anthocyanin colour in fruits such as the strawberry, cherry, blackcurrant and apple as an aid to seed dispersal by animals is self-evident and needs no elaboration here. A less obvious but commercially important function of flavonoids follows from the fact that both fruit and flower colour give immense aesthetic pleasure to man. Conscious selection for colour varieties among garden plants and horticultural crops has been practised for a very long time. Colour "breaks" in the tulip, for example, were known and preserved in European gardens as far back as the sixteenth century (McKay and Warner, 1933). Even today, the production of new flower colour forms remains one of the prime objectives in ornamental-plant breeding. Fruit and vegetable colour play a role in stimulating the appetite for food and there is much emphasis today on breeding fruit varieties for their outward appearance. The production of large, juicy, bright scarlet strawberry varieties which are almost lacking in flavour, however, is one of the mixed blessings of our modern civilisation.

B. Contribution of Flavonoids to Plant Colour

The subject of the flavonoids and flower colour has been reviewed frequently in the past (e.g. Scott-Moncrieff, 1936; Beale, 1941; Blank, 1947; Paech, 1954; Reznik, 1956) and also more recently by the author (Harborne, 1965).

10

Flower and fruit colours have already been discussed in relation to the individual genera and comprehensive lists of anthocyanins in the angiosperms have been given (Chapters 5, 6 and 7). Here, it is only intended to provide a brief summary of the main points.

The contributions of flavonoids to flower colour are summarised in Table 9·1. The descriptions of colour given are rather broad; for more precise notes

TABLE 9·1. Contribution of Flavonoids to Flower Colour

Colour	Pigments[1]	Examples
Ivory and cream	Flavones (e.g. apigenin) and/or flavonols (e.g. quercetin)	Ivory *Antirrhinum mα* white sweet pea
Yellow	(a) Carotenoid alone	Yellow rose
	(b) Flavonol alone (e.g. quercetagetin)	Primrose
	(c) Aurone alone	Yellow *Antirrhinum*
	(d) Carotenoid and flavonol or chalcone	Birdsfoot trefoil, gorse
Orange	(a) Carotenoid alone	*Lilium regale*
	(b) Pelargonidin and aurone	Orange *Antirrhinum*
Scarlet	(a) Pelargonidin	Geranium, *Salvia*
	(b) Cyanidin and carotenoid	Tulip
Brown	Cyanidin on carotenoid background	Wallflower or *Primula polyanthus*
Magenta or crimson	Cyanidin	*Camellia hortense*
Pink	Peonidin	Peony, *Rosa rugosa*
Mauve or violet	Delphinidin	Verbena
Blue	(a) Cyanidin as metal complex	*Centaurea cyanus*
	(b) Delphinidin as metal complex	*Delphinium ajacis*
	(c) Malvidin and co-pigment	*Primula obconica*
Black (purple black)	Delphinidin at high concentration	Tulip "Queen of the Nigh"

[1] For brevity, pigment aglycones are given, but it should be remembered that flavonoids always occur in petal tissue in glycosidic form.

on the colours of individual flowers, the colour charts of Wilson (1938) should be consulted. The colours in Table 9·1 refer to those of the petal or corolla; flavonoids also contribute to other parts of the flower, be it sepal, bract, stamen, style or pollen. Anthocyanins are also responsible for most red and purple pigmentation in leaves, whether it is produced transiently in young or ageing leaves or whether it is permanent coloration. The important contributions of anthocyanin to fruit colours will be discussed in a later section on flavonoids in foods. The specific contributions of anthocyanins, chalcones, aurones and flavones to flower colour will now be outlined.

1. The Anthocyanins

The contribution of anthocyanins to flower colour is basically simple. There are three main pigments: pelargonidin (1), cyanidin (2) and delphinidin (3), which differ in structure only by the number of their hydroxyl groups. Neglecting the complications of glycosylation, methylation and co-pigmentation, these three pigments, either singly or as mixtures, provide the whole range of flower colour from pink and orange to violet and blue. Broadly speaking, all pink, scarlet and orange-red flowers have pelargonidin, all crimson and magenta flowers have cyanidin and mauve and blue flowers have delphinidin.

(1) Pelargonidin, $R = R' = H$
(2) Cyanidin, $R = OH$, $R' = H$
(3) Delphinidin, $R = R' = OH$

Very rarely, alterations in flower colour can be traced to changes in the hydroxylation patterns of the three common anthocyanidins. Loss of the 3-hydroxyl group from cyanidin and pelargonidin give the 3-deoxyantho-cyanidins luteolinidin and apigeninidin, yellower in colour than the parent compounds, which are responsible for the orange-red or orange-yellow hues in the corollas of *Gesneria, Columnea* and *Kohleria* species. Substitution of a hydroxyl in the 6- or 8-position of pelargonidin has a hypsochromic effect on colour and the pigment in the unusual tangerine-coloured flowers of *Impatiens aurantiaca* is of this type.

Other modifications in the structure of anthocyanins have only minor effects on colour. Methylated anthocyanins occur frequently but methylation has only a slight reddening effect on colour and, in fact, the effect of methyl-ation is usually obscured by other factors, particularly co-pigmentation. Most anthocyanidins occur in petals as 3-glucosides or 3,5-diglycosides and there is little difference in colour between these glycosidic classes, although 3,5-di-glycosides do, perhaps, provide the more intense coloration. 3,7-Diglycosides, which occur very rarely, are distinctly different in colour (less blue) than the common glycosides and the unusual orange-yellow colour in petals of the orien-tal poppy, of some forms of Iceland poppy and of *Watsonia* flowers can be traced to the presence of a pelargonidin 3,7-diglycoside.

Variations in the amounts of anthocyanin in the petal have profound effects on colour and large discontinuous differences in anthocyanin content have been noted in the flowers of some plant varieties, e.g. those of the carnation or Chi-nese primrose. At one end of the scale, low pigment concentrations (*ca* 0·01% dry weight of petal) give flowers a faint pinkish or purplish blush (e.g. the rose "Madam Butterfly"). At the other, high concentrations (up to 15% dry weight of petal) produce deep purple-black colours such as can be seen in the tulip "Queen of the Night" or the pansy variety "Jet Black".

No anthocyanin isolated from the petals with acidic methanol is blue in colour and blueness in plants is nearly always due to purple anthocyanins occurring *in vivo* in some complex combination. This combination most frequently involves metal complexing and blue chelates of cyanidin and del-phinidin with magnesium, iron and aluminium have been successfully isolated (by extraction under neutral conditions) from plants such as the cornflower, *Hydrangea* and lupin (Bayer, 1958). Blueness also results when anthocyanins occur in loose association with flavones, the so-called "co-pigmentation" effect, which has been observed in plants such as *Lathyrus odoratus* and *Primula sinensis*. Whether these pigment–copigment complexes are bound together by metal is not yet known, but recent experiences of Asen and Jurd (1966) suggest this is likely. Two more minor factors which may cause purple anthocyanins to appear blue *in vivo* are the pH of the cell sap and the adsorp-tion of anthocyanin on polysaccharide. In some plants, no doubt, blueness is produced by a combination of several of these factors or by other as yet un-known agencies.

2. Chalcones and Aurones

The contribution of chalcones and aurones to flower colour is quite straight-forward. They are bright yellow substances which pigment the flowers of a restricted number of plants, having been found frequently in only one plant family, the Compositae. Carotenoids are much more common as yellow colour-ing matters in flowers and indeed accompany chalcones and aurones in many instances. For example, the flowers of gorse, *Ulex europaeus*, are pigmented jointly by sap-soluble glucosides of 2',4',4-trihydroxychalcone (4) and by the lipid-soluble α- and β-carotene (5), violaxanthin and taraxanthin. Again most

(4) 2',4',4-Trihydroxychalcone   (6) Bracteatin

(5) β-Carotene

composites with chalcones or aurones in their flowers also have carotenoids; this is true, for example, of *Helichrysum bracteatum*, which has the aurone bracteatin (6) as yellow colouring matter.

Chalcones and aurones occur as sole yellow pigments in relatively few plants: yellow *Antirrhinum*, yellow *Dahlia*, *Cosmos sulphureus* and *Petrocosmea kerrii*. In *Antirrhinum*, the yellow aurones aureusidin and bracteatin (6) contribute also to orange flower colour in as much as they combine with the scarlet pelargonidin in some varieties to produce this intermediate shade.

3. Flavonols and Flavones

A large proportion of higher plants have white, ivory or cream flowers; in addition, cyanic-flowered plants in cultivation frequently produce acyanic (white) varieties. The vast majority of such flowers are pigmented by the common flavonols or flavones. Thus, Reznik (1956) found that 86 of some 100 species surveyed with white flowers contained kaempferol and 17 quercetin. Luteolin and apigenin (7) also commonly occur in white petals, the latter, for example, being an abundant constituent of ivory *Antirrhinum* varieties. The function of the flavones and flavonols is to add "body" to petals which would otherwise be translucent. Also, as has already been mentioned, these colourless flavones absorb strongly in the ultra-violet, can be "seen" by bees and possibly other insects and presumably provide the flowers with a satisfactory means of attracting pollinating insects. That flavones serve a useful purpose in white-

(7) Apigenin   (8) Quercetagetin

flowered species may be gauged from the fact that it is almost unknown for white petals to completely lack flavonoids of any kind.

Simple modifications in the hydroxylation, methylation or glycosylation patterns of flavonols may produce substances which are distinctly yellow in colour and flavonols are known to be responsible for yellow flower colour in a number of plants. The most notable is quercetagetin (8), the 6-hydroxy derivative of quercetin, which is the principal yellow substance in the primrose and also contributes, with carotenoid, to yellow colour in many other *Primula* species and cultivars (*see* p. 200). It is the sole yellow colouring matter in yellow flowers of many *Rhododendron* species and also occurs with lutein in the marigold, *Tagetes erecta*. The related 8-hydroxyquercetin or gossypetin is the yellow pigment in cotton flowers (*Gossypium*) and in the corn-marigold, *Chrysanthemum segetum*.

A few methylated flavonols are pale yellow in colour and methylated derivatives of quercetin and myricetin, such as isorhamnetin and syringetin, are minor pigments of the common marigold *Calendula officinalis* and of the meadow sweet pea *Lathyrus pratensis*, respectively. Finally, changes in the glycosylation pattern can cause flavonols to function as yellow pigments. Flavonols always occur in petals in glycosidic form but instead of being present in the more usual 3-glycosidic combination, quercetin occasionally occurs as the 7- or 4'-glycoside. In such instances, e.g. in yellow rose and gorse petals, it probably provides a minor contribution to the colour.

C. FLAVONOIDS OF POPULAR ORNAMENTAL PLANTS

The contributions of flavonoids to flower colour are further illustrated in Table 9·2, which gives the pigments found (as glycosides) in colour varieties of some of the more popular ornamental plants. In some instances, the results shown summarise surveys of many different colour varieties. For example, in tulip, Shibata and Ishikura (1960) looked at the anthocyanidins of 107 varieties and the author and also Arisumi (1963) have examined an equal number of roses for their pigments. The plants given in Table 9·2 are species showing extensive colour variation but similar data are available on practically all other cultivated garden plants.

TABLE 9·2. The Flavonoids in Flowers of Ornamental Plants

Plant, colour form and typical variety	Pigments present[1]
SNAPDRAGON (*Antirrhinum*)	
Magenta	Pure cyanidin
Orange-red	Cyanidin and yellow aurone
Pink	Pure pelargonidin
Orange-yellow	Pelargonidin and yellow aurone
Yellow	Aurone (aureusidin, bracteatin)
Ivory	Flavones (apigenin and luteolin)
SWEET PEA (*Lathyrus odoratus*)	
Deep purple, "Jupiter"	⎫ Malvidin
Pale mauve, "Elizabeth Taylor"	⎭
Deep crimson, "Harrow"	⎫ Peonidin (some cyanidin)
Carmine, "Carlotta"	⎭
Pink, "Mrs R. Bolton"	⎫ Pelargonidin
Scarlet, "Air Warden"	⎭
White, "Swan Lake"	Kaempferol
HYACINTH (*Hyacinthus orientalis*)	
Deep red, "Scarlet O'Hara"	Pelargonidin with 10% cyanidin
Pink, "Pink Perfection"	Pelargonidin and cyanidin (6:4)
Mauve, "Mauve Queen"	Pure cyanidin
Blue, "Delft Blue"	Delphinidin with 10% cyanidin
Pale blue, "Springtime"	Pure delphinidin
CAPE PRIMROSE (*Streptocarpus*)	
Mauve, "Constant Nymph"	⎫ Malvidin, with traces of petudinin
Blue, "Merton Blue"	⎭ and delphinidin
Rose and magenta	Peonidin (+5% cyanidin)
Pink and salmon	Pelargonidin

TABLE 9·2—*continued*

Plant, colour form and typical variety	Pigments present[1]
ROSE (*Rosa*)	
Mauve, "Reine de Violette"	Cyanidin (co-pigmented)
Crimson, "Rose Bourbon"	Cyanidin
Pink, "Roserai de l'Haye"	Peonidin and cyanidin (1:1)
Scarlet, "Will Scarlet"	Pelargonidin and cyanidin (2:1)
Yellow, "All Gold"	Quercetin (as 4'-glucoside) and carotenoids
Cream, "La Perle"	Quercetin (as 3-glycoside)
VERBENA (*Verbena*)	
Purple and blue	Delphinidin, with 10–15% cyanidin
Maroon	Delphinidin, cyanidin and pelargonidin (1:1:1)
Pink	Pelargonidin and cyanidin (4:1)
Scarlet, "Miss Willmott"	Pure pelargonidin
Cream, "Snow Queen"	Apigenin and luteolin
GLADIOLUS (*Gladiolus*)	
Plum, "King of the Black" ⎱	Malvidin, delphinidin with some
Purple, "Argentine" ⎰	cyanidin and peonidin
Claret, "Red Signal"	Malvidin and pelargonidin
Red, "Oriental Red"	Pure pelargonidin
LUPIN (*Lupinus*)	
Purple, mauve and blue	Delphinidin and cyanidin (4:1)
Pink and red	Pelargonidin and cyanidin (1:1)
TULIP (*Tulipa*)	
Black, "Queen of the Night" ⎱	Delphinidin, cyanidin and pelar-
Purple and violet, "Merry Widow" ⎰	gonidin (ratio 6:3:1)
Crimson, "Crimson Cardinal"	As above, but ratio 1:6:3
Orange-red, "Orange Wonder"	As above, but ratio trace:1:1
Yellow, "Golden City"	Carotenoid
White, "Moonbeam"	Kaempferol and quercetin
CHINESE PRIMROSE (*Primula sinensis*)	
Blue, "Reading Blue"	Malvidin (at different pH)
Mauve, "Standard Stellata"	Malvidin (co-pigmented)
Maroon, "Oak Tongue"	Malvidin and petinidin (35%)
Orange, "Dazzler" ⎱	Pelargonidin and trace of cyanidin
Coral, "Pink Star" ⎰	
White, "Beetle"	Kaempferol and quercetin
DAHLIA (*Dahlia*)	
Violet, "Bonny Blue"	Cyanidin (as metal complex)
Vermilion, "Bravissimo"	Cyanidin
Scarlet, "Dr. Webb"	Pelargonidin
Yellow, "Zantine"	Chalcones and aurones
White, "Clare White"	Flavones or flavanones

[1] For brevity, pigment aglycones are given, but it should be remembered that flavonoids occur in petal tissue in glycosidic form. For details of sugar combinations, *see* the Tables in Chapters 5–7.

The main factors concerned in colour production in garden plants may be summarised as follows:

(1) Pure blue strains are rare and are difficult to introduce into a plant unless the "wild type" plant has blue flowers (as in the case of the hyacinth). Hence, the absence of a really true blue rose. This is probably because blue colour is produced by *in vivo* modification of anthocyanin by means of co-pigmentation, metal complexing or a combination of these two factors.

(2) Mauve, violet, crimson, red, scarlet and pink forms are usually present in most species. Pure strains containing a single anthocyanidin are easily obtained (except in the case of tulip) and intermediate shades are produced by judicious mixture of pelargonidin with cyanidin or of cyanidin with delphinidin. Although genetic factors (i.e. epistasy) prevent much mixing of anthocyanidin types in sweet pea, there is nevertheless a fairly extensive range of cyanic colour forms. In some cases (e.g. *Cyclamen*) scarlet forms are absent from present-day varieties, but should be obtainable in the future by inducing the mutation, cyanidin→pelargonidin.

(3) Species with both cyanic and yellow colour forms are infrequent and yellow is not a colour easy to introduce *de novo*. Yellow colour in the modern Hybrid Tea rose was obtained by crossing a cream-flowered cultivar Hybrid Perpetual with a wild yellow-flowered species, *Rosa foetida* (Arisumi, 1963). In the case of the snapdragon, the yellow aurone occurs naturally in the wild species as a lip pigment so that breeding a yellow self-coloured flower simply involved spreading the lip pigment over the rest of the corolla.

(4) White forms are available in practically all ornamental species, the mutation from coloured to white being the easiest of all changes to induce in cultivation.

Although the colour range in most garden plants is already very extensive, considerable efforts are expended in producing new colour forms. In particular, breeders are still searching for a blue rose, a blue snapdragon and a yellow sweet pea. Clearly, knowledge of the pigments already present in the species and those present in related wild species is useful in such breeding work. However, the major barrier is not usually finding a suitable source of the desired pigment but difficulties in incompatibility and ploidy when introducing it into cultivated forms.

Thus, a source of yellow for introducing into the sweet pea is the meadow sweet pea, *Lathyrus pratensis*, the small yellow petals of which are pigmented by carotenoid and the flavonol, syringetin. However, all attempts to hybridise this plant with *L. odoratus* at both diploid and tetraploid levels have so far failed because of the incompatibility system present in this genus. Again, a source of a blue snapdragon is the wild North American species, *Antirrhinum nuttalianum*, the small blue flowers of which contain delphinidin, instead of the cyanidin in the cultivated forms of the European *A. majus*. The two plants however, cannot be crossed and the chances of achieving a fertile hybrid seem remote since the two species involved are in different sections of the *Antirrhinum* genus.

A measure of the difficulties involved in breeding new colour forms may be obtained from Legro's account (1963) of the partly successful attempt to produce orange- and red-flowering *Delphinium*. These were obtained after several years' patient breeding and were produced by doubling up diploid wild species to the tetraploid level and making crosses involving three wild species (*D. cardinale, D. zalil* and *D. nudicaule*) as well as the blue-flowered cultivar (*D. elatum*).

The difficulty with breeding a pure blue rose is that no source of blueness exists in the genus *Rosa* as a whole; cyanidin is universally present whereas delphinidin is absent and cannot be obtained by normal genetic means. The bluer-shaded purple varieties at present in the catalogues contain cyanidin probably co-pigmented with leucoanthocyanidin. There is no evidence of a pure cyanidin–metal complex, which, if it could be achieved, would yield a cornflower-blue colour; there is presumably some factor in *Rosa* (high citric acid content?) which prevents metal ions in the petal forming such a complex.

II. Physiological and Pathological Functions

A. Flavonoids as Growth Regulators

The physiological role of flavonoids in plants has been a subject of much speculation in the past and there is still no clear-cut evidence to show that they play a vital role in growth or metabolism. However, indirect evidence has been obtained in recent years indicating that, as growth regulators, they play a part in dormancy and in root and shoot growth. Certainly *in vitro*, flavonoids in low concentrations can be shown to exert profound effects on isolated systems such as the indoleacetic acid (IAA)–indoleacetic acid oxidase (IAA oxidase) system in peas and beans or the growth of wheat roots in organ culture. How far these experiments have a bearing on the situation in the whole plant is a matter for further study.

Most attention has been given to the indirect action of flavonoids on plant growth via the IAA–IAA oxidase system. The concentration of IAA, one of the major growth hormones in plants, is well known to be regulated by the enzyme indoleacetic acid oxidase, which irreversibly destroys it by oxidation to inactive products. Studies by both Furuya *et al.* (1962) and Mumford *et al.* (1961) on the effect of light on the growth of pea seedlings led to the discovery of low molecular weight dialysable substances in seedlings whose synthesis was light-controlled and which either stimulated or inhibited growth via IAA–IAA oxidase. The major growth inhibitor, which is a co-factor for enzymic destruction of IAA, was isolated and identified as kaempferol 3-(*p*-coumaroyltriglucoside) (9). At the same time, it was noted that quercetin, which also occurs in peas, reverses the effect of kaempferol on growth by inhibiting the enzyme IAA oxidase.

10*

Much effort has been expended subsequently on determining the relation-ship between flavonoid structure and growth activity, but the results obtained to date are still partly contradictory. The most active worker in this field, Stenlid (1963), examined the effect of flavonoids on pea root IAA oxidase, an

(9) Kaempferol 3-(*p*-coumaroyltriglucoside)

(10) Naringenin

enzyme which is highly active and has no requirement (unlike the one in mung bean) for manganese ion or dichlorophenol as co-factor. Stenlid found that all flavonoids were uncoupling agents of oxidative phosphorylation and also all affected the destruction of IAA by the pea root system. In general, all 4′-hydroxyflavonoids were co-factors for the oxidation of IAA and were thus growth inhibitors whereas 3′,4′-dihydroxyflavonoids inhibited the destruction of IAA and were thus growth stimulators. Stenlid's results agree with those of Furuya *et al.* (1962), who noted that quercetin inhibited and kaempferol (at low concentrations) stimulated mung bean IAA oxidase. Furuya *et al.* (1962) found that kaempferol at higher concentrations was capable of inhibiting the enzyme and indeed, Mumford *et al.* (1961) report that kaempferol is inhibitory at both high and low concentrations. Their system, however, as Stenlid points out, is not strictly comparable with that of Furuya *et al.* since they used di-chlorophenol as a co-factor for their IAA oxidase, which was from bean roots. The ambiguous position of kaempferol as a growth regulator is also apparent from the work of Nitsch and Nitsch (1962), who found that quercetin and kaempferol are both synergists of IAA in increasing the growth of oat mesocotyls.

A factor which has not been taken into account in these studies is the pheno-lase activity of IAA oxidase preparations and while it is easy to speculate that quercetin and other catechol-derived flavonoids have a sparing action on IAA by providing an alternative substrate for the enzyme IAA oxidase, it is more difficult to see why flavonoids such as apigenin and naringenin should stimulate enzyme action and thus, in effect, act as growth inhibitors.

Other growth-regulatory studies of flavonoids can be dealt with more briefly. Stenlid (1962) noted that synthetic anthocyanidins at concentrations of 3×10^{-7} M to 10^{-4} M increased root growth in wheat seedlings and also reversed the inhibition in root growth produced by added IAA. This result was confirmed by Reddy et al. (1963) who observed that addition of a cyanidin diglucoside in solution to wheat, oat and maize seedlings increased the growth of the roots. Finally, an association between ability to root and anthocyanin formation in leaf was discovered in Acer rubrum and Eucalyptus camaldulensis by Bachelard and Stowe (1962). These authors reported a significant correlation between total leaf pigment and number of roots formed following treatment of the plants with indolylbutyric acid.

Some work has also been done on flavonoids in relation to dormancy, following the isolation by Hendershott and Walker (1959) of naringenin from dormant buds of peach. Phillips (1961) examined the possibility that naringenin might control dormancy in other systems as well and found that it could be used in place of coumarin to induce a light requirement for germination in lettuce varieties that are not normally light-requiring. He also found that, at concentrations of 40–80 mg/l, naringenin inhibits the germination of lettuce seeds and also successfully competes with gibberellic acid for dormancy control in the same system. There is no evidence as yet that naringenin occurs regularly in dormant buds and seeds so it is not clear that the above experiments have any general validity.

B. Flavonoids and Disease Resistance

The idea that the common flavonoids have a protective function in plants in relation to disease resistance is one that has long been discussed. It is based on the fact that flavonoids are phenolic compounds and it has been known since the days of Joseph Lister that phenols are highly toxic to micro-organisms. The observation of Charles Darwin and other naturalists that anthocyanin-lacking mutants of wild species are sometimes weak, unthrifty plants has fostered the same idea. However, evidence connecting anthocyanins and flavonols with disease resistance is still very tenuous. Two examples may be mentioned to illustrate the present situation.

First, Hulme and Edney (1960) have demonstrated that cyanidin in saturated aqueous solution completely inhibits the germination of Gloeosporium perennans, the fungus responsible for apple fruit rot. It is true that cyanidin occurs (as the 3-galactoside) in apple skins but it is not known whether the pigment prevents the in vivo penetration into the fruit of fungal spores. It certainly does not explain the known resistance of green-skinned apple varieties to fungal infection.

Second, the fact that the flavone tricin occurs in greater amount in "Khapli" wheat, a rust-resistant variety, than in the susceptible "Marquis" suggested to Newton and Anderson (1929) that it might play a role in rust resistance. Subsequent tests (Anderson, 1934) of flavone extracts on the germinating ability of uredospores of Puccinia graminis var. tritici failed to show any clear relationship between flavone toxicity and rust-resistance.

Tannins (including leucoanthocyanidins) have likewise been regarded as beneficial plant substances, promoting resistance to disease in leaf, bark and heartwood. The only firm data on tannins are those regarding their roles in inhibiting virus infection. Cadman (1960) has shown that they inhibit the mechanical transmission of virus in sap of rosaceous species such as the raspberry by causing the virus particles to form clumps and so become uninfective. However, since viruses are usually transmitted from plant to plant by other means, the protective role of tannins in this instance is a very limited one.

The only clear example of a flavonoid being implicated in disease resistance in plants is that of pisatin, the phytoalexin of *Pisum sativum* discovered by Cruickshank and Perrin in 1960. Pisatin was isolated from pea pods infected with *Monolinia fructicola* and identified as the isoflavonoid (11) by Perrin and Bottomley (1962). Subsequently, a closely related substance, phaseollin (12), was obtained from French bean pods inoculated with the same fungus.

(11) Pisatin   (12) Phaseollin

The role of pisatin in disease resistance of pea plants has been discussed at length by Cruickshank and Perrin (1964) (*see* also Cruickshank and Perrin, 1965). The situation may be briefly summarised as follows. Pisatin is absent from healthy plant tissue but is produced in root, leaf, stem and pod of *Pisum sativum* whenever the plant is invaded by any type of fungus. Its significance is that, if produced in sufficient concentration, it is toxic to an invading fungus and prevents its multiplication. Thus, non-pathogenic fungi induce synthesis of pisatin to such an extent that their growth is prevented and pisatin plays a primary role in the plant's resistance. On the other hand, pathogenic fungi are able to develop in pea tissue, since they induce pisatin formation in low, non-lethal concentrations.

These results explain how it is that plants are able to resist some diseases and not others, but more work is required before the phytoalexin theory of disease resistance is fully established. Substances related to pisatin occur in many other legumes; unanswered questions are whether these other iso-flavonoids are also phytoalexins and what substances take the place of iso-flavonoids in non-legumes. Finally, it will be interesting to see whether knowledge of the structure of pisatin helps at all in devising practical methods of protecting plants from pathogenic fungi.

III. Pharmacological Activity of Flavonoids

A. METABOLISM

The vast majority of flavonoids are non-toxic to man and animals, which is just as well since they are widely distributed in foods (cf. Section IV). Known pharmacological activities are limited to a few substances of this group and activities are weak (on a weight basis) when compared with other active plant substances such as alkaloids. Pharmacological activity follows largely from oral ingestion of flavonoids in the diet and knowledge of their metabolism is essential to any understanding of their action *in vivo*. Fortunately the metabolism of most classes of flavonoid in the mammalian body have been studied (Williams, 1964) and the situation can be summarised by considering the fate of rutin, one of the most widely distributed of all flavonoids.

Rutin metabolism has been studied in man, the rat, guinea-pig and rabbit; in all cases, it is hydrolysed to quercetin and some is excreted as a glucuronide conjugate. However, quercetin produced on hydrolysis is largely broken down *in vivo*, with the production of CO_2 from the A-ring and of three phenylacetic acids from the B-ring:

Quercetin   *m*-Hydroxyphenylacetic acid

The dehydroxylation of 3,4-dihydroxyphenylacetic acid to *m*-hydroxyphenyl-acetic acid does not occur in the liver or kidney but is almost certainly a reaction catalysed by the gut bacteria (Booth and Williams, 1963). Flavones and flavanones, e.g. those occurring in *Citrus* fruit, are broken down in a similar fashion, the main products detected in the urine being the related phenyl-propionic, cinnamic and benzoic acids, instead of the phenylacetic acids. An interesting species difference has been observed in the metabolism of naringin; in man, it is largely recovered as naringenin glucuronide whereas in rats, it is almost completely broken down into smaller fragments. Flavone itself, is, not surprisingly, converted in the guinea-pig to 4'-hydroxy- and 3',4'-dihydroxy-flavone but there is little breakdown, since only traces of salicylic acid occur in the urine following its administration (Das and Griffiths, 1966).

The fate of anthocyanins has not been studied but these pigments are almost certainly broken down as the flavonols are; no colour has been observed in the urine following ingestion of fruits with high anthocyanin content such as black-berries. By contrast, it is interesting that the chemically unrelated beta-cyanins, such as betanin the beet root pigment, are occasionally recovered unchanged in the urine. The failure to metabolise betanin is controlled by genetic factors in man and beeturia (betanin in the urine) is found in approximately 14% of the population in this country (Watson, 1964).

Isoflavones differ from other flavonoids in their metabolism by undergoing reduction without the splitting of the central pyran ring. The main product of genistein in the animal body is, apparently, equol, a substance found in fowl's urine in recent feeding experiments (Cayen *et al.*, 1964) but which was detected as a "natural constituent" of mare's urine as long ago as 1932 by Marrian and Haslewood:

Genistein   Equol

B. Pharmacology

Pharmacological activities have been ascribed to rutin and related compounds, to phloridzin and to the isoflavones. Rutin and the related flavanones hesperidin and eriodictyol have attracted much interest owing to their action in decreasing the fragility of blood capillaries in guinea-pigs and these substances were once thought to possess vitamin-like activity in humans. As a result, rutin is widely advertised as a herbal remedy, in the form of buckwheat leaf, and is frequently added to multivitamin tablets. Claims for its general efficacy in treatment of blood pressure, etc. must be considered unfounded since the therapeutic value of rutin in the treatment of diseases associated with capillary fragility, in spite of much research, has never been fully established.

While the claims that rutin and related flavanones are "vitamin-P" substances have never been substantiated, it is true that rutin (in massive doses) has pharmacological activity as an antioxidant towards adrenaline and ascorbic acid; it also relaxes smooth muscle and behaves as a general enzyme inhibitor. For a fuller discussion of what is still very much a controversial subject, the reader should consult the book of Fairbairn (1959) and the articles and references therein.

The physiological activity of phloridzin (17) in the mammalian body is, by contrast, well established. It causes glycosuria by interfering with the tubular

(17) Phloridzin

readsorption of glucose in the kidney and by inhibiting absorption of glucose from the small intestine. Phloridzin is widely used experimentally to study glucose transport across the cell membranes. Phloridzin is highly specific in its action, the related galactoside being inactive. It is also very specific in occurrence, being restricted in the plant kingdom to apple bark; it does not even occur in apple fruit and thus does not present a dietary problem.

The only really important pharmacologically active group of flavonoids are the isoflavones. The isoflavones, genistein (18) and daidzein, were found to be oestrogenic, following their isolation from Australian strains of subterranean clover, as the active factors responsible for interference in the oestral cycle of sheep and subsequent serious reduction in lambing percentages (Bradbury and White, 1951). The structural relationships between isoflavones (e.g. 18) and natural oestrogens such as oestradiol (20) and synthetic oestrogens such as diethylstilboestrol (19) is illustrated below:

(18) Genistein

(19) Diethylstilboestrol

(21) Coumestrol

(20) Oestradiol

Isoflavones are only very weak oestrogens, being only about 5×10^{-6} as active as diethylstilboestrol. They are almost certainly pro-oestrogens (Biggers, 1959) and are converted to much more active substances (e.g. equol) in the animal body. Three of the four major isoflavones of clover, i.e. daidzein, genistein, and biochanin-A, have about the same activity; the fourth, formononetin, has only a third of the activity of the others. Clovers also contain related isoflavonoids which are considerably more active than the isoflavones themselves; the best known of these is coumestrol (21) which is 30 times as active as genistein (Bickoff, 1961).

The presence of oestrogenic isoflavonoids in clover affects fertility in both sheep and cattle and is a considerable agricultural problem, which has been solved in part by restricting the intake of clover during the breeding season. Active steps are also being taken to produce clover strains with little or no isoflavone content (Francis and Millington, 1965). The isoflavone content of clover fodder is by no means always disadvantageous and, in fact, there is little doubt that these substances increase the rate of growth of fattening stock and also have a beneficial effect on lactation in cows, contributing to the so-called "spring flush" in milk yield.

The presence of isoflavones in foods could clearly interfere with oestrus in the human female but although there is a fairly high isoflavone content in soya bean, a staple diet in the Far East, no untoward effects have been recorded. There is no evidence that isoflavonoids are at all toxic to man or animals. Indeed, their very non-toxicity accounts for the continued wide use as insecticide of derris root, the active principle of which is the isoflavonoid, rotenone (22). Synthetic insecticides are, unfortunately, less specific in their action. Rotenone and related rotenoids are also highly toxic to fish and derris and other roots were used for this purpose (i.e. killing fish for human consumption) by primitive Indian communities before being employed as insecticides.

(22) Rotenone

IV. Flavonoids in Foods

A. DISTRIBUTION

Flavonoids are widely distributed in foods, having been recorded in well over half the known food plants (Bate-Smith, 1959; Swain, 1962). Their presence creates important practical problems in the food industry and food scientists are continually concerned with the changes flavonoids undergo during processing and manufacture. Anthocyanins and flavonol glycosides are amongst the most widespread and important flavonoids in foods, adding as they do to colour. Many of the pigments found in common fruits and vegetables are listed in Tables 9·3 and 9·4. These tables could be considerably extended by noting varietal differences and adding quantitative data. Although flavonol glycosides have been fully identified in only a few fruits (Table 9·4), these substances almost certainly occur in many fruits other than those listed; their chemical identification has unfortunately lagged behind that of the anthocyanins.

TABLE 9·3. Anthocyanins of Fruits and Vegetables

Fruit or vegetable	Anthocyanins present[1]
Apple (*Malus pumila*)	Cy 3-galactoside (skin)
Blueberry (*Vaccinium angustifolium*)	Dp 3-glucoside, Dp 3-galactoside and Dp 3-arabinoside[2]
Blackberry (*Rubus fruticosus*)	Cy 3-glucoside and Cy 3-rutinoside
Cherry, sour (*Prunus avium*)	Cy 3-glucosylrutinoside
sweet (*P. cerasus*)	Cy 3-glucoside and Cy 3-rutinoside
Cranberry (*V. macrocarpon*)	Cy and Pn 3-galactosides, Cy and Pn 3-arabinosides
Currant, red (*Ribes rubrum*)	Cy 3-xylosylrutinoside and Cy 3- glucosyl-rutinoside[2]
Currant, black (*R. nigrum*)	Cy and Dp 3-glucoside, Cy 3-rutinoside
Elderberry (*Sambucus nigra*)	Cy 3-glucoside and Cy 3-sambubioside[2]
Gooseberry (*Ribes grossularia*)	Cy 3-glucoside and Cy 3-rutinoside
Grape (*Vitis vinifera*)	Dp, Pt and Mv 3-glucoside and Mv 3-(*p*-coumaroylglucoside) (in "Barlinka")[2]
Mulberry (*Morus nigra*)	Cy 3-glucoside
Orange (*Citrus sinensis*)	Cy and Dp 3-glucoside (juice of "blood" orange)
Passion fruit (*Passiflora edulis*)	Dp 3-glucoside (skin and flesh)
Peach (*Prunus persica*)	Cy 3-glucoside
Pear (*Pyrus communis*)	Cy 3-galactoside (in flesh of some varieties)
Plum (*Prunus domestica*)	Cy and Pn 3-glucoside, Cy and Pn 3-rutinoside
Pomegranate (*Punica granatum*)	Dp 3,5-diglucoside (fruit juice)
Raspberry (*Rubus idaeus*)	Cy 3-glucosylrutinoside[2]
Rhubarb (*Rheum rhaponticum*)	Cy 3-glucoside, Cy 3-rutinoside
Sloe (*Prunus spinosa*)	Cy and Pn 3-glucoside, Cy and Pn 3-rutinoside
Strawberry (*Fragaria × anannasa*)	Pg 3-glucoside
Aubergine (*Solanum melongena*)	Dp 3-rutinoside[2]
Bean, french (*Phaseolus vulgaris*)	Dp 3-glucoside and Dp 3,5-diglucoside[2]
runner (*P. multiflorus*)	Mv 3,5-diglucoside (pod of some varieties)
Cabbage, red (*Brassica oleraceae*)	Cy 3-(diferuloylsophoroside)-5-glucoside[2]
Onion (*Allium cepa*)	Pn 3-arabinoside
Potato (*Solanum tuberosum*)	Pg 3-(*p*-coumaroylrutinoside)-5-glucoside[2]
Radish (*Raphanus sativus*)	Pg 3-(*p*-coumaroylsophoroside)-5-glucoside[2]

[1] Aglycone abbreviations: Pg, pelargonidin; Cy, cyanidin; Pn, peonidin; Dp, delphinidin; Pt, petunidin; Mv, malvidin.
[2] Indicates that only most typical pigments are mentioned. These are accompanied by other related compounds; in particular, Cy 3-glucosylrutinoside is usually accompanied by related 3-sophoroside and 3-rutinoside. There are also varietal differences in pigmentation in species so marked.

Other important classes of flavonoid in foods are leucoanthocyanidins and catechins, substances which contribute to flavour by providing astringency in otherwise insipid foodstuffs. Catechins occur primarily in tea leaf and, during

TABLE 9·4. Flavonol Glycosides of Fruits and Vegetables

Fruit or vegetable	Flavonol glycosides present[1]
Apple (*Malus pumila*)	Qu 3-glucoside, Qu 3-galactoside, Qu 3-rhamnoside, Qu 3-arabinoside and Qu 3-xyloside
Apricot (*Prunus armeniaca*)	Qu 3-glucoside
Bilberry (*Vaccinium myrtillus*)	Qu 3-glucoside and Qu 3-rhamnoside
Cherry, sweet (*Prunus avium*)	Qu 3-rutinoside
Cranberry (*Vaccinium macrocarpon*)	Qu and My 3-arabinosides, Qu 3-galactoside, Qu 3-rhamnoside and My 3-digalactoside
Currant, black (*Ribes nigrum*)	Qu 3-glucoside
Grape (*Vitis vinifera*)	Km, Qu and My 3-glucoside, Qu 3-glucuronide
Pear (*Pyrus communis*)	Isorhamnetin 3-glucoside, 3-rutinoside and 3-rhamnosylgalactoside (in skin)
Plum (*Prunus domestica*)	Qu 3-rutinoside
Onion (*Allium cepa*)	Qu 4'-glucoside, Qu 7,4'-diglucoside and Qu 3,4'-diglucoside
Potato (*Solanum tuberosum*)	Qu 3-glucoside and Qu 3-rutinoside
Tea (*Camellia sinensis*)	Km, Qu and My 3-glucoside, Km, Qu and My 3-rutinoside
Tomato (*Lycopersicon esculentum*)	Qu 3-rhamnoside (skin)

[1] Abbreviations: Km, kaempferol; Qu, quercetin; My, myricetin.

manufacture, undergo important changes in structure and provide much of the colour of the manufactured product. Leucoanthocyanidins occur in many food drinks, especially in wine, cider, tea and cocoa. They are also present in many legumes, i.e. in the seeds or nuts, and occur in considerable amount in fruits such as quince (*Cydonia oblonga*), persimmon (*Diospyros kaki*) and banana (*Musa*). Other flavonoid classes are less abundant. Flavanones are restricted largely to citrus fruits, where they contribute to taste of juice and peel. Flavones are found occasionally in foods; they are present, for example, in many cereal grains, e.g. in wheat (King, 1962).

From the practical aspect, flavonoids are important in foods because of the (usually oxidative) changes they undergo during the storage and processing of the raw plant material. Control of anthocyanin colour during processing is essential, for example, in most fruit preserves. Some foods are the result of fermentation processes, e.g. tea, cocoa and wine, and the changes flavonoids undergo during manufacture have been followed in some detail (for tea manufacture, *see* Roberts, 1962). These practical aspects of flavonoids will be considered further in the following sections on contribution to colour and taste.

B. CONTRIBUTION TO COLOUR

The attractiveness of fruits and vegetables on the table depends considerably on their colour properties. In many instances, colour is due to anthocyanin pigmentation although carotenoids are, of course, important in tomato and carrot and chlorophylls in all green vegetables. The colour of a vegetable or fruit variety is a matter of consumer choice and, in general, cyanic varieties seem to be more popular than others. Thus the red raspberry is preferred to the yellow variety and scarlet radish varieties are more widely grown than white ones. In the potato, the splash of anthocyanin red on the skin around the eye is associated with quality, due to its presence in the variety "King Edward". This association is so strong in the public mind that other varieties with the same colour splash but poorer in quality can be "unloaded" on the market at an inflated price.

As has already been mentioned, the retention of anthocyanin colour is a major problem in the storage and processing of many fruits. Colour is easily lost on heating fruit in hot water; the pelargonidin pigment of strawberry, for example, is notoriously difficult to preserve in jam-making. Fruits also lose their colour prior to thermal treatment and this is probably mainly the result of enzyme action. In the case of anthocyanins, these are of two types. There are the glycosidases or anthocyanases (Huang, 1956), which hydrolyse the pigments to aglycone and sugar; the aglycone then spontaneously and rapidly decomposes to colourless products. Then, there are catecholases, which oxidise anthocyanins based on cyanidin, delphinidin and petunidin again to colourless products. The efficiency of these enzymes depends, of course, on other factors such as pH, and the concentrations of metal ion, ascorbic acid and other catechol-derived compounds in the fruit (van Buren et al., 1960).

Even after destruction of the anthocyanin-attacking enzymes by rapid heating, colour is lost in fruits due to oxidative decomposition, which is accelerated in the presence of oxygen (as peroxide) or free metal ions. Sulphur dioxide, widely used as a preservative, also contributes to this decolorisation. With modern canning procedures, i.e. very rapid heating for short periods of time, fruit colour can be retained satisfactorily, with minimal loss during storage in most cases. Anthocyanin colour is also destroyed by light and this is a particular problem in production of fruit juices. Loss of colour can be minimised by storage in dark bottles, but this procedure, acceptable in the case of red wines, would be difficult to introduce into the soft-fruit industry. One other solution, suggested by Jurd (1964), is the addition to the fruit juices of synthetic anthocyanidins, the structure being based on 3-deoxyanthocyanidins which are much less sensitive to light decomposition. Such substances would, unfortunately, be classified as food additives and would require prior clearance by food and drug officials before they could be used commercially.

The preservation of colour in red wines is accomplished by storage in dark-coloured bottles in cellars in complete darkness and, under such conditions, colour is retained almost indefinitely. Changes in pigment tone occur on storage and while fresh wines show a sharp anthocyanin band at 530 mμ when placed in a spectrophotometer, 10- and 50-year-old vintage wines have maximal absorption at 470 mμ. The mechanisms of the changes that take place are unknown (Ribereau-Gayon, 1964). Colour in wine seems to prevent bacterial decomposition and malvidin is reported to be a bactericidal agent (Masquelier, 1959; cf. Powers *et al.*, 1960).

Anthocyanin colour in foods is occasionally a liability and may require removal. For example, there is a better market in the United States for white rather than red wines and the use of the fungal enzyme preparation anthocyanase, which can be produced commercially from *Aspergillus niger*, has been suggested for converting red wine to white. The same enzyme preparation may also be useful for removing excess pigment deposits, which occur in jam-making in fruits with high anthocyanin concentration (e.g. blackberry, black-currant).

Processing may cause leucoanthocyanidins present in fresh foods to be converted to anthocyanidins and hence give rise to undesirable colours. This happens in the broad bean, and only the variety "Triple White", which lacks leucoanthocyanidin, is suitable for canning (Rowlands and Corner, 1962). Other flavonoids in foods occasionally give trouble in the canning process. Thus, rutin present in asparagus forms a ferrous iron complex if the tin used in the canning has too high an iron content. When the tin is opened, ferrous iron is rapidly oxidised to the ferric state, and the ferric-rutin complex produces a dark discoloration in the vegetable (Dame *et al.*, 1959). Other examples of flavonoids in foods causing undesirable colours after processing are flavonols in onions and anthocyanins in southern peas (Burns and Winzer, 1962).

C. CONTRIBUTION TO TASTE AND FLAVOUR

Two classes of flavonoid, flavanones and leucoanthocyanidins, are known to contribute to flavour and taste in foods. That the flavanone naringin is the major bitter principle of grapefruit and the Seville orange has been known for a long time but it is only recently that the relationship between structure and taste of flavanones has been critically examined (Horowitz, 1964). The bitterness of naringin is very similar to that of quinine and it is a fifth as bitter on a molar basis and is actively bitter in aqueous solution at a concentration of 5×10^{-5} M. Horowitz has shown that naringin and related flavanones, such as neohesperidin of the lemon, are responsible for most of the bitter taste of citrus juices, although terpenoids such as limonin certainly contribute to the bitter taste of citrus peels.

Not all *Citrus* flavanones are bitter (*see* Table 9·5) and an examination of the taste of a series of flavanones from the genus (Horowitz, 1964) showed that bitterness is due to combination of the flavanone structure with the rare di-saccharide, neohesperidose (rhamnosyl-$\alpha(1\rightarrow2)$-glucose), as in naringin (23). The sugars and flavanones, after acid hydrolysis, are all lacking in taste and the replacement of the neohesperidose in the 7-position by the isomeric disaccharide rutinose (rhamnosyl-$\alpha(1\rightarrow6)$-glucose) gives a flavanone glycoside (e.g. naringenin 7-rutinoside) which is completely lacking in taste. The situation is more

TABLE 9.5. The Taste Properties of Citrus Flavanones

Flavanone or chalcone	Relative bitterness or sweetness (on a molar basis)
BITTER	
Hesperitin 7-neohesperidoside (neohesperidin)	2
Naringenin 7-glucoside	6
Phloridzin	10
Naringenin 7-neohesperidoside (naringin)	20
Isosakuranetin 7-neohesperidoside (poncirin)	20
Quinine dihydrochloride	100
NO TASTE	
Naringenin 7-rutinoside	—
and related chalcone	—
and related dihydrochalcone	—
Isosakuranetin 7-rutinoside	—
Eriodictyol 7-rutinoside (eriocitrin)	—
Hesperetin 7-rutinoside (hesperidin)	—
Hesperidin chalcone	—
Naringenin 5-glucoside	—
SWEET	—
Naringin chalcone	—
Naringin dihydrochalcone	1
Neohesperidin dihydrochalcone	20
Saccharin	1

complex in that naringenin 7-glucoside, which should also have no taste, has in fact a third of the bitterness (on a molar basis) of naringin; the dihydrochalcone phloridzin (17) is also bitter. Ring opening of flavanone neohesperi-

Rha—O—Glc—O
α, 2 ←1

HO   O
(23) Naringin   ⟶

Rha—O—Glc—O
α, 2 ←1

HO   O
(24) Naringin dihydrochalcone

dosides to the related chalcones give substances which are, remarkably enough, intensely sweet and the dihydrochalcone (24) of naringin is as sweet on a molar basis as the artificial sweetening agent, saccharin. Just as neohesperidose is more or less necessary for bitterness in the flavanone series, so the same sugar is required for sweetness in chalcones and dihydrochalcones; chalcone rutinosides have no perceptible taste.

That flavanones are mainly responsible for the taste of citrus fruits is further illustrated by the close correlation that exists between occurrence of flavanone rutinosides and neohesperidosides in *Citrus* fruit species and fruit taste. The former glycosides predominate in sweet oranges, whereas the latter occur almost exclusively in the bitter lemon and grapefruit (*see* also Table 5·10, Chapter 5). Again, the concentration of neohesperidosides such as naringin is higher in unripe (bitter tasting) fruit than in the riper (sweeter) fruit.

Knowledge of the flavanone structures have proved useful in the control of bitterness in citrus fruits. Excessive bitterness in the grapefruit, for example, may be reduced by addition of a hydrolytic enzyme, naringinase, which hydrolyses naringin to the non-bitter naringenin (Ting, 1958). Similar debittering may be achieved by adding peroxidase and hydrogen peroxide (Markh and Feldman, 1950).

Astringency, the sensation produced by leucoanthocyanidins, in fruits (e.g. persimmon) is quite distinct from the bitterness of flavanones in *Citrus* or the sourness of organic acids in unripe apples. It is a puckering sensation in the mouth, produced by destruction of the lubricant property of saliva by cross-linking action of polymeric phenol with protein (Swain, 1962). Too much astringency, like excessive bitterness or sourness, is undesirable in foods and varieties have generally been selected for low leucoanthocyanidin content. Some astringency is necessary for avoiding insipidness and suitable blends of leucoanthocyanidin and sugar are essential components of the palatability of wine and cider. Leucoanthocyanidins also contribute to the flavour properties of tea and chocolate and are present, usually in minor amounts, in many fruits, providing a counterbalance to excessive sugar concentrations in some varieties.

Our understanding of the role of leucoanthocyanidins in the palatability of foods has been delayed, largely by confusion over nomenclature and by ignorance of their structures. Following the recent investigations of Swain and his collaborators (*see* especially Goldstein and Swain, 1963) the situation is now much clearer. Leucoanthocyanidins (or condensed tannins) can be classified into three groups: (1) low molecular weight substances, which are probably dimers formed by linkage of a flavan-3,4-diol with a catechin (*see* Chapter 3, p. 98), (2) soluble oligomers, containing 4 to 8 flavan units, and (3) insoluble polymers (flavolans) of 10 or more units. The three groups occur together in plant tissues but changes in amount take place during fruit development, a general conversion of oligomers to polymers being particularly apparent. The property of astringency is limited to the oligomers and hence, the loss of astringency which occurs on ripening in fruits such as the peach, pear, plum, persimmon and banana, is directly attributable to the disappearance of the oligomers. The polymers (or flavolans) are laid down on the cell wall and cannot contribute to taste because of their insolubility.

The ability of leucoanthocyanidin oligomers to tan protein is important in foods in other matters besides palatability. Their presence leads to the formation of unwanted precipitates or hazes in wines and beers and they also remove from foods enzymes which may be beneficial in processing (for fuller discussion, *see* Swain, 1962).

CHAPTER 10

CHEMICAL TAXONOMY OF FLAVONOIDS

I. Flavonoids as Taxonomic Markers

In previous chapters of this book, the many flavonoid pigments have been described and their identification and distribution discussed. Mention has also been made of their inheritance, biosynthesis and function. In conclusion, their role as chemical markers in plant classification and phylogeny will be assessed, since this is a field where flavonoids promise to make a significant contribution.

Chemical plant taxonomy or biochemical systematics is a topic which has developed rapidly within the last 10 years. The pre-eminence of flavonoids as favoured taxonomic markers among the many secondary constituents in plants may be gauged from the fact that in the first book devoted to the subject, i.e. "Chemical Plant Taxonomy" (Swain, 1963), no less than five of the sixteen chapters deal *inter alia* with these substances.

This is not the place for a detailed discussion of the rationale of biochemical systematics. The reader may be referred to the book already mentioned (Swain, 1963), to the proceedings of a more recent symposium (Swain, 1966), to the book of Alston and Turner (1963) entitled "Biochemical Systematics" and the chapter by the author and N. W. Simmonds in "Biochemistry of Phenolic Compounds" (Harborne, 1964). The situation may be summarised by saying that the generally agreed requirements for a chemical character to be of use in plant taxonomy are as follows: chemical complexity and structural variability, physiological stability, widespread distribution and easy and rapid identification. That flavonoids match up well to these various requirements will be apparent from the following pages.

(1) Structural variability. Flavonoids exhibit immense structural variation. Thus, there are at least a dozen classes of flavonoid and each class varies in the degree of hydroxylation, methylation and glycosylation; flavonoids with C-glycosyl, furano, isoprenyl and alkaloidal substituents are also known. In this way flavonoids provide at least as many scorable characters as any other group of secondary substances. The wealth of data available to taxonomists may be assessed by considering the flavonoid pigments that have been fully identified, which must number at least 500. There are certainly over 100 anthocyanins (Chapter 1), perhaps 250 flavones and flavonols (Chapter 2) and another 150 minor flavonoids (Chapter 3). Although the majority of these have been reported from single species, surveys would no doubt indicate that many have a significant distribution pattern among related plants.

(2) Widespread distribution. Flavonoids occur universally in angiosperms, gymnosperms and pteridophytes (Chapter 4). They have the advantage that they are more widely distributed than most other secondary substances. Thus, they have been found in every family, and practically every species, examined for their presence. By contrast, alkaloids are restricted to between 10 and 20% of angiosperm families and essential oils are abundant in only 5 to 10% of these families.

(3) Stability. All the indications are that flavonoids are amongst the most stable chemical characters in plants. Qualitative variation at the species level is very limited. Leaf constituents naturally show some quantitative variation with environmental factors (cf. Chapter 8) but petal flavonoids are very little affected by physiological conditions. Genetic variation is limited to a relatively few (mainly ornamental) plants and is not a serious handicap in taxonomic work, since it mainly affects the anthocyanins, the other flavonoids being more constant (Parks, 1965).

(4) Ease and speed of identification. Flavonoids again rate highly here. Plants can be rapidly scored for flavonoids by one- or two-dimensional paper chromatography. A great advantage is that they can be visualised directly on paper chromatograms in ultra-violet light, without requiring the chromogenic sprays necessary for substances such as alkaloids, non-protein amino acids, sugars, etc. The only other major tool required for identification is the ultra-violet spectrophotometer, but this is widely available in scientific laboratories and the results obtained with this instrument are simpler to interpret than those provided by i.r. or n.m.r. spectroscopy.

II. Contribution to Plant Systematics

A. Introduction

From a consideration of their known distribution in higher plants (outlined in Chapters 5–7) it is abundantly clear that flavonoids have considerable potentialities as taxonomic markers in plant classification. Their present contribution to the study of plant relationships is naturally slight, since surveys have so far been limited. The chemical data, a product of only the last decade, cannot hope to compete at the present time with the large corpus of morphological data accumulated over the centuries since the time of Linnaeus. Taxonomists cannot be expected to accept chemical results when, as in most cases, plant groups have been sampled for flavonoids at the level of ascertainment of 1% or even less, and where the bias has been in favour of those species most frequently grown in cultivation or those of economic importance.

Clearly, the solution is to have unbiased and more extensive surveys. The major difficulty facing the flavonoid chemist with this in mind is access to collections of living plant material, complete at the generic, tribal or family level. Such collections are few and far between, being mainly limited to plants of outstanding ornamental value, and it is doubtful whether they will ever be available for the majority of plant groups. The taxonomist largely works from herbarium sheets, of which there are much more extensive collections, and the chemist will have to follow suit to fill the gaps when fresh plants are not available. Fortunately, in the case of the flavonoids, much can be done with dried leaves taken from herbarium sheets; the Umbelliferae, for example, has been successfully surveyed for flavonol and flavone aglycones using such material (see p. 181).

Turning to the results that are already available, it may be noted that significant correlations between flavonoid distribution patterns and morphological features have been established in several instances. The most striking examples of these will be discussed briefly. Since the actual data on distribution have been presented earlier in this volume, the emphasis in the seven chosen cases will be on assessing the coverage achieved and the value to the systematist.

B. FLAVANONES IN PINES

The genus *Pinus* is unusually rich in flavonoids and a range of flavanones, flavones and flavanonols occur in the heartwood (see p. 121). *Pinus* species can be divided into two groups according to whether the pattern of heartwood flavanones is simple or complex (Table 10·1). In fact, these groups correspond precisely with the generally accepted division of the genus into Haploxylon and Diploxylon, the species of which are morphologically differentiated by the number (one or two respectively) of bundles there are in the pine needles.

Fifty-two of some ninety species were surveyed (Erdtman, 1956); ascertainment is thus well over 50%. This correlation between chemistry and subgeneric classification is obviously a satisfactory one to the chemist but it solves no outstanding taxonomic problem. The taxonomist working with pines is more concerned with the species concept and it is the smaller variations in flavonoid composition between species of the subgenus Haploxylon (Table 10·1) that are of most interest to him. The fact that species of the sub-sections Gerardianae and Strobi are very distinctive by reason of having C-methylated flavonoids (e.g. strobopinin) found nowhere else in the genus is one of considerable systematic importance.

TABLE 10·1. Distribution of Flavanones in *Pinus* Heartwoods

Subgenus and subsection (no. of species surveyed in parenthesis)	Flavones			Flavanones			Flavanonols	
	1	2	3	4	5	6	7	8
HAPLOXYLON								
Cembrae (3)	+	+	−	+	+	−	+	−
Flexiles (1)	+	+	−	+	+	−	+	−
Strobi (8)	+	(+)	(+)	+	(+)	(+)	+	(+)
Cembroides (1)	+	+	−	+	+	−	+	−
Gerardianae (2)	−	−	−	+	+	(+)	+	+
Balfourianae (2)	+	+	−	+	+	−	+	−
DIPLOXYLON								
Leiophyllae (2)	−	−	−	+	−	−	+	−
Longifoliae (2)	−	−	−	+	−	−	+	−
Pineae (1)	−	−	−	+	−	−	+	−
Lariciones (7)	−	−	−	+	−	−	(+)	−
Australes (9)	−	−	−	+	−	−	+	−
Insignes (12)	−	−	−	+	−	−	+	−
Macropae (2)	−	−	−	+	−	−	+	−

Key: 1, chrysin; 2, tectochrysin; 3, strobochrysin; 4, pinocembrin; 5, pinostrobin; 6, strobopinin; 7, pinobanksin; 8, strobobanksin (for structures, *see* Chapter 4, p. 121). +, uniformly present; (+) not present in all species; −, absent. Data adapted from Erdtman (1956). Subsectional names are derived from type species within the subsection, e.g. Cembrae after *P. cembra* L., Flexiles after *P. flexilis* Jam., etc.

C. BIFLAVONYLS IN GYMNOSPERMS

The almost complete restriction of biflavonyls in their occurrence to the Gymnospermae is one of the most interesting correlations that have emerged from chemical plant taxonomy (Baker and Ollis, 1961; *see* also p. 118). This correlation is based on a sampling of the leaves of about 14% of the gymnosperms (i.e. some 100 species of the 700 described) (Sawada, 1958) and their general absence from angiosperms has been assumed and is not based on actual surveys.

The value of the biflavonyls for defining the gymnosperms in chemical terms is lessened by at least three factors: (1) their absence (44 species surveyed) from one of the largest gymnosperm families, the Pinaceae; (2) the recent discovery of their presence in two widely different angiosperms, the relatively primitive *Casuarina stricta* and the highly advanced *Viburnum opulus*; (3) their occurrence in two lower plants, *Selaginella tamariscina* (Hsu, 1959) and *Psilotum triquetrum* (Voirin and Lebreton, 1966).

Clearly, taxonomists will not find the biflavonyl character a useful one for distinguishing gymnosperms from other plants; this is not a matter, in any case, which presents any difficulty. Rather, it is the distribution of these substances within the gymnosperms, e.g. their absence from Pinaceae, which is of interest to plant systematists and should prompt more extensive surveys.

D. 3-DEOXYANTHOCYANINS IN GESNERADS

These rare pigments occur in flowers and/or leaves of 18 of 21 species of the sub-family Gesnerioideae of the Gesneriaceae but are absent from all of 25 species of the sub-family Cyrtandroideae. Only about 1% of all the species in the family have been studied, but those surveyed represent about a quarter of the genera described.

This result provides a positive contribution to plant taxonomy in that it supports Burtt's recent reclassification (1962) of the two sub-families on the basis of presence or absence of anisocotyly and geographical distribution. By contrast, it does not agree with Fritsch's earlier division of the family (1893–4) based on whether the ovary is superior or inferior. More extensive sampling of gesnerads would obviously be valuable, particularly since there are indications that several other flavonoid types (O-methylated flavones, chalcones and aurones) are restricted to one or other of the two sub-families (Harborne, 1966c). The 3-deoxyanthocyanin character might also illuminate family relationships within the order Tubiflorae, to which the Gesneriaceae belong. 3-Deoxyanthocyanins have also been found once in the closely related Bignoniaceae but are almost certainly absent from more distantly related families, i.e. the Labiatae, Solanaceae and Hydrophyllaceae.

E. REPLACEMENT OF ANTHOCYANINS BY BETACYANINS IN THE CENTROSPERMAE

The discovery that anthocyanins are absent from all but one of the 10 families comprising the order Centrospermae and are replaced by purple pigments of quite different structures, the betacyanins, is usually regarded (cf. Mabry, 1966) as one of the most successful correlations to have emerged from phytochemical studies. Betacyanins (or the related yellow betaxanthins) have been detected in about 83 (25%) of the 350 genera involved (Table 4·3, Chapter 4) but at the species level, positive results are recorded for only 2% of an approximate total of 5500. Betacyanins are particularly valuable since by replacing anthocyanins they provide more positive information than simple presence/absence characters; the two types of pigment are mutually exclusive of each other in occurrence.

From the taxonomic point of view, the betacyanin/anthocyanin criterion has been useful in supporting the inclusion of the Cactaceae and the Dideriaceae in the order and for indicating that the affinities of the Caryophyllaceae, the one family to retain anthocyanin pigmentation, should be re-examined. To the author, it seems premature to exclude the Caryophyllaceae completely from the Centrospermae, as has been suggested by Mabry (1966), unless there are other (biological or chemical) reasons for doing this. Most families of the Centrospermae, although lacking anthocyanin, have a range of other structurally interesting flavonoids and it would be valuable to see if their distribution supported, or otherwise, the removal of the Caryophyllaceae from the order.

F. AzaLEATIN IN *Plumbago, Eucryphia* AND *Rhododendron*

Azaleatin (5-*O*-methylquercetin) is not a substance which is easily overlooked during chromatographic surveys because it has an intense yellow fluorescence in u.v. light. That it is relatively rare in plants (it has only been found in four families) seems well established and it is unlikely that this fact will be altered by more extensive surveys. The four families containing it (Ericaceae, Eucryphiaceae, Juglandaceae and Plumbaginaceae) are all predominantly woody in character but they are not normally thought of as being particularly closely related (two are in the Archichlamydeae and two in the Sympetalae). It is the distribution of azaleatin at the generic level which is of most taxonomic interest.

Some 65 species (out of a total of 300) of the Plumbaginaceae have been surveyed for azaleatin and it has been detected in the leaf or flower of the majority (6 out of 9) of members of the tribe Plumbagineae but is absent, from all species surveyed (56) of the tribe Staticeae. Its distribution thus is satisfactorily correlated with the natural division of the family into two tribes, based on the position of the stamens on the ovary.

In the Ericaceae, azaleatin occurs in both *Erica* and *Rhododendron* (*see* p. 194). In *Rhododendron* it occurs in the petals of 44 of 83 species surveyed (total number of species 250) and its distribution is correlated with the lepidote character, azaleatin being found mainly in species with scurfy scales underneath the leaves. In *Eucryphia* (Eucryphiaceae) azaleatin occurs (together with the 3,5-dimethyl ether, caryatin) in leaves of two of the five known species but is absent from the leaves of the other three. The correlation here is with plant geography, since the first two species (*E. glutinosa, E. cordifolia*) are South American plants whereas the latter three (*E. lucida, E. moorei* and *E. milliganii*) are Australasian in origin (Bate-Smith *et al.*, 1966)

Thus, azaleatin is a good example of a flavonoid of value in studying species relationships. Because it is so easily detected (even in low concentrations in herbarium material) it is the type of substance which could be used immediately for purposes of plant classification.

G. Isoflavones in the Legume Sub-family Papilionatae

Isoflavones occur in rich profusion in only one group of plants, the legume sub-family Papilionatae (*see* p. 168) and are found indiscriminately in flower, leaf, seed, root and heartwood. As taxonomic markers, they appear to be very characteristic of this group; unfortunately they also occur sporadically in two related families (Rosaceae and Moraceae) and in three unrelated groups (Amarantaceae, Iridaceae and Podocarpaceae). Even within the Papilionatae, isoflavone distribution seems to be erratic. Although they occur in 10 of the 11 tribes, they are found in all species of some genera (e.g. *Cytisus, Ulex*), in most species of others (e.g. *Trifolium*) and in single species of yet others (e.g. *Lathyrus*).

The isoflavone is thus a chemical character which is clearly useful in determining whether a legume of doubtful affinities, cf. the case of *Amphimas* (p. 169), should be placed in this sub-family or be left in one of the two sub-families which lack isoflavone. On the other hand, its value at other levels of classification seems limited; it is difficult to correlate isoflavone occurrence with the tribal and generic divisions of the sub-family. It is, perhaps, the type of character, like that of the amino acid canavanine (which also occurs exclusively but sporadically in the Papilionatae), which is best dealt with by means of numerical taxonomy. Certainly, some valuable correlations may well emerge when the sub-family has been more deliberately searched for isoflavones and related flavonoids.

H. Dihydrochalcones in Apples

That there is a distinction between the apple genus *Malus* and the pear genus *Pyrus* is a view that taxonomists have not always subscribed to (*see* e.g. Willis, 1960) although these taxa are recognised today as separate genera of the Rosaceae. The situation is one is which chemical data can usefully contribute in retrospect and it is satisfying to find that the distribution of flavonoids in the leaves of these fruit trees completely supports this division.

All 25 known species of *Malus* have been surveyed by Williams (1966) and all contain dihydrochalcones, substances which are completely absent from *Pyrus* and from all other members of the Rosaceae studied. Furthermore in *Pyrus* dihydrochalcones are replaced, as major leaf constituents, by a simpler phenol, the hydroquinone glucoside arbutin. Flavonoids are of further interest, since the type of dihydrochalcone (and flavone) found within the genus *Malus* varies and appears to be correlated with plant geography (for details, *see* p. 160).

I. Conclusion

The seven chemotaxonomic correlations quoted above are the results of deliberate, if incomplete, phytochemical surveys but there are many other known instances of structurally unusual flavonoids occurring in related taxa; unfortunately, it is not usually possible to draw systematic conclusions because the data are incomplete in some respect. When a chemist discovers a new flavonoid in a particular species, he may look for it in a few other species of the same genus but usually fails (1) to record those that are negative, (2) to look at a statistically significant sample of the genus and (3) to extend the survey far enough to be sure of its absence from related genera. The examples already quoted do at least indicate the potentialities of the method, show that results can be obtained at all levels (or ranks) of systematic classification and demonstrate that analyses should be carried out on as many different parts of the plant as possible.

Some taxa are particularly rich in flavonoids (e.g. members of the Leguminosae and the Compositae) and in these instances individual species can be characterised by the pattern of flavonoids as revealed by two-dimensional chromatography of leaf and petal extracts. This is true of some species of *Baptisia* (Leguminosae) and a number of species-specific flavonoids have been detected in this genus (Alston and Turner, 1963). Much use has been made of two-dimensional chromatography for determining the origin of *Baptisia* species hybrids collected in the field. In some instances, conventional morphological and cytological criteria fail completely to indicate the species involved in the crosses and chromatography of the flavonoids is the only means of identification.

To separate the contribution of the flavonoids from those of other secondary constituents to the practice of chemotaxonomy, as has been done here, naturally does less than justice to this novel approach to systematics. One great advantage of this method is that a host of *structurally different* chemical characters may be applied to classification. The flavonoid results in *Pinus*, for example, should be combined with those obtained from studying stilbenes, essential oils, alkaloids and cyclitols, all of which are known to occur in these plants. In the past, little attempt has been made to mount a co-ordinated chemical attack on a group of plants, the ordering of which is a matter of taxonomic dispute; however, programmes with this objective in mind are under way in several laboratories.

III. Contribution to Phylogeny

Phylogeny is the study of the origin and evolution of taxa. In the case of the angiosperms, the largest group of plants (some 250,000 species), it is doubtful whether a complete understanding of their evolution will ever be achieved; there is so little fossil evidence to go on. However, many phylogenetic trees have been constructed, based on subjective evaluations of present morphological relationships, and provide the experimental taxonomist with material for critical evaluation.

In spite of much conflict of opinion over matters of detail, the main trends of plant evolution are widely accepted, i.e. that gymnosperms are less advanced than angiosperms or that the Magnoliaceae are a less specialised family than the Compositae (Davis and Heywood, 1963). There seems no reason why this broad picture of plant development should not be used to relate chemistry to phylogeny, comparisons of the type of chemical constituents of primitive taxa with those of advanced taxa being a valid exercise.

In the case of the flavonoids (Fig. 10·1), there is a general evolutionary trend towards complex structures and the wealth of highly substituted flavonoids are concentrated in highly specialised families like the Leguminosae, Gesneriaceae and Compositae. A simpler pattern, by contrast, is found in all the more primitive angiosperm groups such as the Magnoliaceae and the Ranunculaceae

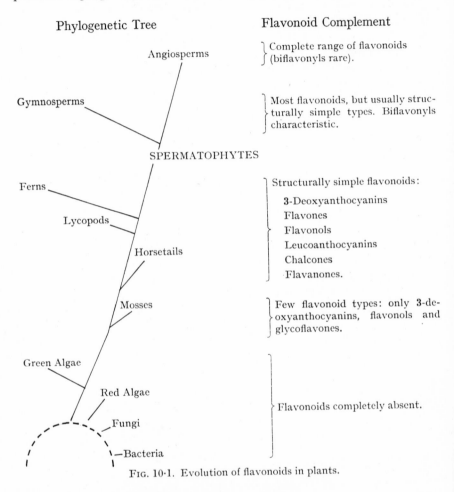

FIG. 10·1. Evolution of flavonoids in plants.

(cf. p. 146). Flavonoids with simple hydroxylation and glycosylation patterns again characterise the less highly developed plant orders—the gymnosperms, ferns, horsetails, and mosses. Flavonoids are indeed only found in the more complex cryptogams; these structures are apparently absent from the simpler organisms such as the bacteria, fungi and algae.

The discovery that flavonoid complexity is related to morphological advancement does not, of itself, further knowledge of plant evolution. Chemistry makes an unambiguous contribution to phylogeny because the biosynthetic origin of most plant constituents is known or can be determined. Different flavonoids may be placed in sequence on a biosynthetic pathway, A→B→C→ D→E, and it may be accepted that a plant making only A has fewer enzymes at its disposal than a plant containing E, i.e., A is the simpler organism. The difficulty in assessing whether one morphological character is "advanced" over another (e.g. free petals over united petals) is the lack of information regarding the development of one from the other; this difficulty vanishes in the case of chemical features which are biosynthetically related.

Table 10·2. Evolutionary Status of Flavonoid Characters

PRIMITIVE CHARACTERS	1. 3-Deoxyanthocyanidins 2. Flavonols 3. Leucoanthocyanidins 4. Chalcones, flavanones and dihydrochalcones 5. C-substitution (C-methylation, C-prenylation, C-glycosylation, biflavonyl formation)

ADVANCED CHARACTERS	Gain Mutations
	  1. Complex O-glycosylation
	  2. 6- and 8-Hydroxylation
	  3. 2′-Hydroxylation
	  4. O-Methylation
	  5. Oxidation of chalcones to aurones
	  6. Trihydroxylation of anthocyanidin B-ring (in flowers)
	Loss Mutations
	  1. Replacement of flavonols by flavones
	  2. Elimination of leucoanthocyanins
	  3. Elimination of trihydroxylation of flavonol B-ring (in leaves)

ISOLATED CHARACTERS	1. Replacement of anthocyanin by betacyanin 2. Shift of flavonoid B-ring from 2- to 3-position: isoflavone formation 3. Elimination of 5-hydroxyl group

Although knowledge of flavonoid biosynthesis is still limited (see Chapter 8), it is possible to draw some conclusions about the relative advancement of most flavonoid characters. The general scheme outlined in Table 10·2 is based on this knowledge and on present information of distribution. Some of the assignments in this table must be regarded as speculative. It is, for example, difficult to assess the position of isoflavones—it remains to be seen whether their production is a biosynthetic cul-de-sac, as is suggested here, or whether they have a more important evolutionary role.

Two general trends are indisputable: the change in flavonoid pattern in leaf (e.g. flavonol→flavone) with the replacement of woody by herbaceous habit

11

and the correlation of anthocyanin type with natural selection for flower colour. Exceptions to these general trends may, of course, be found. That 3-deoxyanthocyanins, for example, are primitive characters is based on biosynthetic arguments and also on their occurrence in some 10 ferns and a few mosses. The exceptional presence in one fern, *Davallia divaricata*, of ordinary anthocyanin does not invalidate this conclusion; nor does the occasional occurrence of 3-deoxyanthocyanins in angiosperms if their synthesis can be related to evolutionary factors (*see* p. 280).

The study of the comparative biochemistry of flavonoids is worthwhile in its own right and can only enrich our knowledge of these fascinating pigments. The possibility of using the information so obtained to solve biological problems is a considerable incentive to such study. Nowhere is the need for such biochemical data greater than in the field of plant taxonomy, which in its broadest sense embraces every aspect of plant science and which has the task of arranging in order over a quarter of a million plants.

References

Abe, Y. and Gotoh, K. (1956). *App. Rep. natn. Inst. Genet. (Japan)* **6**, 75.

Abe, Y. and Gotoh, K. (1959). *Bot. Mag. (Tokyo)* **72**, 432.

Acheson, R. M. (1956). *Proc. R. Soc.* B **145**, 549.

Acheson, R. M., Harper, J. L. and McNaughton, I. H. (1956). *Nature, Lond.* **178**, 1283.

Acheson, R. M., Harper, J. L. and McNaughton, I. H. (1962). *New Phytol.* **61**, 256.

Adinarayana, D. and Seshadri, T. R. (1965). *Tetrahedron* **21**, 3727.

Akai, S. (1955). *J. pharm. Soc. Japan* **55**, 537.

Albach, R. F., Kepner, R. E. and Webb, A. D. (1965). *Fd. Res.* **30**, 69.

Albach, R. F., Webb, A. D. and Kepner, R. E. (1965). *Fd. Res.* **30**, 620.

Alston, R. E. (1958). *Am. J. Bot.* **45**, 689.

Alston, R. E. (1964). *In* "Biochemistry of Phenolic Compounds" (J. B. Harborne, ed.) pp. 171–204. Academic Press, London and New York.

Alston, R. E. (1966). *In* "Comparative Phytochemistry" (T. Swain, ed.), pp. 33–56. Academic Press, London and New York.

Alston, R. E. and Hagen, C. W. (1957). *Genetics, Princeton,* **43**, 35.

Alston, R. E., Rosler, H., Naifeh, K. and Mabry, T. J. (1965). *Proc. natn. Acad. Sci. U.S.A.* **54**, 1458.

Alston, R. E. and Turner, B. L. (1962). *Proc. natn. Acad. Sci. U.S.A.* **48**, 130.

Alston, R. E. and Turner, B. L. (1963). "Biochemical Systematics." Prentice-Hall, New Jersey.

Anderson, J. A. (1934). *Can. J. Res.* **11**, 667.

Anyos, T. and Steelink, C. (1960). *Archs Biochem. Biophys.* **90**, 63.

ApSimon, J. W., Haynes, N. B., Sim, K. Y. and Whalley, W. B. (1963). *J. chem. Soc.* 3780.

Arcoleo, A. (1958). *Annali Chim.* **51**, 751.

Ardenne, R. (1965). *Z. Naturf,* **20b**, 186.

Arisumi, K. (1963). *Sci. Bull. Fac. Agric. Kyushu Univ.* **20**, 131.

Aritomi, M. (1962). *J. pharm. Soc. Japan.* **82**, 771.

Aritomi, M. (1963). *J. pharm. Soc. Japan* **83**, 737.

Aritomi, M. (1964). *Chem. pharm. Bull., Tokyo* **12**, 841.

Aritomi, M. (1965). *Chem. Abstr.* **62**, 1966.

Aritomi, M., Shimojo, M. and Mazaki, T. (1964). *J. pharm. Soc. Japan* **84**, 895.

Arthur, H. R., Hui, W. H. and Ma, C. N. (1956). *J. chem. Soc.* 632.

Arthur, H. R. and Tam, S. W. (1960). *J. chem. Soc.* 3197.

Asahina, Y. (1908). *Arch. Pharm. Berl.* **246**, 260.

Asen, S. (1958). *Pl. Physiol. Lancaster* **33**, 14.

Asen, S. (1961). *Proc. Am. Soc. hort. Sci.* **78**, 586.

Asen, S. (1965). *J. Chromat.* **18**, 602.

Asen, S. and Budin, P. S. (1966). *Phytochemistry* **5**, 1263.

Asen, S. and Emsweller, S. L. (1962). *Proc. Am. Soc. hort. Sci.* **81**, 530.

Asen, S., Jansen, L. L. and Hilton, J. L. (1963). *Nature, Lond.* **198**, 185.

Asen, S. and Jurd, L. (1966). *Phytochemistry,* **5**, 1257.

Asen, S., Siegelman, H. W. and Stuart, N. W. (1957). *Proc. Am. Soc. hort. Sci.* **69**, 561.

Avadhani, P. N. and Lim, G. (1964). *Abstr. 10th Int. Bot. Congr.* 326

Awe, W., Schaller, J. F. and Kummell, H. J. (1959). *Naturwissenschaften* **46**, 458.

Bachelard, E. P. and Stowe, B. B. (1962). *Nature, Lond.* **194**, 209.

Baker, H. G. (1948). *Ann. Bot.* N.S. **12**, 207.

Baker, W. and Ollis, W. D. (1961). *In* "Recent Developments in the Chemistry of Natural Phenolic Compounds" (W. D. Ollis, ed.), pp. 152–184. Pergamon Press, Oxford.

Baker, W., Hemming, R. and Ollis, W. D. (1951). *J. chem. Soc.* 691.

Balakrishna, S., Rao, M. M. and Seshadri, T. R. (1962). *Tetrahedron* **18**, 1503.

Balakrishna, K. J. and Seshadri, T. R. (1947). *Proc. Indian Acad. Sci.* **25A**, 449.

Banerji, A., Murti, W. S., Seshadri, T. R. and Thakur, R. S. (1963). *Indian J. Chem.* **1**, 25.

Baraud, J., Genevois, L. and Ponart, J. P. (1964). *J. Agric. trop. Bot. appl.* **11**, 55.

Barber, G. A. (1962a). *Biochemistry* **1**, 463.

Barber, G. A. (1962b). *Archs Biochem. Biophys.* **7**, 204.

Barger, G. (1906). *J. chem. Soc.* **89**, 1210.

Barger, G. and White, F. D. (1923). *Biochem. J.* **17**, 836.

Barloy, J. (1963). *Annali. Physio. Veget.* **5**, 141.

Barnes, R. A. and Gerber, N. N. (1955). *J. Am. chem. Soc.* **77**, 3259.

Barua, P. K. (1956). *Camellian* **7**, 18.

Baruah, S. and Swain, T. (1959). *J. Sci. Fd. Agric.* **10**, 125.

Bate-Smith, E. C. (1948). *Nature, Lond.* **161**, 835.

Bate-Smith, E. C. (1954). *Biochem. J.* **58**, 122.

Bate-Smith, E. C. (1956). *Sci. Proc. R. Dublin Soc.* **27**, 165.

Bate-Smith, E. C. (1958). *J. Linn. Soc. (Bot.)* **58**, 39.

Bate-Smith, E. C. (1959). *In* "Pharmacology of Plant Phenolics" (J. W. Fairbairn, ed.), pp. 133–147. Academic Press, London and New York.

Bate-Smith, E. C. (1961). *J. Linn Soc. (Bot.)* **58**, 39.

Bate-Smith, E. C. (1962). *J. Linn. Soc. (Bot.)* **58**, 39.

Bate-Smith, E. C. and Harborne, J. B. (1963). *Nature, Lond.* **198** 1307.

Bate-Smith, E. C., Harborne, J. B. and Davenport, S. M. (1966). *Nature, Lond.* **212**, 1065.

Bate-Smith, E. C., Nordström, C. G. and Swain, T. (1955). *Nature, Lond.* **176**, 1016.

Bate-Smith, E. C. and Swain, T. (1960). *Chemy Ind.* 1132.

Bate-Smith, E. C. and Swain, T. (1966). *In* "Comparative Phytochemistry" (T. Swain, ed.), pp. 159–174. Academic Press, London and New York.

Battersby, A. R. (1961). *Q. Rev. Chem. Soc.* **15**, 259.

Batyuk, V. S., Prokopenko, A. P. and Kolesnikov, D. G. (1965). *Chem. Abstr.* **63**, 16297.

Bauer, K. H. and Dietrich, H. (1933). *Chem. Ber.* **66**, 1053.

Bayer, E. (1958). *Chem. Ber.* **91**, 1115.

Bayer, E. (1959). *Chem. Ber.* **92**, 1062.

Beadle, G. W. (1945). *Chem. Rev.* **37**, 15.

Beale, G. H. (1941). *J. Genet.* **42**, 197.

Beale, G. H., Price, J. R. and Sturgess, V. C. (1941). *Proc. R. Soc.* B **130**, 113.

Beale, G. H., Robinson, G. M., Robinson, R. and Scott-Moncrieff, R. (1939). *J. Genet.* **37**, 375.

Beck, A. B. (1964). *Aust. J. agric. Res.* **15**, 223.

Beck, E., Merxmuller, H. and Wagner, H. (1962). *Planta* **58**, 220.

Beckmann, S. and Geiger, H. (1963). *Phytochemistry* **2**, 281.

Belic, I., Bergant-Dolar, J. and Morton, R. A. (1961). *J. chem. Soc.* 2523.

Bell, J. C. and Robinson, R. (1934). *J. chem. Soc.* 813.

Bendz, G. and Martensson, O. (1961). *Acta. chem. scand.* **15**, 1185.

Bendz, G., Martensson, O. and Nilsson, E. (1966). *Acta chem. scand.* **20**, 277.
Bendz, G., Martensson, O. and Terenius, L. (1962). *Acta chem. scand.* **16**, 1183.
Bentham, G. and Hooker, J. D. (1862–1883). "Genera Plantarum." London.
Bevan, C. W. L., Ekong, D. E. U., Obasi, M. E. and Powell, J. W. (1966). *J. chem. Soc.* (C) 509.
Bhandari, P. R. (1964). *J. Chromat.* **16**, 130.
Bhatia, V. K., Gupta, S. R. and Seshadri, T. R. (1966). *Tetrahedron* **22**, 1147.
Bickoff, E. M. (1961). "Symposium on the Biochemistry of Plant Phenolic Substances, Fort Collins, Colo.", pp. 125–169.
Bickoff, E. M., Livingston, A. L. and Witt, S. C. (1965). *Phytochemistry* **4**, 523.
Biggers, J. D. (1959). *In* "Pharmacology of Plant Phenolics" (J. W. Fairbairn, ed.), pp. 51–68. Academic Press, London and New York.
Billot, J. (1964). *C.r. hebd. Séanc. Acad. Sci., Paris* **258**, 2386.
Birch, A. J. and Hextall, P. (1955). *Aust. J. Chem.* **8**, 263.
Birkofer, L. and Kaiser, C. (1962). *Z. Naturf.* **17b**, 359.
Birkofer, L., Kaiser, C., Koch, W. and Lange, H. W. (1962). *Z. Naturf.* **17b**, 352.
Birkofer, L., Kaiser, C., Koch, W. and Lange, H. W. (1963). *Z. Naturf.* **18b**, 367.
Birkofer, L., Kaiser, C. and Kosmol, H. (1965). *Z. Naturf.* **20b**, 605, 923.
Bischoff, H. (1876). "Inaug. Diss." Tübingen.
Bjorkman, O. and Holmgren, P. (1958). *Physiologia. Pl.* **11**, 254.
Blasdale, W. C. (1945). *J. Am. chem. Soc.* **67**, 491.
Blank, F. (1947). *Bot. Rev.* **13**, 241.
Blank, F. (1958). *In* "Encyclopedia of Plant Physiology" (K. Paech and M. V. Tracey, eds.), Vol. 10, p. 300. Springer Verlag, Berlin.
Blunden, G. and Challen, S. B. (1965). *Nature, Lond.* **208**, 388.
Bockain, S. H., Kepner, R. E. and Webb, A. D. (1955). *J. Agric. Fd. Chem.* **3**, 695.
Bockova, H., Holubek, J. and Cekan, Z. (1964). *Colln. Czech. chem. Commun.* **29**, 1484.
Bogs, H. U. and Bogs, U. (1965). *Pharmazie* **20**, 706.
Boichinov, A., Panova, D. and Asenov, I. (1965). *Farmatsiya Mosk.* **15**, 15.
Bolley, P. (1860). *Jber. Chem.* 889.
Booth, A. N. and Williams, S. R. T. (1963). *Biochem. J.* **88**, 66P.
Bopp, M. (1957). *Planta* **48**, 631.
Bopp, M. (1958). *Z. Naturf.* **13b**, 699.
Bopp, M. (1959). *Z. Bot.* **47**, 197.
Bopp, M. and Boll, M. (1960). *Naturwissenschaften* **47**, 159.
Bose, P. K. and Bhattachanya, S. N. (1938). *J. Indian Chem. Soc.* **15**, 311.
Bose, P. K. and Bose, J. (1939). *J. Indian Chem. Soc.* **16**, 183.
Bottomley, W., Smith, H. and Galston, A. W. (1965). *Nature, Lond.* **207**, 1311.
Bradbury, R. B. and White, D. E. (1951). *J. chem. Soc.* 3447.
Bragt, J. van (1962). *Meded. Landbouwhogeschool Opzoekingstat. Staat. Cent.* **62**, 1.
Brandwein, B. J. (1965). *Fd. Res.* **30**, 680.
Brass, K. and Krantz, H. (1932). *Liebig's Ann. Chem.* **499**, 175.
Brewerton, H. W. (1958). *N.Z. Jl. Sci. Technol.* **1**, 220.
Bridel, M. and Charaux, C. (1925). *C.r. hebd. Séanc. Acad. Sci., Paris* **180**, 387.
Brieskorn, C. H. and Meister, G. (1965). *Arch. Pharm., Berl.* **298**, 435.
Briggs, L. H., Cain, B. F. and Cebalo, T. P. (1959). *Tetrahedron* **6**, 143; **7**, 262.
Briggs, L. H., Cambie, R. C. and Hoare, J. L. (1961). *J. chem. Soc.* 4645.
Briggs, L. H. and Locker, R. H. (1950). *J. chem. Soc.* 2376, 2379.
Brown, A. G., Falshaw, C. P., Haslam, E., Holmes, A. and Ollis, W. D. (1966). *Tetrahedron Letters* 1193.

Brune, W. and Geissman, T. A. (1965). *Aust. J. Chem.* **18**, 1649.

Buren, J. P. van, Scheiner, D. M. and Wagenknecht, A. C. (1960). *Nature, Lond.* **185**, 165.

Burns, E. F. and Winzer, J. W. (1962). *Proc. Am. Soc. hort. Sci.* **80**, 449.

Burrows, B. F., Ollis, W. D. and Jackman, L. M. (1960). *Proc. chem. Soc.* 177.

Burtt, B. L. (1962). *Notes R. Bot. Gdn. Edinb.* **14**, 205.

Butler, W. L., Hendricks, S. B. and Siegelman, H. W. (1965). *In* "Chemistry and Biochemistry of Plant Pigments" (T. W. Goodwin, ed.), pp. 197–210. Academic Press, London and New York.

Cambie, R. C. and Seelye, R. N. (1961). *N.Z. Jl. Sci. Technol.* **4**, 189.

Cadman, C. H. (1960). *In* "Phenolics in Plants in Health and Disease" (J. B. Pridham, ed.), pp. 101–105. Pergamon Press, Oxford.

Carles, J. (1935). *Revue gén. bot.* 363.

Carthy, J. D. (1964). *In* "Colour and Life" (W. B. Broughton, ed.), pp. 69–78. Institute of Biology, London.

Cayen, M. N., Carter, A. L. and Common, R. H. (1964). *Biochim. biophys. Acta* **86**, 56.

Casparis, P. and Steinegger, E. (1945). *Pharm. Acta Helv.* **20**, 174.

Casparis, P., Spracher, P. and Muller, H. J. (1946). *Pharm. Acta Helv.* **21**, 341.

Chadha, J. S. and Seshadri, T. R. (1962). *Curr. Sci.* **31**, 235.

Chandler, B. V. (1958). *Nature, Lond.* **182**, 933.

Chandler, B. V. and Harper, K. A. (1958). *Nature, Lond.* **181**, 131.

Chandler, B. V. and Harper, K. A. (1961). *Aust. J. Chem.* **14**, 586.

Chandler, B. V. and Harper, K. A. (1962). *Aust. J. Chem.* **15**, 114.

Chandrashakar, V., Krishnamurti, M. and Seshadri, T. R. (1965). *Curr. Sci.* **34**, 609.

Chapman, E., Perkins, A. G. and Robinson, R. (1927). *J. chem. Soc.* 3015.

Charaux, C. (1924). *Bull. Soc. chim. biol.* **6**, 641.

Charaux, C. (1925). *C.r. hebd. Séanc. Acad. Sci., Paris* **180**, 1419.

Charaux, C. and Rabate, J. (1931). *Bull. Soc. chim. biol.* **13**, 814.

Charaux, C. and Rabate, J. (1940). *J. Pharm. Chim., Paris* **9**, 155.

Chmielewska, I. (1936). *Roczn. fenol.* **16**, 385.

Chmielewska, I., Kawowska, I. and Lipinski, B. (1955). *Bull. acad. pol. Sci. Cl.* III, **3**, 527.

Chopin, J., Bouillant, M. L. and Lebreton, P. (1964). *Bull. Soc. Chim.* 1038.

Chopin, J., Dellamonica, G. and Lebreton, P. (1963). *C.r. hebd. Séanc. Acad. Sci., Paris* **257**, 534.

Chopin, J., Roux, B. and Durix, A. (1964). *C.r. hebd. Séanc. Acad. Sci., Paris* **259**, 3111.

Chumbalov, T. K. and Kil, T. A. (1962). *Chem. Abstr.* **57**, 5124.

Claisen, L. and Claparede, A. (1881). *Ber. dt. chem. Ges.* **14**, 2463.

Clevenger, S. (1959). *Archs Biochem. Biophys.* **76**, 131.

Clevenger, S. (1964). *Can. J. Biochem.* **42**, 154.

Clifford, H. T. and Harborne, J. B. (1966). *J. Linn. Soc. (Bot.)* (In press.)

Collot, A. M. and Charaux, C. (1939). *Bull. Soc. Chim. biol.* **21**, 455.

Cooke, R. G. and Haynes, H. F. (1960). *Aust. J. Chem.* **13**, 150.

Cooper, R. L. and Elliott, F. C. (1964). *Co-op. Sci.* **4**, 367.

Crabbe, P., Leeming, P. R. and Djerassi, C. (1958). *J. Am. chem. Soc.* **80**, 5258.

Craigie, J. S. and McLachlan, J. (1964). *Can. J. Bot.* **42**, 23.

Creasy, L. L. and Swain, T. (1965). *Nature, Lond.* **207**, 150.

Creasy, L. L., Maxie, E. C. and Chichester, C. O. (1965). *Phytochemistry* **4**, 517.

Crombie, L. and Whiting, D. A. (1962). *Tetrahedron Letters* 801.

Cruickshank, I. A. M. and Perrin, D. R. (1960). *Nature, Lond.* **187**, 799.

Cruickshank, I. A. M. and Perrin, D. R. (1964). *In* "Biochemistry of Phenolic Compounds" (J. B. Harborne, ed.), pp. 511–544. Academic Press, London and New York.

Cruickshank, I. A. M. and Perrin, D. R. (1965). *Aust. J. biol. Sci.* **18**, 803, 817, 829.

Culvenor, C. C. J., Bon, R. D. and Smith, L. W. (1964). *Aust. J. chem.* **17**, 1301.

Dame, C., Chichester, C. O. and Marsh, G. L. (1959). *Fd. Res.* **24**, 20, 28.

Darwin, C. (1876). "The Effects of Cross and Self Fertilisation in the Vegetable Kingdom." Murray, London.

Das, N. P. and Griffiths, L. A. (1966). *Biochem. J.* **98**, 488.

Dasa Rao, C. J., Walawalkar, D. G. and Srikantan, B. S. (1938). *J. Indian chem. Soc.* **15**, 27.

Dave, K. G., Telang, S. A. and Venkataraman, K. (1962). *Tetrahedron Letters* 9.

Davis, P. H. and Heywood, V. H. (1963) ."Principles of Angiosperm Taxonomy." Oliver and Boyd, Edinburgh.

Dean, F. M. (1963). "Naturally Occurring Oxygen Ring Compounds." Butterworths, London.

Denliev, P. K., Pakudina, Z. P. and Sadykov, A. S. (1963). *Dokl. Akad. Nauk. USSR* **20**, 19.

Dietrichs, H. H. and Schaich, E. (1962). *Naturwissenschaften* **50**, 478.

Di Modica, G. and Rivero, A. M. (1962). *J. Chromat.* **7**, 133.

Di Modica, G., Rossi, P. F. and Rivero, A. M. (1959). *Atti Accad. Naz. Lincei Rc.* VIII, **27**, 127.

Dipalma, J. R. (1965). Unpublished results.

Dodds, K. S. and Harborne, J. B. (1964). *A. Rep. John Innes Inst.* 34.

Dodds, K. S. and Long, D. H. (1955). *J. Genet.* **53**, 136.

Dodds, K. S. and Long, D. H. (1956). *J. Genet.* **54**, 27.

Doherty, G. O. P., Haynes, N. B. and Whalley, W. B. (1963). *J. chem. Soc.* 5577.

Doporto, M. L., Gallagher, K. M., Gowan, J. E., Hughes, A. C., Philbin, E. M., Swain, T. and Wheeler, T. S. (1955). *J. chem. Soc.* 4249.

Douglass, C. D., Howard, W. L. and Wender, S. H. (1949). *J. Am. chem. Soc.* **71**, 2658.

Dreiding, A. S. (1961). *In* "Recent Developments in the Chemistry of Natural Phenolic Compounds" (W. D. Ollis, ed.), p. 194. Pergamon Press, Oxford.

Drewes, S. E. and Roux, D. G. (1963). *Biochem. J.* **87**, 167.

Duewell, H. (1954). *J. chem. Soc.* 2562.

Dunstan, W. R. and Henry, T. A. (1901). *Chem. Zent Bl. II* 593.

Eade, R. A., Salasoo, I. and Simes, J. J. H. (1962). *Chemy Ind.* 1720.

Eberhardt, F. (1959). *Planta* **53**, 334.

Eddy, B. P. and Mapson, L. W. (1951). *Biochem. J.* **49**, 694.

Edmondson, Y. H. and Thimann, K. V. (1950). *Archs Biochem.* **25**, 79.

Egger, K. (1959). *Z. Naturf.* **14b**, 401.

Egger, K. (1961a). *Z. analyt. Chem.* **182**, 161.

Egger, K. (1961b). *Z. Naturf.* **16b**, 430.

Egger, K. (1962). *Z. Naturf.* **17b**, 489.

Egger, K. and Keil, M. (1965). *Ber. dt Bot. Ges.* **78**, 153.

Egger, K. and Reznik, H. (1961). *Planta* 57, 239.

El-Khadem, H. and Mohammed, Y. S. (1958). *J. chem. Soc.* 3320.

Endo, T. (1954). *Jap. J. Bot.* **14**, 187.

Endo, T. (1957). *Nature, Lond.* **179**, 378.

Endo, T. (1959). *Bot Mag. (Tokyo)* **72**, 10.

Endo, T. (1959). *Jap. J. Genet.* **34**, 116.

Endo, T. (1962). *Jap. J. Genet.* **37**, 284.

Engler, A. (1936). "Syllabus der Pflanzenfamilien", 11th Ed. (L. Diels, ed.). Borntraeger, Berlin.

Engler, A. (1964). "Syllabus der Pflanzenfamilien", 12th Ed. (H. Melchior, ed.), Vol. II. Borntraeger, Berlin.

Erikson, D., Oxford, A. E. and Robinson, R. (1938). *Nature, Lond.* 142, 211.

Erdtman, H. (1956). *In* "Perspectives in Organic Chemistry" (A. R. Todd, ed.), p. 430, Interscience, New York.

Eyton, W. B., Ollis, W. D., Sutherland, I. O., Magalhaes M. T. and Jackman, L. M. (1965). *Tetrahedron* 21, 2683.

Fairbairn, J. W. (ed.) (1959). "The Pharmacology of Plant Phenolics." Academic Press, London and New York.

Falco, M. R. and de Vries, J. (1964). *Naturwissenschaften* 51, 462.

Farkas, L., Horhammer, L. and Wagner, H. (1963). *Tetrahedron Letters* 727.

Farkas, L. and Nogradi, M. (1965). *Chem. Ber.* 98, 164.

Farkas, L., Nogradi, M. and Pallos, L. (1963a). *Chem. Ber.* 96, 1865.

Farkas, L., Nogradi, M. and Pallos, L. (1963b). *Tetrahedron Letters* 1999.

Farkas, L., Nogradi, M. and Pallos, L. (1964). *Chem. Ber.* 97, 1044.

Farooq, M. O., Gupta, S. R., Kiamuddin, M., Rahman, W. and Seshadri, T. R. (1953). *J. Sci. Indian Res.* 12B, 400.

Faust, M. (1965). *Proc. Am. Soc. hort. Sci.* 87, 1, 10.

Fedde, F. (1909). *Pflanzenreich* 4, 366.

Feenstra, W. J. (1960). *Meded. Landbouwhogeschool Opzoekingstat Staat. Cent.* 60, 1.

Feenstra, W. J., Johnson, B. L., Ribereau-Gayon, P. and Geissman, T. A. (1963). *Phytochemistry* 2, 273.

Fiedler, U. (1955). *Arnzneimittel-Forsch.* 5, 609.

Fincham, J. R. S. (1963). *A. Rep. John Innes Institute* 20.

Finnemore, H. (1910). *Pharm. Bull.* 31, 604.

Fisel, J. (1965). *Naturwissenschaften* 52, 592.

Flores, S. E. and Herran, J. (1958). *Tetrahedron* 2, 308.

Flores, S. E. and Herran, J. (1960). *Chemy Ind.* 29.

Ford, E. B. (1941). *Proc. R. Ent. Soc. Lond.* 16, 65.

Forsyth, W. G. C. and Quesnel, V. C. (1957). *Biochem. J.* 65, 177.

Forsyth, W. G. C. and Simmonds, N. W. (1954). *Proc. R. Soc.* 142, 549.

Forsyth, W. G. C. and Simmonds, N. W. (1957). *Nature, Lond.* 180, 247.

Francis, C. M. and Millington, A. J. (1965). *Aust. J. Agric. Res.* 16, 565.

Francis, F. J. and Harborne, J. B. (1966a). *Proc. Am. hort. Sci.* (In press.)

Francis, F. J. and Harborne, J. B. (1966b). *J. Fd. Sci.* 31, 524.

Francis, F. J., Harborne, J. B. and Barker, W. G. (1966). *J. Fd. Sci.* 31, 583.

Francois, M. T. and Chaix, L. (1961). *Chem. Abstr.* 55, 3744.

Frankel, S. and David, E. (1927). *Biochem. J.* 187, 146.

Freudenberg, K. and Hartmann, A. (1954). *Liebig's Ann. Chem.* 587, 207.

Friedrich, H. C. (1956). *Phyton* 6, 220.

Friedrich, H. (1962a). *Arch. Pharm., Berl.* 295, 59.

Friedrich, H. (1962b). *Naturwissenschaften* 49, 541.

Fritsch, K. (1893–4). *In* "Die Naturliche Pflanzenfamilien" (Engler and Prantl, eds.), IV (3B), p. 133.

Fritsch, K. von (1950). "Bees, their Vision, Chemical Senses and Language." Cornell, Ithaca, New York.

Fuchs, L. (1949). *Sci. Pharm. (Vienna)* 17, 128.

Fukuchi, G. and Imai, K. (1951). *Chem. Abstr.* 45, 3999.

Fujise, S. (1929). *Sci. Papers Inst. Phys. Chem. Res. (Tokyo).* 11, 111.

Furuya, M. and Thimann, K. V. (1964). *Archs Biochem. Biophys.* 198, 109.

Furuya, M., Galston, A. W. and Stowe, B. B. (1962). *Nature, Lond.* 193, 456.

Fujikawa, F. and Nakajima, K. (1948). *J. Pharm. Soc. Japan* **68**, 175.
Gairdner, A. E. (1936). *A. Rep. John Innes hort. Inst.* 16.
Gallop, R. A. (1965). "Variety, Composition and Colour in Canned Fruits, particularly Rhubarb." Fruit and Veg. Canning Res. Assoc., Chipping Campden, Glos.
Ganguly, A. K. and Seshadri, T. R. (1959). *Tetrahedron* **6**, 21.
Gascoigne, R. M., Ritchie, E. and White, D. R. (1948). *J. R. Soc. N.S.W.* **82**, 44.
Gehrmann, H. J., Endres, L., Cobet, R. and Fiedler, V. (1955). *Naturwissenschaften* **42**, 181.
Geiger, H. and Beckmann, S. (1965). *Z. Naturf.* **20b**, 1139.
Geissman, T. A. (1941). *J. Am. chem. Soc.* **63**, 656.
Geissman, T. A. (1956). *Archs Biochem. Biophys.* **60**, 21.
Geissman, T. A. (1958), *Aust. J. Chem.* **2**, 376.
Geissman, T. A. (ed.) (1962). "Chemistry of the Flavonoid Compounds." Pergamon Press, Oxford.
Geissman, T. A. and Dittmar, H. F. K. (1965). *Phytochemistry* **4**, 359.
Geissman, T. A. and Harborne, J. B. (1956). *J. Am. chem. Soc.* **78**, 832.
Geissman, T. A., Harborne, J. B. and Seikel, M. K. (1956). *J. Am. chem. Soc.* **78**, 825.
Geissman, T. A., Hinreiner, E. and Jorgensen, E. (1956). *Genetics* **41**, 93.
Geissman, T. A. and Heaton, C. D. (1943). *J. Am. chem. Soc.* **65**, 677.
Geissman, T. A. and Heaton, C. D. (1944). *J. Am. chem. Soc.* **66**, 486.
Geissman, T. A. and Hinreiner, E. (1952). *Bot. Rev.* **18**, 77.
Geissman, T. A., Jorgensen, E. C. and Johnson, B. L. (1954). *Archs Biochem. Biophys.* **49**, 368.
Geissman, T. A. and Jurd, L. (1955). *Archs Biochem. Biophys.* **56**, 259.
Geissman, T. A. and Melhquist, G. A. L. (1947). *Genetics* **32**, 410.
Geissman, T. A. and Steelink, C. (1957). *J. org. Chem.* **22**, 946.
Gentili, B. and Horowitz, R. M. (1964). *Tetrahedron* **20**, 2313.
Gertz, O. (1906). "Studier of ver Anthocyan." Lund, Sweden.
Gertz, O. (1938). *Kgl. Fhysiograf. Sallskap Lund Forh* **8**, 62.
Ghanim, A., Prakash, L., Zaman, A. and Kidwai, A. R. (1963a). *Indian J. Chem.* **1**, 230.
Ghanim, A., Prakash, L. and Kidwai, A. R. (1963b). *Indian J. Chem.* **1**, 320.
Gintl, W. (1868). *Jahresber. Chem.* 800.
Glyzin, V. I. and Senov, P. L. (1965). *Chem. Abstr.* **64**, 3308.
Goldstein, J. L. and Swain, T. (1963). *Phytochemistry* **2**, 371.
Goodwin, T. W. (1965). *In* "Biosynthetic Pathways in Higher Plants" (J. B. Pridham and T. Swain, eds.), pp. 37–71. Academic Press, London and New York.
Gopinath, K. W., Kidwai, A. R. and Prakash, L. (1961). *Tetrahedron* **16**, 201.
Gopinath, K. W., Prakash, L. and Kidwai, A. R. (1963). *Indian J. Chem.* **1**, 187.
Gorin, P. A. J. and Perlin, A. S. (1959). *Can. J. Chem.* **37**, 1930.
Goto, R. and Taki, M. (1938). *J. Pharm. Soc. Japan* **58**, 933.
Govindachari, T. R., Parthasarthy, P. C., Pai, B. R. and Subramanian, P. S. (1965). *Tetrahedron* **21**, 2633.
Gowing, D. P. and Lange, A. H. (1962). *Proc. Am. Soc. hort. Sci.* **80**, 645.
Grill, R. and Vince, D. (1964). *Planta* **63**, 1.
Gripenberg, J., Honkanen, E. and Silander, K. (1956). *Acta chem. scand.* **10**, 393.
Grisebach, H. (1957). *Z. Naturf.* **12b**, 227.
Grisebach, H. (1965). *In* "Chemistry and Biochemistry of Plant Pigments" (T. W. Goodwin, ed.), pp. 279–308. Academic Press, London and New York.
Grisebach, H. and Bopp, M. (1959). *Z. Naturf.* **14b**, 485.
Grisebach, H. and Brandner, G. (1962). *Biochim. biophys. Acta* **60**, 51.
Grisebach, H. and Kellner, S. (1964). *Z. Naturf.* **19b**, 125.

11*

Grisebach, H. and Patschke, L. (1961). Z. Naturf. **16b**, 645.
Grove, K. E. and Robinson, R. (1931). Biochem. J. **25**, 1706.
Grzybowska, J. and Jerzmanowska, Z. (1954). Roczn. Chem. **28**, 197.
Gupta, S. R., Pankajamini, K. S. and Seshadri, T. R. (1957). J. Sci. Indian Res. B. **16**, 154.
Gyorgy, P., Murata, K. and Ikehata, H. (1964). Nature, Lond. **203**, 870.
Habich, O. (1891). Stellin. ent. Ztg. **52**, 36.
Hagen, C. W. (1966). Am. J. Bot. **53**, 54.
Halevy, A. H. (1962). Biochem. J. **83**, 637.
Halevy, A. H. and Asen, S. (1959). Pl. Physiol. **34**, 494.
Hall, D. (1940). "The Genus Tulipa." Roy. Hort. Soc., London.
Hallier, H. (1912). Arch. Neerl. Sci. Exact. Nat. Ser. IIIB, **1**, 146.
Hamamura, Y., Hayashiya, K., Naito, K., Matsuura, K. and Nishida, J. (1962). Nature, Lond. **194**, 754.
Hansel, R., Langhammer, L. and Albrecht, A. G. (1962). Tetrahedron Letters, 599.
Hansel, R., Langhammer, L., Frenzl, J. and Ranft, G. (1963). J. Chromat. **11**, 369.
Hansel, R., Leuchert, C., Rimpler, H. and Schaaf, K. D. (1965). Phytochemistry **4**, 19.
Hansel, R. and Klaffenbach, J. (1961). Arch. Pharm., Berl. **294**, 158.
Hansel, R., Ranft, G. and Bahr, P. (1963). Z. Naturf. **18b**, 370.
Hansel, R., Rimpler, H. and Walther, K. (1966). Naturwissenschaften **53**, 19.
Harada, T. and Saiki, Y. (1955). Pharm. Bull. Tokyo **3**, 469.
Harada, T. and Saiki, Y. (1956). Seikatsu Kagaka **3**, 163.
Harborne, J. B. (1957). Biochem. J. **70**, 22.
Harborne, J. B. (1958a). Chromat. Rev. **1**, 209.
Harborne, J. B. (1958b). Nature, Lond. **181**, 25.
Harborne, J. B. (1958c). Chemy Ind. 1590.
Harborne, J. B. (1959). Chromat. Rev. **2**, 105.
Harborne, J. B. (1960a). Nature, Lond. **187**, 140.
Harborne, J. B. (1960b). Biochem. J. **74**, 262.
Harborne, J. B. (1960c). Chemy Ind. 229.
Harborne, J. B. (1961). Experientia **17**, 72.
Harborne, J. B. (1962a). Archs Biochem. Biophys. **96**, 171.
Harborne, J. B. (1962b). Chemy Ind. 222.
Harborne, J. B. (1962c). Biochem. J. **84**, 100.
Harborne, J. B. (1962d). Phytochemistry **1**, 203.
Harborne, J. B. (1962e). Fortschr. Chem. org. Naturst. **20**, 165.
Harborne, J. B. (1962f). In "Chemistry of Flavonoid Compounds" (T. A. Geissman, ed.), pp. 593–617. Pergamon Press, Oxford.
Harborne, J. B. (1963a). In "Chemical Plant Taxonomy" (T. Swain, ed.), pp. 359–388. Academic Press, London and New York.
Harborne, J. B. (1963b). Phytochemistry **2**, 85.
Harborne, J. B. (1963c). Phytochemistry **2**, 327.
Harborne, J. B. (1963d). Experientia **19**, 7.
Harborne, J. B. (1964a). Phytochemistry **3**, 151.
Harborne, J. B. (1964b). A. Rep. John Innes Inst. 45.
Harborne, J. B. (ed.) (1964c). "Biochemistry of Phenolic Compounds." Academic Press, London and New York.
Harborne, J. B. (1965a). Phytochemistry **4**, 107.
Harborne, J. B. (1965b). Phytochemistry **4**, 647.
Harborne, J. B. (1965c). In "Chemistry and Biochemistry of Plant Pigments" (T. W. Goodwin, ed.), pp. 247–278. Academic Press, London and New York.
Harborne, J. B. (1965d). Nature, Lond. **207**, 984.

Harborne, J. B. (1966a). *In* "Comparative Phytochemistry" (T. Swain, ed.), pp. 271–295. Academic Press, London and New York.

Harborne, J. B. (1966b). *Phytochemistry* **5**, 11.

Harborne, J. B. (1966c). *Phytochemistry* **5**, 589.

Harborne, J. B. (1966d). *Z. Naturf.* **216**, 604.

Harborne, J. B. and Hall, E. (1964a). *Phytochemistry* **3**, 421.

Harborne, J. B. and Hall, E. (1964b). *Phytochemistry* **3**, 453.

Harborne, J. B. and Hurst, H. M. (1966). *Phytochemistry*, (In preparation.)

Harborne, J. B. and Paxman, G. J. (1964). *Heredity* **19**, 505.

Harborne, J. B. and Sherratt, H. S. A. (1957). *Experientia* **13**, 486.

Harborne, J. B. and Sherratt, H. S. A. (1961). *Biochem. J.* **78**, 298.

Harper, S. H., Kemp, A. D. and Underwood, W. G. E. (1965). *Chemy Ind.* 562.

Harper, S. H., Kemp, A. D. and Underwood, W. G. E. (1965). *Chem. Commun.* 309.

Harrison, B. J. and Fincham, J. R. S. (1964). *Heredity* **19**, 237.

Hasegawa, M. (1958). *J. Jap. For. Soc.* **40**, 111.

Hasegawa, M. (1959). *J. org. Chem.* **24**, 408.

Hasegawa, M. and Shirato, T. (1953). *J. Am. chem. Soc.* **75**, 5507.

Hathway, D. E. and Seakin, J. W. T. (1957). *Biochem. J.* **65**, 32.

Hattori, S. (1951). *Nature, Lond.* **168**, 788.

Hattori, S. (1962). "Chemistry of the Flavonoid Compounds" (T. A. Geissman, ed.), pp. 317–352. Pergamon Press, Oxford.

Hattori, S. and Hasegawa, M. (1940). *Proc. Imp. Acad. Tokyo* **16**, 9.

Hattori, S. and Hasegawa, M. (1943). *Acta Phytochim. Japan* **13**, 99.

Hattori, S. and Hasegawa, M. (1952). *Chem. Abstr.* **46**, 2541.

Hattori, S., Hasegawa, M. and Hayashi, K. (1938). *J. chem. Soc. Japan* **58**, 844.

Hattori, S., Hasegawa, M. and Shimokoriyama, M. (1944). *Acta Phytochim. Japan* **14**, 1.

Hattori, S. and Hayashi, K. (1937). *Acta Phytochim. Japan* **10**, 129.

Hattori, S. and Matsuda, H. (1949). *Acta Phytochim. Japan* **15**, 233.

Hattori, S. and Matsuda, M. (1952). *Archs Biochem. Biophys.* **37**, 85.

Hattori, S. and Matsuda, H. (1954). *J. Am. chem. Soc.* **76**, 5792.

Hattori, S. and Shimokoriyama, M. (1956). *Bull. Soc. Chin. Biol.* **38**, 912.

Hattori, S., Shimokoriyama, M. and Kanao, M. (1952). *J. Am. chem. Soc.* **74**, 3614.

Hawthorne, B. J. and Morgan, J. W. W. (1962). *Chemy Ind.* 1504.

Hayashi, K. (1933). *Bot. Mag. Tokyo* **47**, 394.

Hayashi, K. (1939). *Acta Phytochim. Japan* **11**, 81.

Hayashi, K. (1940). *Proc. Imp. Acad. Tokyo* **16**, 478.

Hayashi, K. (1941). *Acta Phytochim. Japan* **12**, 65.

Hayashi, K. (1942). *Acta Phytochim. Japan* **13**, 25.

Hayashi, K. (1943). *Acta Phytochim. Japan* **13**, 85.

Hayashi, K. (1944). *Acta Phytochim. Japan* **14**, 39, 55.

Hayashi, K. (1949). *Acta Phytochim. Japan* **15**, 35.

Hayashi, K. (1960). *Proc. Japan. Acad.* **36**, 340.

Hayashi, K. (1962). *In* "Chemistry of Flavonoid Compounds" (T. A. Geissman, ed.), pp. 248–285. Pergamon Press, Oxford.

Hayashi, K. and Abe, Y. (1953). *Misc. Rep. Res. Inst. nat. Resour. Tokyo* **29**, 1.

Hayashi, K. and Abe, Y. (1955). *Bot. Mag. Tokyo* **68**, 299.

Hayashi, K. and Noguchi, T. (1952). *Proc. Japan. Acad.* **28**, 429.

Hayashi, K., Noguchi, T. and Abe, Y. (1954). *Pharm. Bull. (Japan)* **2**, 41.

Hayashi, K., Noguchi, T. and Abe, Y. (1955). *Bot. Mag. Tokyo* **68**, 129.

Hayashi, K. and Ouchi, K. (1946). *Proc. Japan. Acad.* **22**, 251.

Hayashi, K., Suzushinu, G. and Ouchi, K. (1951). *Proc. Japan. Acad.* **27**, 430.

Hayashi, K., Abe, Y., Noguchi, T. and Suzushinu, G. (1953). *Pharm. Bull. (Japan)* **1**, 130.

Hayashiya, K. (1959). *J. agric. Sci. Japan* **33**, 977.

Haynes, L. J. (1965). *Adv. Carbohyd. Chem.* **20**, 357.

Heap, T. and Robinson, R. (1926). *J. chem. Soc.* **2336**.

Hegnauer, R. (1962). "Chemotaxonomie der Pflanzen", Vol. I. Birkhauser Verlag, Basel.

Hegnauer, R. (1963). "Chemotaxonomie der Pflanzen", Vol. II. Birkhauser Verlag, Basel.

Hegnauer, R. (1964). "Chemotaxonomie der Pflanzen", Vol. III. Birkhauser Verlag, Basel.

Hehl, F. W. (1919). *J. Am. chem. Soc.* **41**, 1285.

Hein, S. (1959). *Planta med.* **7**, 185.

Hendershott, C. H. and Walker, D. R. (1959). *Science N.Y.* **130**, 798.

Hendricks, S. B. and Borthwick, H. A. (1965). *In* "Chemistry and Biochemistry of Plant Pigments" (T. W. Goodwin, ed.), pp. 405–436. Academic Press, London and New York.

Henrick, C. A. and Jefferies, P. R. (1964). *Aust. J. Chem.* **17**, 934.

Henrick, C. A. and Jefferies, P. R. (1965). *Tetrahedron* **21**, 3219.

Hergert, H. L. (1956). *J. org. Chem.* **21**, 534.

Hergert, H. L. (1962). *In* "Chemistry of Flavonoid Compounds" (T. A. Geissman, ed.), pp. 553–592. Pergamon Press, Oxford.

Herrmann, K. (1955). *Arch. Pharm., Berl.* **288**, 362.

Herrmann, K. (1958). *Arch. Pharm., Berl.* **291**, 238.

Herz, W. (1961). *J. org. Chem.* **26**, 3014.

Herz, W. and Sumi, Y. (1964). *J. org. Chem.* **29**, 3438.

Hess, D. (1963a). *Z. Bot.* **51**, 142.

Hess, D. (1963b). *Planta* **59**, 567.

Hess, D. (1964a). *Planta* **61**, 73.

Hess, D. (1964b). *Z. Naturf.* **19b**, 148, 447.

Hess, D. (1965). *Z. Pflanzenphysiol.* **53**, 1.

Hess, D. and Meyer, C. (1962). *Z. Naturf.* **17b**, 853.

Hillis, W. E. (1956). *Aust. J. Chem.* **9**, 544.

Hillis, W. E. and Horn, D. H. S. (1965). *Aust. J. Chem.* **18**, 531.

Hillis, W. E. and Horn, D. H. S. (1966). *Aust. J. Chem.* **19**, 705.

Hillis, W. E. and Isoi, K. (1965). *Phytochemistry* **4**, 54.

Hirao, S. (1935). *J. Agric. chem. Soc. Japan* **11**, 921.

Hirose, Y. (1962). *Kumamoto Pharm. Bull.* **5**, 44.

Hisamichi, S. (1961). *Pharm. Soc. Japan. J.* **81**, 446.

Hodges, R. (1965). *Aust. J. Chem.* **18**, 1491.

Hoffman,, E. (1876). *Ber. dt. Chem. Ges.* **9**, 685.

Hollande, A. C. (1913). *Archs Zool. exp. gén.* **51**, 53.

Horhammer, L. and Griesinger, R. (1959). *Naturwissenschaften* **46**, 427.

Horhammer, L., Hansel, R. and Endres, L. (1956a). *Arch. Pharm., Berl.* **289**, 213.

Horhammer, L., Hansel, R. and Endres, W. (1956b). *Arch. Pharm., Berl.* **289**, 133.

Horhammer, L., Hansel, R., Kriesmair, G. and Endres, W. (1955). *Arch. Pharm., Berl.* **288**, 419.

Horhammer, L. and Muller, K. (1954). *Arch. Pharm., Berl.* **287**, 126.

Horhammer, L., Stich, L. and Wagner, H. (1961). *Arch. Pharm., Berl.* **294**, 685.

Horhammer, L. and Votz, E. (1955). *Arch. Pharm., Berl.* **288**, 58.

Horhammer, L. and Wagner, H. (1961). *In* "Chemistry of Natural Phenolic Compounds" (W. D. Ollis, ed.), pp. 185–193. Pergamon Press, Oxford.

Horhammer, L. and Wagner, H. (1962). *In* "Chemistry of Natural and Synthetic Colour Matters" (T. S. Gore, ed.), pp. 315–330. Academic Press, London and New York.

Horhammer, L., Wagner, H. and Dhingra, H. S. (1959). *Arch. Pharm., Berl.* **292**, 82.

Horhammer, L., Wagner, H. and Gloggengiesser, F. (1958). *Arch. Pharm., Berl.* **291**, 126.

Horhammer, L., Wagner, H., Graf, E. and Farkas, L. (1965). *Chem. Ber.* **98**, 548.

Horhammer, L., Wagner, H. and Leeb, W. (1964). *Arch. Pharm., Berl.* **293**, 264.

Horhammer, L., Wagner, H. and Luck, R. (1956). *Arch. Pharm., Berl.* **289**, 613.

Horhammer, L., Wagner, H. and Luck, R. (1957). *Arch. Pharm., Berl.* **290**, 338, 342.

Horhammer, L., Wagner, H. and Probst, W. (1960). *Naturwissenschaften* **47**, 63.

Horhammer, L., Wagner, H. and Reinhardt, H. (1965). *Naturwissenschaften* **52**, 161.

Horhammer, L., Wagner, H., Rosprim, L., Mabry, T. and Rosler, H. (1965). *Tetrahedron Letters*, 1707.

Horhammer, L., Wagner, H. and Salfer, B. (1963). *Arzneimittell-Forsch.* **13**, 33.

Horhammer, L., Wagner, H. and Schilcher, H. (1962). *Arzneimittell-Forsch.* **12**, 1.

Horhammer, L., Wagner, H., Arndt, H. G., Kraemer, H. and Farkas, L. (1966). *Tetrahedron Letters*, 567.

Horowitz, R. M. (1957). *J. org. Chem.* **22**, 1733.

Horowitz, R. M. (1964). *In* "Biochemistry of Phenolic Compounds" (J. B. Harborne, ed.), pp. 545–572. Academic Press, London and New York.

Horowitz, R. M. and Gentili, B. (1960). *Nature, Lond.* **185**, 319.

Horowitz, R. M. and Gentili, B. (1964). *Chemy Ind.* 498.

Horowitz, R. M. and Gentili, B. (1966). *Chemy Ind.* 625.

Hoshi, T., Takamura, E. and Hayashi, K. (1963). *Bot. Mag. Tokyo* **76**, 431.

Hsia, C. L., Luh, B. S. and Chichester, C. O. (1965). *J. Fd. Sci.* **30**, 5.

Hsu, H. Y. (1959). *Bull. Taiwan Provincial Hyg. Lab.*, 1.

Hsu, K. K. and Wong, W. H. (1965). *Chem. Abstr.* **62**, 12159.

Huang, H. T. (1955). *J. Agric. Fd. Chem.* **3**, 141.

Huang, H. T. (1956). *J. Am. chem. Soc.* **78**, 2390.

Hulme, A. C. and Edney, K. L. (1960). *In* "Phenolics in Plants in Health and Disease" (J. B. Pridham, ed.), pp. 87–94. Pergamon Press, Oxford.

Hukuti, G. (1936). *J. Pharm. Soc. Japan* **56**, 569.

Hutchinson, J. (1959). "Families of Flowering Plants." Macmillan & Co., London.

Hutchinson, J. (1964). "The Genera of Flowering Plants", Vol. I. Oxford University Press.

Ice, C. H. and Wender, S. H. (1953). *J. Am. chem. Soc.* **75**, 50.

Imai, K. and Mayama, S. (1953). *A. Rep. Takamiro Lab. (Japan)* **5**, 23.

Inglett, G. E. (1956). *Nature, Lond.* **178**, 1346.

Inglett, G. E. (1957). *J. org. Chem.* **22**, 189.

Ishikura, N. and Hayashi, K. (1962). *Bot. Mag. Tokyo* **75**, 28.

Ishikura, N. and Hayashi, K. (1965). *Bot. Mag. Tokyo* **78**, 91.

Jacobsen, J. S. (1961). *Archs Biochem. Biophys.* **93**, 580.

Jain, A. C., Mathur, S. K. and Seshadri, T. R. (1965). *Indian J. Chem.* **3**, 418.

Janes, N. F., King, F. E. and Morgan, J. W. W. (1963). *J. chem. Soc.* 1356.

Jefferies, P. R., Knox, J. R. and Middleton, E. J. (1962). *Aust. J. Chem.* **15**, 532.

Jefferies, P. R. and Payne, T. G. (1965). *Aust. J. Chem.* **18**, 1441.

Jentzch, K., Spoegel, P. and Fuchs, L. (1962). *Planta Med.* **10**, 1.

Jermstad, A. and Jensen, K. B. (1951). *Bull. soc. chim. biol.* **33**, 258.

Jerzmanowska, Z. (1937). *Widomssci farmac.* **64**, 527.

Jerzmanowska, Z. and Grzybowska, J. T. (1960). *Nature, Lond.* **186**, 807.

Johns, S. R., Russel, J. H. and Hefferman, M. L. (1965). *Tetrahedron Letters*, 1987.

Johnson, A. P., Pelter, A. and Stainton, P. (1966). *J. chem. Soc.* (C) 192.

Jorgensen, E. C. and Geissman, T. A. (1955). *Archs Biochem. Biophys.* **55**, 389.

Jurd, L. (1962). In "Chemistry of the Flavonoid Compounds" (T. A. Geissman, ed.), pp. 107–155. Pergamon Press, Oxford.
Jurd, L. (1964). Fd. Techn. 18, 157.
Jurd, L., Geissman, T. A. and Seikel, M. K. (1957). Archs Biochem. Biophys. 67, 284.
Jurd, L. and Horowitz, R. M. (1961). J. org. Chem. 26, 2561.
Kaizmarck, F. and Ostrowska, B. (1962). Chem. Abstr. 59, 2756.
Kanao, M. and Shimokoriyama, M. (1949). Acta Phytochim. Japan 15, 229.
Karrer, P. and de Meuron, G. (1932). Helv. chim. Acta 15, 507, 1212.
Karrer, P. and Schwab, G. (1941). Helv. chim. Acta 24, 297.
Karrer, P. and Widmer, R. (1927). Helv. chim. Acta 10, 5, 67, 758.
Karrer, P. and Widmer, R. (1928). Helv. chim. Acta 11, 837.
Karrer, P. and Widmer, R. (1929). Helv. chim. Acta 12, 292.
Karrer, W. (1958). "Konstitution und Vorkommen der organischen Pflanzenstoffe", p. 673. Birkhauser Verlag, Basel.
Karsten, U. (1965). Naturwissenschaften 52, 83.
Kataoka, T. (1936). Acta Phytochim. Japan 9, 35.
Kawaguchi, R., Kim, K. and Matsushita, K. (1939). J. Pharm. Soc. Japan 59, 44.
Kawano, N., Miura, H. and Waiss, A. C. (1964). Chemy Ind. 2020.
Kawase, Y. and Shibata, M. (1963). Bot. Mag. Tokyo 76, 89.
Khanna, R. N. and Seshadri, T. R. (1963). Tetrahedron 19, 219.
Kiang, A. K., Sim, K. Y. and Goh, J. (1965). J. chem. Soc. 6371.
King, F. E. and Acheson, R. M. (1950). J. chem. Soc. 168.
King, F. E. and King, T. J. (1951). J. chem. Soc. 569.
King, F. E., King, T. J. and Neill, K. G. (1953). J. chem. Soc. 1055.
King, F. E., King, T. J. and Rustidge, D. W. (1962). Chem. Abstr. 57, 764.
King, F. E., King, T. J. and Sellars, K. (1952). J. chem. Soc. 92.
King, F. E., King, T. J. and Stokes, P. J. (1954). J. chem. Soc. 4587.
King, H. G. C. (1962). J. Sci. Fd. Agric. 27, 446.
King, H. G. C. (1966). Nature, Lond. 211, 944.
King, H. G. C. and White, T. (1957). J. chem. Soc. 3901.
King, H. G. C. and White, T. (1961). J. chem. Soc. 3538.
Kirby, K. S. and White, T. (1955). Biochem. J. 60, 582.
Kishimoto, Y. (1956). Chem. Abstr. 50, 13894.
Kjaer, A. (1963). In "Chemical Plant Taxonomy" (T. Swain, ed.). Academic Press, London and New York.
Klein, O. (1922). Ber. Akad. Wiss. Wien. I, 131, 221.
Klein, A. O. and Hagen, C. W. (1961). Plant Physiol. 36, 1.
Kobayashi, K. and Hayashi, K. (1952). J. Pharm. Soc. Japan, 72, 3.
Koeppen, B. H. and Basson, D. S. (1966). Phytochemistry 5, 183.
Koeppen, B. H. and Roux, D. G. (1965). Tetrahedron Letters, 3497.
Koeppen, B. H., Smit, C. J. B. and Roux, D. G. (1962). Biochem. J. 83, 507.
Kohlmuenzer, S. (1965). Dissnes pharm., Warz. 17, 357.
Kolos-Pethes, E. (1965). Acta Pharm. Hung. 35, 225.
Komatsu, M. and Tomimori, T. (1966). Tetrahedron Letters, 1611.
Komissarenko, N. F., Chernobai, V. T. and Kolesnikov, D. G. (1964). Dokl. Akad. Nauk. SSSR. 158, 904.
Komissarenko, N. F., Chernobai, V. T. and Kolesnikov, D. G. (1965). Chem. Abstr. 62, 14987.
Kondo, T. and Ito, H. (1956). Chem. Abstr. 50, 7959.
Kondo, T., Imamura, H. and Suda, M. (1959). J. Pharm. Soc. Japan 79, 1298.
Koukol, J. and Conn, E. E. (1961). J. biol. Chem. 236, 2692.
Kranen-Fiedler, U. (1956), Arzneimittel-Forsch. 6, 475.
Krebs, K. G. and Matern, J. (1957). Naturwissenschaften 44, 422.

Krebs, K. G. and Matern, J. (1958). *Arch. Pharm., Berl.* **291**, 163.

Krishnamoorthy, V. and Seshadri, T. R. (1962). *J. Sci. Indian Res.* **21b**, 561.

Krishnamoorthy, V., Krishnaswamy, N. R. and Seshadri, T. R. (1963). *Curr. Sci.* **32**, 16.

Krishnamurti, M., Ramanathan, J. D., Seshadri, T. R. and Shankaran, P. R. (1965). *Indian J. Chem.* **3**, 270.

Krishnamurty, H. G., Krishnamoorthy, V. and Seshadri, T. R. (1963). *Phytochemistry* **2**, 47.

Krogmann, D. W. and Stiller, M. L. (1962). *Biochem. biophys. Res. Commun.* **7**, 46.

Krug, H. and Borkowski, B. (1965). *Pharmazie* **20**, 692.

Krug, H. and Olechnowiez-Stepien, W. (1965). *Dissnes Pharm., Warz.* **17**, 213.

Kubista, V. (1950). *Experientia* **6**, 100.

Kubota, T. and Hase, T. (1956). *J. chem. Soc. Japan* **77**, 1059.

Kuhn, R. and Low, I. (1944). *Chem. Ber.* **77b**, 196.

Kuroda, C. and Wada, M. (1935). *Proc. Imp. Acad. Tokyo* **11**, 189.

Kuwada, H. (1964). *Chem. Abstr.* **61**, 7359.

Kuwatsuka, S. and Oshima, Y. (1964). *Nippon NogeiKagaku Kaishi* **38**, 351.

Lamberton, J. A. (1964). *Aust. J. Chem.* **17**, 692.

Lamprecht, H. (1957). *Agric. hort. genet., Landskrona* **15**, 155.

Latour and de la Source, M. (1877). *Bull. Soc. Chim. Paris* **228**, 337.

Lawrence, G. H. M. (1951). "Taxonomy of Vascular Plants." Macmillan, New York.

Lawrence, W. J. C. (1957). *Heredity* **11**, 337.

Lawrence, W. J. C., Price, J. R., Robinson, G. M. and Robinson, R. (1938). *Biochem. J.* **32**, 1661.

Lawrence, W. J. C., Price, J. R., Robinson, G. M. and Robinson, R. (1939). *Phil. Trans. R. Soc.* **230**, 149.

Lawrence, W. J. C. and Scott-Moncrieff, R. (1935). *J. Genet.* **30**, 155.

Lawrence, W. J. C. and Sturgess, V. C. (1957). *Heredity* **11**, 303.

Lee, K., Huang, F. and Yang, T. (1961). *J. Taiwan Pharm. Assoc.* **13**, 25.

Lee, H. H. and Tan, C. H. (1965). *J. chem. Soc.* 2743.

Legro, R. A. H. (1963). *J. R. hort. Soc.* **88**, 13.

Lele, S. S. (1959). *J. Sci. Indian Res.* **18b**, 243.

Leon, A., Robertson, A. and Robinson, R. (1931). *J. chem. Soc.* 2672.

Lewis, Y. S. and Johar, D. S. (1956). *Curr. Sci.* **25**, 325.

Lewis, Y. S. and Neelakantan, S. (1964). *Curr. Sci.* **15**, 460.

Li, K. C. and Wagenknecht, A. C. (1956). *J. Am. chem. Soc.* **78**, 1979.

Li, K. C. and Wagenknecht, A. C. (1958). *Nature, Lond.* **182**, 657.

Litvinenko, V. I. (1964). *Chem. Abstr.* **60**, 6700.

Litvinenko, V. I. and Sergienko, T. A. (1965). *Chem. Abstr.* **63**, 10233.

Livingston, A. L. and Bickoff, E. M. (1964). *J. Pharm. Sci.* **53**, 1557.

Mabry, T. J. (1966). *In* "Comparative Phytochemistry" (T. Swain, ed.), pp. 231–244. Academic Press, London and New York.

Mabry, T. J., Kagan, J. and Rosler, H. (1965). *Phytochemistry* **4**, 177, 487.

Mabry, T. J., Taylor, A. and Turner, B. L. (1963). *Phytochemistry* **2**, 61.

Mabry, T. J., Wyler, H., Sassou, G., Mercier, M., Perikh, T. and Dreiding, A. S. (1962). *Helv. chim. Acta.* **45** 640.

McClure, J. W. and Alston, R. E. (1964). *Nature, Lond.* **201**, 311.

McKay, M. B. and Warner, M. F. (1933). *Nat. Hort. Mag.* **178**.

McLeod, A. M. and MacCorquodale, H. (1958). *New Phytol.* **57**, 168.

McMurry, T. B. H. and Theng, C. Y. (1960). *J. chem. Soc.* 1491.

Mahesh, V. B. and Seshadri, T. R. (1954). *J. Sci. Indian Res.* **13B**, 835.

Mainx, F. (1923). *Lotos* **71**, 183.

Maksyntina, N. P. (1965). *Chem. Abstr.* **63**, 1858.

Malhotra, A., Murti, V. V. S. and Seshadri, T. R. (1965). *Tetrahedron Letters* 3191.
Manning, A. (1956). *New Biol.* 21, 59.
Marini-Bettolo, G. B., Deulofeu, V. and Hug, E. (1950). *Gazz. Chim. ital* 80, 63.
Marini-Bettolo, G. B., Chiavarelli, S. and Casinovi, C. G. (1957). *Gazz. Chim. ital.* 87, 1185.
Markakis, P. (1960). *Nature, Lond.* 187, 1092.
Markh, A. T. and Feldman, A. L. (1950). *Biokhimiya* 15, 230.
Marks, G. E. (1958). *New Phytol.* 57, 300.
Marks, G. E. (1965). *New Phytol.* 64, 293.
Marks, G. E., McKee, R. K. and Harborne, J. B. (1965). *Nature, Lond.* 208, 359.
Maroto, A. L. (1950). *Chem. Abstr.* 46, 581.
Marquart, L. C. (1835). "Die Farben der Bluthen, eine chemisch-physiologische Abhandlung." Bonn.
Marrian, G. F. and Haslewood, G. A. D. (1932). *Biochem. J.* 26, 1227.
Marsh, C. A. (1955). *Biochem. J.* 59, 58.
Masquelier, J. and Michaud, J. (1958). *J. Bull. Soc. Pharm. Bordeaux* 97, 77.
Masquelier, J. and Ricci, R. (1962). 1° Sym. "Les methodes d'analyse des aliments." Bordeaux.
Masquelier, J. (1959). In "Pharmacology of Plant Phenolics" (J. W. Fairbairn, ed.), pp. 123–131. Academic Press, London and New York.
Matsuno, T. (1958). *J. Pharm. Soc. Japan* 78, 1311.
Matsuno, T. and Amano, Y. (1962). *Kyoto Yakka Yaigaku Gakku* 10, 17.
Matsuno, T. and Shintano, A. (1962). *Kyoto Yakku Yaigaku Gakku* 10, 19.
Matsuno, T., Amano, Y., Shintano, A. and Umeda, H. (1963). *Kyoto Yakku Yaigaku Gakku* 11, 15.
Matsushita, A. and Iseda, S. (1965). *Nippon NogeiKagaku Kaishi.* 39, 317.
Maxyutin, N. P. and Litvinenko, V. I. (1964). *Dokl. Akad. Nauk. SSSR.* 154, 1123.
Mazur, L. Y. and Meisels, A. (1955). *Bull. Res. Coun. Israel* 5A, 67.
Mehta, C. R. and Mehta, T. P. (1954). *Chem. Abstr.* 48, 7008.
Melchert, T. E. and Alston, R. E. (1965). *Science, N.Y.* 150, 11701.
Meier, H. and Zenk, M. H. (1965). *Z. Pflanzenphysiol.* 53, 415.
Meier, W. and Furst, A. (1962). *Helv. chim. Acta* 45, 232.
Merz, J. W. and Wu, Y. H. (1936). *Arch. Pharm.* 274, 126.
Metche, M. and Gay, R. (1964). *Branntweinwirtschaft* 17, 347.
Metche, M. and Urion, E. (1961). *C. r. hebd. Séanc. Acad. Sci., Paris* 252, 356.
Meyer, C. (1964). *Z. VererbLehre* 95, 171.
Minaeva, V. G. and Volkhonskaya, T. A. (1964). *Dokl. Akad. Nauk. SSSR* 154, 956.
Mittal, O. P. and Seshadri, T. R. (1956). *J. Chem. Soc.* 2176.
Mitsui, S., Hayashi, K. and Hattori, S. (1959). *Proc. Japan. Acad.* 35, 169.
Moewus, F. (1950). *Biol. Zbl.* 69, 181.
Mohr, H. (1962). *A. Rev. Pl. Physiol.* 13, 465.
Moldenhauer, F. (1856). *Liebig's Ann.* 100, 180.
Molisch, H. (1911). *Ber. dt. Bot. Ges.* 29, 487.
Mongkolsuk, S. and Dean, F. M. (1964). *J. chem Soc.* 4654.
Mongkolsuk, S., Dean, F. M. and Houghton, F. M. (1966). *J. chem. Soc.* (C) 125.
Morita, N. (1957). *J. Pharm. Soc. Japan* 77, 31.
Morita, N. (1960). *Chem. Pharm. Bull Tokyo* 8, 59.
Morita, N. and Shimizu, M. (1963). *Chem. Abstr.* 59, 15374.
Morita, N., Shimizu, M. and Fukuta, M. (1965). *Yakugaku Zasshi* 85, 374.
Morot, F. S. (1849). *Annls Sci. Nat.* 13, 160.
Morris, S. J. and Thomson, R. H. (1963). *J. Insect Physiol.* 9, 391.
Morris, S. J. and Thomson, R. H. (1964). *J. Insect Physiol.* 10, 377.
Muller, H. (1915). *J. chem. Soc.* 107, 872.

Mullick, D. B., Faris, D. G., Brink, V. C. and Acheson, R. M. (1958). *Can. J. Pl. Sci.* **38**, 445.

Mumford, F. E., Smith, D. H. and Castle, J. E. (1961). *Pl. Physiol. Lancaster* **36**, 752.

Murti, V. V. S., Raman, P. V. and Seshadri, T. R. (1964). *Tetrahedron Letters*, 2995.

Murti, V. V. S., Raman, P. V., Seshadri, T. R. and Thakur, R. S. (1962). *J. Sci. Indian Res.* **21B**, 80.

Nagai, W. and Hattori, S. (1930). *Acta Phytochim. Japan* **5**, 1.

Nagai, I., Suzushino, G. and Suzuki, Y. (1960). *Japan. J. Breed.* **10**, 247.

Nagarajan, G. R. and Seshadri, T. R. (1964). *Phytochemistry* **3**, 477.

Nagasi, M. (1942). *Chem. Abstr.* **36**, 3625.

Nair, A. G. R., Subramanian, S. S. and Sridharan, K. (1963). *Curr. Sci.* **32**, 115.

Nair, A. G. R., Nagarajan, S. and Subramanian, S. S. (1964). *Curr. Sci.* **33**, 14, 431.

Nair, A. G. R., Nagarajan, S. and Subramanian, S. S. (1965). *Curr. Sci.* **34**, 79.

Nair, A. G. R. and Subramanian, S. S. (1962). *Curr. Sci.* **31**, 155, 504.

Nair, A. G. R. and Subramanian, S. S. (1964). *Curr. Sci.* **33**, 211.

Nakabayashi, T. (1952). *J. Agric. chem. Soc. Japan* **26**, 140, 331.

Nakabayashi, T. (1953). *J. Agric. chem. Soc. Japan* **27**, 469.

Nakabayashi, T. (1955). *Bull. Agric. chem. Soc. Japan* **19**, 104.

Nakabayashi, T. (1956). *Chem. Abstr.* **50**, 14185.

Nakabayashi, T. (1957). *Chem. Abstr.* **51**, 8905.

Nakamura, H. and Hukuti, G. (1940). *Chem. Abstr.* **34**, 7910.

Nakamura, H., Ohta, T. and Hukuti, G. (1936). *J. Pharm. Soc. Japan* **56**, 531.

Nakaoki, T. (1936). *Chem. Abstr.* **30**, 725.

Nakaoki, T. (1941). *Chem. Zbl.* II, 892.

Nakaoki, T. (1944). *J. Pharm. Soc. Japan* **64** (11A), 57.

Nakaoki, T., Morita, N. and Asaki, M. (1961). *J. pharm. Soc. Japan* **81**, 1697.

Nakaoki, T. and Morita, N. (1955). *J. Pharm. Soc. Japan* **75**, 112, 173.

Nakaoki, T. and Morita, N. (1956a). *Chem. Abstr.* **50**, 9687.

Nakaoki, T. and Morita, N. (1956b). *J. Pharm. Soc. Japan* **76**, 320, 350.

Nakaoki, T. and Morita, N. (1957). *J. Pharm. Soc. Japan* **77**, 108, 110.

Nakaoki, T. and Morita, N. (1958). *J. Pharm. Soc. Japan* **78**, 521, 558.

Nakaoki, T. and Morita, N. (1959). *J. Pharm. Soc. Japan* **79**, 1338.

Nakaoki, T. and Morita, N. (1960). *J. Pharm. Soc. Japan* **80**, 1298–9.

Nakaoki, T., Morita, N., Hiraki, A. and Kurokawa, Y. (1956). *J. Pharm. Soc. Japan* **76**, 347.

Nakaoki, T., Morita, N. and Isetani, A. (1961). *J. Pharm. Soc. Japan* **81**, 558, 1158.

Nakaoki, T., Morita, N. and Yoshida, Y. (1957). *J. Pharm. Soc. Japan* **77**, 112.

Neelakantan, K. and Seshadri, T. R. (1939). *Proc. Indian Acad. Sci.* **9**A, 365.

Neelakantan, K., Rao, P. S. and Seshadri, T. R. (1943). *Proc. Indian Acad. Sci.* **17**A, 26.

Needham, A. E. (1966). *Life Sci.* **5**, 33.

Noguchi, M. (1958). *Chem. Abstr.* **52**, 13998.

Neish, A. C. (1961). *Phytochemistry* **1**, 1.

Neish, A. C. (1964). *In* "Biochemistry of Phenolic Compounds" (J. B. Harborne, ed.), pp. 295–360. Academic Press, London and New York.

Newton, R. and Anderson, J. A. (1929). *Can. J. Res.* **1**, 86.

Neyland, M., Ng, Y. L. and Thimann, K. V. (1963). *Pl. Physiol. Lancaster* **38**, 44.

Ng, Y. L. and Thimann, K. V. (1962). *Archs Biochem. Biophys.* **96**, 336.

Nilsson, M. (1961). *Acta chem. scand.* **15**, 154, 211.

Nishida, K. and Funaoka, K. (1956). *Chem. Abstr.* **50**, 14729.

Nitsch, J. B. and Nitsch, C. (1962). *Ann. Physiol. Vegetale* **4**, 211.

Nordström, C. G. (1956). *Acta chem. scand.* **10**, 491.

Nordström, C. G. and Swain, T. (1953). *J. chem. Soc.* 2764.

Nordström, C. G. and Swain, T. (1956). *Archs Biochem. Biophys.* **60**, 329.

Nortje, B. K. and Koeppen, B. H. (1965). *Biochem. J.* **97**, 209.

Nybom, N. (1964). *Physiol. Plant.* **17**, 157.

Nystrom, C. W., Howard, W. L. and Wender, S. H. (1957). *J. org. Chem.* **22**, 1272.

Obara, H. (1964). *J. chem. Soc. Japan* **85**, 514.

Obdulio, F. and Lobete, M. P. (1944). *Chem. Zbl.* **I**, 32.

Ockendon, D. J., Alston, R. E. and Naifeh, K. (1966). *Phytochemistry* **5**, 601.

Oesterle, O. A. and Wander, G. (1925). *Helv. chim. Acta* **8**, 519.

Ohta, T. (1940). *Z. Physiol.* **263**, 221.

Ohta, T. and Miyazaki, T. (1956). *Jap. J. Pharmac.* **10**, 7.

Ohta, T. and Miyazaki, T. (1958). *A. Rep. Tokyo Coll. Pharm.* **8**, 151.

Ohta, T. and Miyazaki, T. (1959). *J. Pharm. Soc. Japan* **76**, 323.

Ohta, T. Miyazaki, T. and Mihashi, S. (1960). *Chem. Pharm. Bull. Tokyo* **8**, 647.

Oku, M. (1934). *J. Agric. Chem. Soc. Japan,* **10**, 1029.

Olechowska-Baranska, K. and Lamer, E. (1962). *Acta polon. Pharm.* **19**, 199.

Ollis, W. D. (1962). *In* "The Chemistry of Flavonoid Compounds" (T. A. Geissman, ed.), pp. 352–405. Pergamon Press, Oxford.

Onslow, M. W. (1925). "The Anthocyanin Pigments of Plants", 2nd Ed. Cambridge University Press.

Ootani, S. and Miura, T. (1961). *Inst. Breed. Res. Tokyo Univ. Agric. Bull.* **2**, 16.

Ouchi, K. (1953). *Misc. Rep. Res. Inst. Nat. Resour. Tokyo* **32**, 1.

Oshima, Y. and Nakabayashi, T. (1954). *J. agric. Chem. Soc. Japan* **27**, 754.

Pacheco, H. (1955). *Bull. Soc. chim. Biol.* **37**, 723.

Pacheco, H. (1966). *In* "Actualites de Phytochimie Fondamentale" (C. Mentzer, ed.), Vol. II, pp. 73–84. Masson et Cie, Paris.

Paech, K. (1954). *A. Rev. Pl. Physiol.* **6**, 273.

Pakudina, Z. P. and Sodykov, A. S. (1963). *Chem. Abstr.* **62**, 9457.

Pallares, E. S. and Garza, H. M. (1949). *Arch. Biochem.* **21**, 377.

Palmer, L. S. and Knight, H. H. (1924). *J. Biol. Chem.* **59**, 451.

Paris, C. D., Haney, W. J. and Wilson, G. B. (1960). *Techn. Bull.* 281. *Agric. Exp. Station*, Michigan State Univ.

Paris, R. R. (1953). *Bull. Soc. chim. biol.* **35**, 655.

Paris, R. R. (1957). *C. r. hebd. Séanc. Acad. Sci., Paris* **245**, 443.

Paris, R. R. and Charles, A. (1962). *C. r. hebd. Séanc. Acad. Sci., Paris* **254**, 325.

Paris, R. R. and Delaveau, P. (1961). *C. r. hebd. Séanc. Acad. Sci., Paris* **252**, 1510.

Paris, R. R. and Delaveau, P. (1962). *C. r. hebd. Séanc. Acad. Sci., Paris* **254**, 928; *Lloydia* **25**, 151.

Paris, R. R. and Etchepare, S. (1965). *Ann. Pharm. Franc.* **23**, 627.

Paris, R. R. and Foucaud, A. (1959). *C. r. hebd. Séanc. Acad. Sci., Paris* **248**, 2634.

Paris, R. R. and Fougeras, G. (1965). *C. r. hebd. Séanc. Acad. Sci., Paris* **261**, 1761.

Paris, R. R. and Paris, M. (1964). *C. r. hebd. Séanc. Acad. Sci., Paris* **258**, 361.

Parks, C. R. (1965). *Am. J. Bot.* **52**, 309.

Parry, W. D. and Smithson, F. (1964). *Ann. Bot. N.S.* **28**, 169.

Parthasarathy, M. R. and Seshadri, T. R. (1964). *Indian J. Chem.* **2**, 130.

Partridge, S. M. and Westall, R. G. (1948). *Biochem. J.* **42**, 249, 251.

Patschke, L., Barz, W. and Grisebach, H. (1966). *Z. Naturf.* **21b**, 201.

Patschke, L., Hess, D. and Grisebach, H. (1964). *Z. Naturf.* **19b**, 1114.

Patschke, L. and Grisebach, H. (1965). *Z. Naturf.* **20b**, 1039.

Paul, H. (1908). *Mitt. bayer. bot. Ges. (MoorkAnst)* **2**, 63.

Payne, N. M. (1931). *J. mar. biol. Ass. U.K.* **17**, 742.

Pearl, I. A. and Darling, S. F. (1963). *J. org. Chem.* **28**, 1442.

Pecket, R. C. (1960). *New Phytol.* **59**, 138.

Pelter, A. and Johnson, A. P. (1966). *J. chem. Soc.* (C), 701.

Perkin, A. G. (1900). *J. chem. Soc.* **77**, 422.

Perkin, A. G. (1902). *J. chem. Soc.* **81**, 478, 479.

Perkin, A. G. (1910). *J. chem. Soc.* **97**, 1776.

Perkin, A. G. and Everest, A. E. (1918). "The Natural Organic Colouring Matters." Longmans, Green and Co., London.

Perkin, A. G. and Hummel, J. J. (1896). *Chem. News.* **74**, 278.

Perrin, D. R. and Bottomley, W. (1962). *J. Am. chem. Soc.* **84**, 1919.

Petersen, G. E., Livesay, R. and Futch, H. (1961). *Chem. Abstr.* **55**, 17738.

Pew, J. C. (1948). *J. Am. chem. Soc.* **70**, 3031.

Phillips, D. J. (1961). *Nature, Lond.* **192**, 240.

Piattelli, M. and Minale, L. (1964). *Phytochemistry* **3**, 547.

Piattelli, M., Minale, L. and Proto, G. (1964). *Ann. Chim. (Rome)* **50**, 9638.

Piattelli, M., Minale, L. and Nicolaus, R. A. (1965). *Phytochemistry*, **4**, 817.

Piattelli, M. and Nicolaus, R. A. (1965). *Rc. Accad. Sci. Fis. mat., Napoli* **4**, 32.

Pijl, L. van der (1961). *Evolution* **15**, 44.

Power, F. B. and Tutin, F. (1907). *J. chem. Soc.* **91**, 887.

Powers, J. J., Somaatmadja, D., Pratt, D. E. and Hamoy, M. K. (1960). *Fd. Techn.* **14**, 626.

Prakash, L., Zaman, A. and Kidwai, A. R. (1965). *J. org. Chem.* **30**, 3561.

Prakken, R. (1940). *Genetica* **22**, 331.

Pramer, D. and Wright, J. M. (1955). *Pl. Dis. Reptr.* **39**, 118.

Prat, H. (1962). *Bull. Soc. bot. Fr.* **107**, 32.

Price, J. R. (1963). *In* "Chemical Plant Taxonomy" (T. Swain, ed.), pp. 429–452. Academic Press, London and New York.

Price, J. R. and Robinson, R. (1937). *J. chem. Soc.* 449.

Price, J. R. and Sturgess, V. C. (1938). *Biochem. J.* **32**, 1658.

Price, J. R., Sturgess, V. C., Robinson, G. M. and Robinson, R. (1938). *Nature, Lond.* **142**, 356.

Price, J. R., Robinson, G. M. and Robinson, R. (1938). *J. chem. Soc.* 281.

Price, J. R., Robinson, R. and Scott-Moncrieff, R. (1939). *J. chem. Soc.* 1465.

Pridham, J. B. (1965). *Adv. Carbohyd. Chem.* **20**, 371.

Pruthi, J. S., Susheela, R. and Lal, G. (1961). *Fd. Res.* **26**, 385.

Puri, B. and Seshadri, T. R. (1955). *J. chem. Soc.* 1589.

Puski, C. and Francis, F. J. (1966). *Fd. Res.* (In press.)

Rabate, J. (1928). *Bull. Soc. Pharm.* **35**, 70.

Rabate, J. (1930). *Bull. Soc. chim. biol.* **12**, 974.

Rabate, J. (1933). *Bull. Soc. chim. biol.* **15**, 130.

Rabate, J. (1938). *J. Pharm. chim.* **8**, 28,478.

Rabate, J. and Dussy, J. (1938). *Bull. Soc. chim. biol.* **20**, 459.

Radhakrishnan, P. V., Rao, A. V. R. and Venkataraman, K. (1965). *Tetrahedron Letters*, 663.

Radner, B. S. and Thimann, K. V. (1963). *Archs Biochem. Biophys.* **102**, 92.

Rahman, W. and Ilyas, M. (1962). *J. org. Chem.* **27**, 153.

Rahman, A. U. and Khan, M. S. (1962). *Z. Naturf.* **17b**, 9.

Rahman, W., Ilyas, M. and Khan, A. W. (1963). *Naturwissenschaften* **50**, 477.

Ralha, A. J. C. (1954). *Chem. Abstr.* **48**, 2696.

Rangaswami, S. and Rao, K. H. (1959). *Proc. Indian Acad. Sci.* **49A**, 241.

Rangaswami, S. and Sambamurthy, K. (1960). *Proc. Indian Acad. Sci.* **51A**, 322.

Rao, C. B. and Venkateswarlu, V. (1962). *J. Sci. Indian Res.* **21B**, 313.

Rao, P. S. (1965). *Naturwissenschaften* **34**, 262.

Rao, P. S. and Seshadri, T. R. (1939). *Proc. Indian Acad. Sci.* **9A**, 177, 365.

Rao, P. S. and Seshadri, T. R. (1941). *Proc. Indian Acad. Sci.* **14A**, 265, 643.

Rao, P. S. and Seshadri, T. R. (1942). *Proc. Indian Acad. Sci.* **15A**, 148; **16A**, 323.

Rapson, W. S. (1938). *J. chem. Soc.* 282.

Rauh, W. and Reznik, H. (1961). *Bot. Jb.* **81**, 94.

Rayndu, G. V. N. and Rajadurai, S. (1965). *Leather Sci.* **12**, 301.
Reddy, G. M. (1963). *Maize Genet. Coop News Lett.* **37**, 14.
Reddy, G. M. and Coe, E. H. (1962). *Science N.Y.* **138**, 149.
Reddy, G. M., Katsumi, M. and Phinney, B. O. (1963). *Maize Genet. Coop News Lett.* **37**, 14.
Reichel, L. and Reichwald, W. (1960). *Naturwissenschaften* **47**, 41.
Reichel, L., Stroh, H. and Reichwald, W. (1957). *Naturwissenschaften* **44**, 468.
Reznik, H. (1956). *Sbr. heidelb. Akad. Wiss.* 125.
Reznik, H. (1957). *Planta* **49**, 406.
Reznik, H. (1961). *Flora* **150**, 454.
Reznik, H. and Neuhausel, R. (1959). *Z. Bot.* **47**, 471.
Ribereau-Gayon, P. (1959). *Rev. gen. Bot.* **66**, 581.
Ribereau-Gayon, P. (1964). "Les Composés Phenoliques du Raisin et du Vin". Institut National de la Recherche Agronomique, Paris.
Rimpler, H. and Hansel, R. (1965). *Arch. Pharm., Berl.* **298**, 838.
Rimpler, H., Langhammer, L. and Frenzel, H. J. (1963). *Planta med.* **11**, 325.
Ritchie, E., Taylor, W. C. and Vautin, S. T. K. (1965). *Aust. J. Chem.* **18**, 2021.
Roberts, E. A. H. (1962). *In* "Chemistry of the Flavonoid Compounds" (T. A. Geissman, ed.), pp. 468–512. Pergamon Press, Oxford.
Roberts, E. A. H., Cartwright, R. A. and Wood, D. J. (1956). *J. Sci. Fd. Agric.* **7**, 637.
Roberts, E. A. H. and Williams, D. M. (1958). *J. Sci. Fd. Agric.* **9**, 217.
Robinson, G. M. and Robinson, R. (1931). *Biochem. J.* **25**, 1687.
Robinson, G. M. and Robinson, R. (1932). *Biochem. J.* **26**, 1647.
Robinson, G. M. and Robinson, R. (1933). *Biochem. J.* **27**, 206.
Robinson, G. M. and Robinson, R. (1934). *Biochem. J.* **28**, 1712.
Robinson, G. M., Robinson, R. and Todd, A. R. (1934). *J. chem. Soc.* 809.
Robinson, R. (1934). *Ber. dt. chem. Ges.* **67A**, 85.
Robinson, R. (1935). *Nature, Lond.* **135**, 732.
Robinson, R. and Smith, H. (1955). *Nature, Lond.* **175**, 634.
Rochleder, F. and Hlasiwetz, H. (1852). *Liebig's Ann.* **82**, 196.
Roller, K. (1956). *Z. Bot.* **44**, 477.
Roots, B. I. and Johnston, P. V. (1966). *Comp. Biochem. Biophys.* **17**, 285.
Rosler, H., Mabry, T. J., Cranmer, M. F. and Kagan, J. (1965). *J. org. Chem.* **30**, 4346.
Rosler, H., Rosler, V., Mabry, T. J. and Kagan, J. (1966). *Phytochemistry* **5**, 189.
Roux, D. G. and de Bruyn, G. C. (1963). *Biochem. J.* **87**, 439.
Roux, D. G. and Paulus, E. (1962). *Biochem. J.* **82**, 324.
Row, L. R. and Viswanadham, N. (1954). *Proc. Indian Acad. Sci.* **39A**, 240.
Rowlands, D. G. and Corner, J. J. (1962). *Proc. 3rd. Congress Eucarpia (Paris)*, 229.
Rowley, G. D. (1957). *J. R. hort. Soc.* **82**, 484.
Rudolph, H. (1965). *Planta* **64**, 178.
Ryan, F. J. (1955). *Science N.Y.* **122**, 470.
Rzadkowska-Bodalska, H. and Bodalski, T. (1965). *Chem. Abstr.* **63**, 7346.
Sachindrak, D. S., and Chakraborti, T. B. (1958). *Trans. Bose. Res. Inst.* **21**, 61.
Sakamoto, S. (1956). *Chem. Abstr.* **50**, 15619.
Sakamura, S. and Francis, F. J. (1961). *Fd. Res.* **26**, 318.
Sambamurthy, K., Rangaswami, S. and Verraswamy, P. (1962). *Planta Med.* **10**, 173.
Sando, C. E. (1926). *J. biol. chem.* **64**, 74.
Sando, C. E. (1937). *J. biol. chem.* **117**, 45.
Sando, C. E. and Bartlett, H. H. (1920). *J. biol. Chem.* **41**, 295.

Sando, C. E. and Lloyd, J. U. (1924). *J. biol. chem.* **58**, 737.
Sannie, C. and Sauvain, H. (1952). *Mém. Mus. natn. Hist. nat.*, *Paris* serie B, Botanique **2**, 1.
Santavy, F., Potesilova, H. and Kubicek, R. (1957). *Chem. Listy* **51**, 1767.
Santamour, F. S. (1965). *Morris Arboretum Bull.* **16**, 43.
Santamour, F. S. (1966). *Forest Science* (In press.)
Sasaki, T. and Mikami, M. (1963). *Yakugaku Zasshi* **83**, 879.
Sasaki, T. and Watanabe, Y. (1956). *J. Pharm. Soc. Japan* **76**, 1893.
Sawada, T. (1958). *J. Pharm. Soc. Japan* **78**, 1023.
Sawhney, P. L. and Seshadri, T. R. (1956). *J. Sci. Indian Research* **15C**, 154.
Scarpati, M. L. and Oriente, G. (1958). *Tetrahedron* **4**, 43.
Scott-Moncrieff, R. (1930). *Biochem. J.* **24**, 753.
Scott-Moncrieff, R. (1936). *J. Genet.* **32**, 117.
Scott-Moncrieff, R. and Sturgess, V. C. (1940). *Biochem. J.* **34**, 268.
Schilcher, H. (1964). *Z. Naturf.* **19b**, 857.
Schindler, O. (1945). *Helv. Chim. Acta* **28**, 1157.
Schmidt, H. (1962). *Biol. Zbl.* **81**, 213.
Schunck, E. (1858). *Manchester Memoirs* **155** (2), 122.
Schwarze, P. (1959). *Planta* **54**, 152.
Seikel, M. K. (1955). *J. Am. chem. Soc.* **77**, 5685.
Seikel, M. K. and Bushnell, A. J. (1959). *J. org. Chem.* **24**, 1995.
Seikel, M. K., Bushnell, A. J. and Birzgalis, R. (1962). *Archs Biochem. Biophys.* **99**, 451.
Seikel, M. K., Chow, J. H. S. and Feldman, L. (1966). *Phytochemistry*, **5**, 439.
Sen, S., Siegelman, H. W. and Stuart, N. W. (1957). *Proc. Am. Soc. hort. Sci.* **69**, 561.
Seshadri, T. R. (1951). *Rev. Pure Appl. Chem.* **1**, 186.
Seshadri, T. R. (1956). *Sci. Proc. R. Dublin Soc.* **27**, 77.
Seshadri, T. R. and Thakur, R. S. (1960). *Curr. Sci.* **29**, 57.
Seshadri, T. R. and Vasishta, K. (1965). *Phytochemistry* **4**, 989.
Seshadri, T. R. and Venkateswarlu, V. (1946). *Proc. Indian Acad. Sci.* **23A**, 192.
Seyffert, W. (1955). *Z. indukt. Abstamm.-u-VererbLehre* **87**, 311.
Seyffert, W. (1959). *Naturwissenschaften* **46**, 271.
Seyffert, W. (1960). *Z. PflZucht.* **44**, 4.
Sharma, R. C., Zaman, A. and Kidwai, A. R. (1964). *Indian J. Chem.* **2**, 83.
Sherratt, H. S. A. (1958). *J. Genet.* **56**, 1.
Shibata, M. (1958). *Sci. Rep. Tohoku Univ.* **24**, 89.
Shibata, M. and Ishikura, N. (1960). *Jap. J. Bot.* **17**, 230.
Shibata, M. and Ishikura, N. (1964). *Bot. Mag. Tokyo* **77**, 277.
Shibata, M. and Nozaka, K. (1963). *Bot. Mag. Tokyo* **76**, 317.
Shibata, M. and Sakai, E. (1958). *Bot. Mag. Tokyo* **71**, 6, 193.
Shibata, M., Takakuwa, N. and Ishikura, N. (1962). *Bot. Mag. Tokyo* **75**, 413.
Shibata, S., Murata, T. and Fujita, M. (1963). *Chem. Pharm. Bull. Tokyo* **11**, 372.
Shimokoriyama, M. (1949). *Acta Phytochim. Japan* **16**, 63.
Shimokoriyama, M. (1957). *J. Am. chem. Soc.* **79**, 214.
Shimokoriyama, M. (1962). *In* "Chemistry of Flavonoid Compounds" (T. A. Geissman, ed.), pp. 286–316. Pergamon Press, Oxford.
Shimokoriyama, M. and Geissman, T. A. (1960). *J. org. Chem.* **25**, 1956.
Shimokoriyama, M. and Geissman, T. A. (1962). *In* "Chemistry of Natural and Synthetic Colouring Matters" (T. Gore *et al.*, eds.), p. 255. Academic Press, New York and London.
Shimokoriyama, M. and Hattori, S. (1953). *J. Am. chem. Soc.* **75**, 2277.
Shriner, R. L. and Anderson, R. J. (1928). *J. biol. Chem.* **80**, 743.
Sieburth, J. M. and Conover, J. T. (1965). *Nature, Lond.* **208**, 52.

COMPARATIVE BIOCHEMISTRY OF FLAVONOIDS

Siegelman, H. W. (1955). *J. biol. Chem.* **213**, 647.
Siegelman, H. W. (1964). *In* "Biochemistry of Phenolic Compounds" (J. B. Harborne, ed.), pp. 437–456. Academic Press, London and New York.
Siegelman, H. W. and Hendricks, S. B. (1957). *Pl. Physiol. Lancaster* **32**, 393.
Simmonds, N. W. (1954). *Nature, Lond.* **173**, 402.
Simmonds, N. W. and Harborne, J. B. (1965). *Heredity* **20**, 315.
Skrzypczakowa, L. (1964). *Chem. Abstr.* **60**, 16208.
Smith, A. C. (1943). *J. Arnold Arboretum* **24**, 119.
Smith, H. G. (1898). *J. chem. Soc.* **73**, 697.
Somaatmadja, D. and Powers, J. J. (1963). *Fd. Res.* **28**, 617.
Somers, T. C. (1965). *Nature, Lond.* **209**, 368.
Sondheimer, E. (1958). *Archs Biochem. Biophys.* **74**, 131.
Sondheimer, E. and Karash, C. B. (1956). *Nature, Lond.* **178**, 648.
Sondheimer, E. and Kertesz, Z. I. (1948). *J. Am. chem. Soc.* **70**, 3476.
Sondheimer, F. and Meisels, A. (1960). *Tetrahedron* **9**, 139.
Sosa, A. (1949). *Bull. Soc. Chim. Biol.* **31**, 57.
Sosa, A. and Sosa, C. (1966). *C. r. hebd. Séanc. Acad. Sci., Paris* **262**, 1144.
Sosa, F. and Percheron, F. (1965). *C. r. hebd. Séanc. Acad. Sci., Paris* **261**, 4544.
Spada, A. and Cameroni, R. (1956). *Gazz. Chim. ital.* **86**, 965.
Spada, A. and Cameroni, R. (1958). *Gazz. Chim. ital.* **88**, 204.
Spiridonov, V. N., Prokopenko, A. P. and Kolesnikov, D. G. (*Zh. obshch. Khim.* **34**, 4128.
Sprengel, C. K. (1793). "Das entdeckte Geheimnis der Natur im Bau und in der Befruchtung der Blumen."
Stafford, H. A. (1965). *Pl. Physiol. Lancaster* **40**, 130.
Stambouli, A. and Paris, R. R. (1961). *Annls Pharm. Franc.* **19**, 732.
Stanko, S. A. and Bardinskaya, M. S. (1962). *Dokl. Akad. Nauk. S.S.S.R.* **146**, 956.
Stearn, W. T. (1965). *J. R. hort. Soc.* **90**, 279.
Stearn, W. T. (1966). *Lloydia* **29**, 196.
Stebbins, G. L. (1959). *In* "Vistas in Botany" (W. B. Turrill, ed.), pp. 258–290. Pergamon Press, London.
Steinegger, E. and Sonanini, D. (1960a). *Pharmazie* **15**, 643.
Steinegger, E. and Sonanini, D. (1960b). *Pharm. Acta Helv.* **36**, 662.
Steinegger, E., Sonanini, D. and Tsingaridas, K. (1963). *Pharm. Acta Helv.* **38**, 119.
Stenlid, G. (1962). *Physiol. Pl.* **15**, 598.
Stenlid, G. (1963). *Physiol. Pl.* **16**, 110.
Stepien, W. and Krug, H. (1965). *Dissnes. Pharm., Warz.* **17**, 389.
Stout, G. H. and Stout, V. F. (1961). *Tetrahedron* **14**, 296.
Straus, J. (1959). *Pl. Physiol. Lancaster* **34**, 536.
Straus, J. (1960). *Pl. Physiol. Lancaster* **35**, 645.
Stroh, H. H. (1959). *Z. Naturf.* **14b**, 699.
Stroh, H. H. and Siedel, H. (1965) *Z. Naturf.* **20b**, 39.
Strohl, M. J. and Seikel, M. K. (1965). *Phytochemistry* **4**, 383.
Subramanian, S. S. (1963). *Curr. Sci.* **32**, 308.
Subramanian, S. S. and Nair, A. G. R. (1963). *Indian J. Chem.* **1**, 501.
Subramanian, S. S., Nair, A. G. R., and Nagarajan, S. (1965). *Curr. Sci.* **34**, 246.
Subramanian, S. S. and Swamy, M. N. (1964). *Curr. Sci.* **33**, 112.
Sugano, N. and Hayashi, K. (1960). *Bot. Mag. Tokyo* **73**, 231.
Suginome, H. (1959). *J. org. Chem.* **24**, 1655.
Suomalainen, H. and Keranen, A. J. A. (1961). *Nature, Lond.* **191**, 498.
Swain, T. (1962). *In* "Chemistry of Flavonoid Compounds" (T. A. Geissman, ed.), pp. 513–552. Pergamon Press, Oxford.
Swain, T. (ed.) (1963). "Chemical Plant Taxonomy." Academic Press, London and New York.

Swain, T. (1965). *In* "Biosynthetic Pathways in Higher Plants" (J. B. Pridham and T. Swain, eds.), pp. 9–36. Academic Press, London and New York.

Swain, T. (ed.) (1966). "Comparative Phytochemistry." Academic Press, London and New York.

Swain, T. and Bate-Smith, E. C. (1962). *In* "Comparative Biochemistry", Vol. III (M. Florkin and H. S. Mason, eds.), p. 755. Academic Press, New York and London.

Takahashi, M., Ito, T. and Mizutami, A. (1960). *J. Pharm. Soc. Japan* **80**, 1557.

Takino, Y., Ferretti, A., Flanagan, Y., Gianturco, M. and Vogel, M. (1965). *Tetrahedron Letters* 4019.

Tanret, C. and Tanret, G. (1899). *C. r. hebd. Séanc. Acad. Sci., Paris* **129**, 725.

Tappi, G. and Monzani, A. (1955). *Gazz. Chim. ital.* **85**, 732.

Tatsuta, H. (1957). *Chem. Abstr.* **51**, 5918.

Taylor, A. O. and Wong, E. (1965). *Tetrahedron Letters* 3675.

Taylor, T. W. J. (1940). *Proc. R. Soc.* B. **129**, 230.

Thimann, K. V. and Edmondson, Y. H. (1949). *Archs Biochem. Biophys.* **22**, 33.

Thimann, K. V. and Radner, B. S. (1955). *Archs Biochem. Biophys.* **58**, 484; **59**,511.

Thimann, K. V. and Radner, B. S. (1958). *Archs Biochem. Biophys.* **74**, 209.

Thomson, D. L. (1926). *Biochem. J.* **20**, 73.

Ting, S. V. (1958). *J. Agric. Fd. Chem.* **6**, 546.

Tominaga, T. (1956). *Chem. Abstr.* **50**, 9396.

Tseng, K. F. and Chen, C. H. (1961). *Chem. Abstr.* **55**, 24734.

Tunmann, R. and Isaac, O. (1957). *Arch. Pharm.* **290**, 37.

Tutin, F. (1913). *J. chem. Soc.* **103**, 2006.

Ueno, A. (1963). *Chem. Abstr.* **59**, 736.

Ueno, A., Oguri, N., Hori, K., Saiki, Y. and Harada, T. (1963). *Yakugaku Zasshi* **83**, 420.

Uphof, J. C. T. (1924). *Genetics* **9**, 292.

Valentine, J. and Wagner, G. (1953). *Pharm. Zentralhalle Dtl.* **92**, 354.

Vega, F. A. and Martin, C. (1962). *Nature, Lond.* **197**, 382.

Venkataraman, K. (1956). *Sci. Proc. R. Dublin. Soc.,* **27**, 93.

Voirin, B. and Lebreton, P. (1966). *C. r. hebd. Séanc. Acad. Sci., Paris* **262**, 707.

Volkhonskaya, T. A. and Minaeva, V. G. (1964). *Byull. glavn. Bot. Sada, Leningrad* **56**, 57.

Von Gerichten, E. (1901). *Liebig's Ann.* **318**, 121.

Vrkoc, J., Herout, V. and Sorm, F. (1959). *Coll. Czech. Chem. Commun.* **24**, 3938.

Vuataz, L., Brandenberger, H. and Egli, R. H. (1959). *J. Chromat.* **2**, 173.

Wada, M. (1950). *Misc. Rep. Res. Inst. Nat. Resour. Tokyo* **17–18**, 197.

Wada, E. (1952). *J. Agric. chem. Soc. Japan* **26**, 103.

Wada, M. (1956). *J. Am. chem. Soc.* **78**, 4725.

Wagner, H. (1965). *In* "Methods in Polyphenol Chemistry" (J. B. Pridham, ed.), pp. 37–48. Pergamon Press, Oxford.

Wagner, H., Horhammer, L. and Kirchner, W. (1960). *Arch. Pharm., Berl.* **293**, 1053.

Wagner, H., Horhammer, L. and Munster, R. (1965). *Naturwissenschaften* **52**, 305.

Wagner, R. (1850). *J. Prakt. Chem.* I, **51**, 82.

Wagner, J. (1964). *Z. Physiol. Chem.* **335**, 232.

Watanabe, R. and Wender, S. H. (1965). *Archs Biochem. Biophys.* **112**, 111.

Watkin, J. E. (1960). *Proc. Plant Phenolics Group N. America* (Fort Collins, Colo.), pp. 39–51.

Watkin, J. E., Magrill, D. S. and Steck, W. (1965). *Proc. Plant Phenolics Group N. America* (Albany, Calif.), p. 21.

Watson, W. C. (1964). *Biochem. J.* **90**, 3P.

Wawzonek, S. (1951). *In* "Heterocyclic Compounds", Vol. II (R. C. Elderfield, ed.), p. 329. John Wiley, New York.

Weinges, K. and Freudenberg, K. (1965). *Chemy Comm.* 220.

Weiss, A. (1842). *Chem. Zbl.* 305.

Weiss, U. and Altland, H. W. (1965). *Nature, Lond.* **207**, 1295.

Werckmeister, P. (1960). *Bull. Am. Iris Soc.* 25.

Wettstein, R. von (1935). "Handbuch der Systematischen Botanik", 2nd Ed. Leipzig.

Whalley, W. B. (1962). *In* "The Chemistry of the Flavonoid Compounds" (T. A. Geissman, ed.), pp. 441–467. Pergamon Press, Oxford.

Will, W. (1885). *Chem. Ber.* **18**, 1311.

Williams, A. H. (1964). *Chemy Ind.* 1318.

Williams, A. H. (1960). *In* "Phenolics in Plants in Health and Disease" (J. B. Pridham, ed.), pp. 3–7. Pergamon Press, Oxford.

Williams, A. H. (1966). *In* "Comparative Phytochemistry" (T. Swain, ed.), pp. 297–307. Academic Press, London and New York.

Williams, B. L. and Wender, S. H. (1953). *J. Am. chem. Soc.* **75**, 4363.

Williams, B. L. and Wender, S. H. (1952). *J. Am. chem. Soc.* **74**, 4566.

Williams, R. T. (1964). *In* "Biochemistry of Phenolic Compounds" (J. B. Harborne, ed.), pp. 205–248. Academic Press, London.

Willis, J. C. (1960). "A Dictionary of Flowering Plants and Ferns", 6th Ed. Cambridge University Press.

Willstater, R. and Bolton, E. K. (1914). *Liebig's Ann.* **408**, 42.

Willstater, R. and Bolton, E. K. (1916). *Liebig's Ann.* **412**, 136.

Willstater, R. and Burdick, C. L. (1916). *Liebig's Ann.* **412**, 149.

Willstater, R. and Burdick, C. L. (1917). *Liebig's Ann.* **412**, 217.

Willstater, R. and Everest, A. E. (1913). *Liebig's Ann.* **401**, 189.

Willstater, R. and Mieg, W. (1915). *Liebig's Ann.* **408**, 61, 122.

Willstater, R. and Nolan, T. J. (1915). *Liebig's Ann.* **408**, 1, 136.

Willstater, R. and Weil, F. (1917). *Liebig's Ann.* **412**, 231.

Willstater, R. and Zollinger, E. H. (1915). *Liebig's Ann.* **408**, 83.

Willstater, R. and Zollinger, E. H. (1917). *Liebig's Ann.* **412**, 195.

Wilson, R. F. (1938). "Horticultural Colour Chart." Roy. Hort. Sci., London.

Wolf, F. T. (1956). *Physiol. Plant.* **9**, 559.

Wolfrom, M. L. and Mahan, J. (1942). *J. Am. chem. Soc.* **64**, 308.

Wong, E. (1963). *Tetrahedron Letters* 159.

Wong, E. (1966a). *Phytochemistry* **5**, 463.

Wong, E. (1966b). *Chemy Ind.* 598.

Wong, E., Mortimer, P. I. and Geissman, T. A. (1965). *Phytochemistry* **4**, 89.

Wu, M. A. (1957). *Diss. Abstr.* **17**, 2147.

Wu, M. A. and Burrell, R. C. (1958). *Archs Biochem. Biophys.* **74**, 114.

Yamamoto, R. and Oshima, Y. (1931). *Agric. chem. Soc. Japan J.* **7**, 312.

Yang, C. H., Braymer, H. D., Murphy, E. L., Chorney, W., Scully, N. and Wender, S. H. (1960). *J. org. Chem.* **25**, 2063.

Yap, F. and Reichardt, A. (1964). *Züchter* **34**, 143.

Yasue, M. and Hasegawa, M. (1962). *Nippon Kagaku zasshi* **44**, 170.

Yeh, P. Y. and Huang, P. K. (1961). *Tetrahedron* **12**, 181.

Young, M. R., Towers, G. H. N. and Neish, A. C. (1966). *Can. J. Bot.* **44**, 341.

Yu, M. H. (1933). *Bull. Soc. chim. biol.* **15**, 482.

Zane, A. and Wender, S. H. (1961). *J. org. Chem.* **26**, 4718.

Zapsalis, C. and Francis, F. J. (1965). *Fd. Res.* **30**, 396.

Zemplen, G. and Gerecs, A. (1935). *Ber.* **68**, 2054.

Zemplen, G. and Bognar, R. (1941). *Ber.* **74**, 1783.

Zwenger, C. and Dronke, F. (1861). *A. Auppl.* **1**, 263.

ADDENDUM

Literature survey from April to October, 1966

I. NEW ANTHOCYANIN RECORDS

Species	Pigments present[1]	Organ	References
ARCHICHLAMYDEAE (cf. Chapter 5, p. 127)			
ROSALES Leguminosae			
Cassia marginata	Pg 3-diglucoside with C$_5$-residue at C-6 or C-8[2]	Petal	Adinarayana and Seshadri, 1966
Glycine hispida	Cy 3-glucoside	Seedcoat	Manabe *et al.*, 1966
SYMPETALAE (cf. Chapter 6, p. 186)			
TUBIFLORAE Labiatae			
Coleus blumei cvs.	Cy 3,5-diglucoside Pg 3-glucoside and acylated derivs.	Petal	Palmieri and Landi, 1966
Perilla ocimoides var. *crispa*	Cy 3-(6-*p*-coumaroyl-glucoside)-5-gluco-side [complete structure]	Leaf	Watanabe *et al.*, 1966
Solanaceae			
Solanum melongena	Dp 3-(4'-*p*-coumaroyl-rutinoside)-5-glucoside [complete structure]	Fruit skin	Watanabe *et al.*, 1966
Gesneriaceae Sub-family Gesnerioideae			
Chrysothemis pulchella	Gesnerin Cy 3-rutinoside	Sepal Petal	
Episcia mellittifolia	Lt 5-glucoside Cy 3-rutinoside	Stem Petal	J. B. Harborne, unpublished
Kohleria bogotensis	Pg 3-rutinoside Columnin	Petal Leaf	
Sarmientia repens	Lt glycoside	Petal	

---

[1] Abbreviations: Pg, pelargonidin; Cy, cyanidin; Pn, peonidin; Dp, delphinidin; Lt, luteolinidin.

[2] The first *C*-alkylated anthocyanin to be reported; the anthocyanidin has R_f 0·81 in Forestal and $\lambda_{\mathrm{max}}^{\mathrm{MeOH\text{-}HCl}}$ at 525 mμ

Species	Pigments present[1]	Organ	References

Sub-family Cyrtandroideae

Aeschynanthus marmoratus	Cy 3-sambubioside	Leaf ⎫	
A. tricolor	Cy and Pg 3-sambubiosides	Petal ⎬	J. B. Harborne, unpublished
Didymocarpus humboldtianus	Dp glycoside	Petal ⎭	

Bignoniaceae

Catalpa bignonioides	Cy 3-rutinoside	Petal ⎫	J. B. Harborne,
Eccremocarpus scaber	Cy 3-rutinoside	Petal ⎬	unpublished
Tecoma garrocha	Cy 3-rutinoside	Petal ⎭	

MONOCOTYLEDONEAE

(cf. Chapter 7, p. 234)

LILIFLORAE

Liliaceae

Asparagus officinalis	Cy and Pn 3-rutinosides Cy and Pn 3-rhamnosyl diglucosides	Shoot	F. J. Francis, unpublished

FARINOSAE

Pontederiaceae

Eichhornia crassipes (Water hyacinth)	Dp 3-diglucoside	Petal	Shibata *et al.*, 1965

II. NEW FLAVONOL AND FLAVONE GLYCOSIDE RECORDS

Species	Pigments present[3]	Organ	References

MOSSES

(cf. Chapter 4, p. 114)

Marchantiales

Corsinia coriandrina	Qu and Km	Thallus	Reznik and Wierman, 1966

FERNS

(cf. Chapter 4, p. 117)

Psilotales

Psilotum triquestrum	Amentoflavone and hinokiflavone	Whole plant	Voirin and Lebreton, 1966

[3] Abbreviations: My, myricetin; Qu, quercetin; Km, kaempferol; Lu, Luteolin; Ap, apigenin.

Species	Pigments present[3]	Organ	References

GYMNOSPERMS
(cf. Chapter 4, p. 117)
Pinaceae
Ducampopinus 6,8-dimethyldihydro- Heartwood[4] Erdtman *et al.*,
krempfii chrysin 1966

ANGIOSPERMS—ARCHICHLAMYDEAE
(cf. Chapter 5, p. 133)

FAGALES

Fagaceae
Nothofagus fusca Km 3-rhamnoside Heartwood Hillis and Inoue,
1966
Quercus incana Qu 3-galactosyl- Leaf Kalra *et al.*, 1966
arabinoside

CENTROSPERMAE

Chenopodiaceae
Beta vulgaris Vitexin xyloside Leaf Gardner *et al.*,
1966

CACTALES

Cactaceae
Cactus grandiflora Isorhamnetin 3- Whole Hörhammer *et al.*,
galactoside and plant 1966a
isorhamnetin
3-rutinoside

MAGNOLIALES

Calycanthaceae
Calycanthus Km 3-rutinoside Leaf Plouvier, 1966
occidentalis

RANUNCULALES

Nymphaceae
Nelumbium Lu 7-glucoside Petal Nagarajan *et al.*,
speciosum Qu 7-glucoside 1966

GUTTIFERALES

Guttiferae
Ficus carica Qu 3-rutinoside Leaf El-Kholy and
Shaban, 1966
Garcinia livingstonei Eriodictyol Seed Srivastara and
Sharma, 1966
Hypericum Qu 3-galactoside Whole plant Hargreaves, 1966
androsaemum

[4] The heartwood also contains typical flavonoids of *Pinus*, see p. 121.

Species	Pigments present[3]	Organ	References
SARRACENIALES			
Droseraceae			
Drosera rotundifolia	Qu 3-galactoside Qu 3-digalactoside	Root	Bienenfeld and Katzlmeier, 1966
PAPAVERALES			
Papaveraceae			
Romneya coulteri	Km 3-rutinoside	Petal	Plouvier, 1966
ROSALES			
Leguminosae			
Albizzia lebek	Melanoxetin (3,5,8,3′,4′- pentahydroxy- flavone) and the chalcone, okanin	Heartwood	Gupta *et al.*, 1966
Bauhinia variegata	Km 3-galactoside Km 3-rutinoside	Petal	Rahman and Begum, 1966
Cassia javanica	Km 3-rhamnoside	Leaf	Bhutani *et al.*, 1966
Gliricidia maculata	Km 3-rutinoside	Leaf	Rangaswami and Iyer, 1966
Intsia bijuga	Myricetin and robinetin	Wood	Wojcicka and Massicot, 1965
Trifolium repens	Myricetin	Seed	Fottrell *et al.*, 1964
Rosaceae			
Crataegus monogyna	Qu 3-rutinoside and Qu 3-rhamnosyl- galactoside	Leaf	Fisel, 1966
C. oxycantha	Vitexin 4′-rutinoside, vitexin 4′-rhamno- side and vitexin 7,4′-diglucoside	Leaf	Lewak, 1966
Neviusia alabamensis	Acacetin 7-rutinoside	Leaf	Plouvier, 1966
Sorbus aucuparia	Qu 3-glucoside, Qu 3-sophoroside	Petal	Krolikowska and Kamecki, 1965
S. pendula	Qu 3-galactoside	Petal	Pavlii *et al.*, 1966
GERANIALES			
Euphorbiaceae			
Euphorbia *cyparissias*	Km and Qu 3-glucuronides	Leaf	Stadtmann and Pohl, 1966
E. palustris	Qu 3-galactoside	Leaf	Sotnikova and Chagorets, 1966

Species	Pigments present[3]	Organ	References
SAPINDALES			
Sapindaceae			
Xanthoceras sorbifolia	My 3-rhamnoside	Leaf	Plouvier, 1966
MALVALES			
Malvaceae			
Eriodendron anfractuosum Tiliaceae *Tilia cordata*	Acacetin 7-rutinoside	Leaf	Plouvier, 1966
CUCURBITALES			
Cucurbitaceae			
Bryonia dioica	Di-C-glycosylapigenin (alliaroside)	Leaf	Paris *et al.*, 1966
MYRTIFLORAE			
Myrtaceae			
Eucalyptus citriodora *E. globulosus* *E. rostrata*	Qu 3-rutinoside Qu 3-rhamnoside	Leaf	Elkeiy *et al.*, 1964
Onagraceae			
Oenothera lavandulaefolia	My 3-galactoside	Leaf and petal	Kagan, 1966
Eleagnaceae			
Hippophae rhamnoides	Isorhamnetin 3-glucoside Isorhamnetin 3-rutinoside	Berry	Hörhammer *et al.*, 1966c
UMBELLIFLORAE			
Araliaceae			
Acanthopanax pentaphyllus *A. henryii*	Qu and Km 3-rutinosides Qu 3-rutinoside	Flower	Plouvier, 1966
Umbelliferae			
Ammi visnaga	Qu and Km 3-rutinosides Qu and Km 3-glucosides	Whole plant	Akacic and Kustrak, 1965
Pastinaca sativa	Qu 3-glucoside Qu 3-rutinoside Isorhamnetin 3-glucoside-4'-rhamnoside	Fruit	Maksyutina and Litvinenko, 1966

Species	Pigments present[3]	Organ	References

ANGIOSPERMS—SYMPETALAE
(cf. Chapter 6, p. 191)

GENTIANALES
Rubiaceae

| *Crusea calocephala* | Qu 3-rhamnoside | Leaf | Brooker and Eble, 1966 |

TUBIFLORAE

Boraginaceae

| *Lithospermum arvense* | Qu 3-rutinoside | Whole plant | Krolikowska and Swiatek, 1966 |

Labiatae

| *Coleus blumei* cvs. | 5,6,7-Trihydroxy-flavone 5,6,7,4'-Tetra-hydroxyflavone | Petal | Palmieri and Landi, 1966 |
| *Leonuris quinquelobatus* | Qu 3-rutinoside Ap 7-glucoside-4'-p-coumarate | Whole plant | Petrenko, 1965 |

DIPSACALES

Caprifoliaceae

| *Kolkwitzia amabilis* | Acacetin 7-rutinoside | Leaf | Plouvier, 1966 |

Dipsacaceae

| *Dipsacus sylvestris* *D. fullonum* *Scabiosa succisa* | Isovitexin 7-glucoside | Leaf | |

CAMPANULALES

Compositae

Chrysanthemum arcticum	Acacetin 7-rutino-side	Leaf	Plouvier, 1966
Cynara scolymus	Lu 7-rutinoside-4'-glucoside	Leaf	Dronik and Chernobai, 1966
Iva axillaris	Quercetagetin 3,6-dimethyl ether	Whole plant	Hörhammer et al., 1966b
I. nevadensis	Pectolinaringenin, hispidulin and nevadensin (5,7-diOH-6,8,4'-tri-OMeflavone)		Farkas et al., 1966
Tanacetum vulgare	Acacetin		Khvorost et al., 1966
Xanthium spinosum	8-(γ,γ-dimethylallyl) apigenin		Pashchenko et al., 1966

REFERENCES TO THE ADDENDUM

Adinaraya, D. and Seshadri, T. R. (1966). *Indian J. Chem.* **4**, 73.
Akacic, B. and Kustrak, D. (1965). *Acta pharm. jugosl.* **15**, 171.
Bhutani, S. P., Chiber, S. S. and Seshadri, T. R. (1966). *Curr. Sci.* **35**, 363.
Bienenfeld, W. and Katzlmeier, H. (1966). *Arch. Pharm.* **299**, 598.
Brooker, R. M. and Eble, J. N. (1966). *Lloydia* **29**, 230.
Dranik, L. I. and Chernobai, V. T. (1966). *Chem. Abstr.* **65**, 3947.
El-Kholy, I. E. and Shaban, M. A. M. (1966). *J. chem. Soc.* (C), 1140.
Elkeiy, M. A., Darwish, M., Hashim, F. M. and Khadega, A. A. (1966). *Chem. Abstr.* **64**, 16277.
Erdtman, H., Kimland, B. and Norin, T. (1966). *Phytochemistry* **5**, 927.
Farkas, L., Nogradi, M., Herz, W. and Sudarsanam, V. (1966). *Abstr. 4th Int. Natural Products Symp.* (*Stockholm*) 44.
Fisel, J. (1966). *Arzneimittel-Forsch.* **16**, 80.
Fottrell, P. F., O'Connor, S. and Masterson, C. L. (1964). *Ir. J. agric. Res.* **3**, 246.
Gardner, R. L., Kerst, A. F., Wilson, D. M. and Payne, M. G. (1966). *Phytochemistry* (In press.)
Gupta, S. R., Malik, K. K. and Seshadri, T. R. (1966). *Indian J. Chem.* **4**, 139.
Hargreaves, K. R. (1966). *Nature, Lond.* **211**, 417.
Hillis, W. E. and Inoue, T. (1966). *Phytochemistry* (In press.)
Hörhammer, L., Wagner, H., Arndt, H. and Farkas, L. (1966a). *Chem. Ber.* **99**, 1384.
Hörhammer, L., Wagner, H., Farkas, L., Ruger, R., Herz, W. and Sudarsanam, V. (1966b). *Abstr. 4th Int. Natural Products Symp.* (*Stockholm*), 43
Hörhammer, L., Wagner, H. and Khalil, E. (1966c). *Lloydia* **29**, 225.
Kagan, J. (1966). *Phytochemistry* (In press.)
Kalra, V. K., Kukla, A. S. and Seshadri, T. R. (1966). *Curr. Sci.* **35**, 204.
Khvorost, P. P., Chernobai, V. T. and Kolesnikov, D. G. (1966). *Chem. Abstr.* **64**, 19545.
Krolikowska, M. and Kamecki, J. (1965). *Roczn. Chem.* **39**, 1937.
Krolikowska, M. and Swiatek, L. (1966). *Diss. Pharm. Pharmacol.* **18**, 157.
Lewak, S. (1966). *Roczn. Chem.* **40**, 443.
Maksyutina, N. P. and Litvinenko, V. I. (1966). *Chem. Abstr.* **65**, 788.
Manabe, T., Kubo, S., Kodama, M. and Bessho, Y. (1966). *Chem. Abstr.* **64**, 20068.
Nagarajan, S., Nair, A. G. R., Ramakrishnan, S. and Subramanian, S. S. (1966). *Curr. Sci.* **35**, 176.
Palmieri, F. and Landi, A. (1966). *Chem. Abstr.* **65**, 4258.
Paris, R. R., Delaveau, P. G. and Leiba, S. (1966). *C. r. hebd. Séanc. Acad. Sci. Paris* **262**, ser. D, 1372.
Pashchenko, M. M., Pivnenko, G. P. and Litvinenko, V. T. (1966). *Farmatsevt. Zh.* (*Kiev*) **21**, 44.
Pavlii, O. I., Makorova, G. Y. and Borisyuk, Y. G. (1966). *Farmatsevt. Zh.* (*Kiev*) **21**, 55.
Petrenko, V. V. (1966). *Chem. Abstr.* **64**, 15827.
Plouvier, V. (1966). *C. r. hebd. Séanc. Acad. Sci. Paris* **262**, ser. D, 1369.
Rahman, W. and Begum, S. J. (1966). *Naturwissenschaften* **53**, 385.
Rangaswami, S. and Iyer, V. S. (1966). *Curr. Sci.* **35**, 364.
Reznik, H. and Wierman, R. (1966). *Naturwissenschaften* **53**, 230.
Shibata, M., Yamazaki, K. and Ishikura, N. (1965). *Bot. Mag.* (*Tokyo*) **78**, 299.

Sotnikova, O. M. and Chagovets, R. K. (1966). *Farmatsevt. Zh.* (*Kiev*) **21**, 49.
Srivastava, S. N. and Sharma, V. N. (1966). *Curr. Sci.* **35**, 290.
Stadtmann, H. and Pohl, R. (1966). *Naturwissenschaften* **53**, 362.
Watanabe, S., Sakamura, S. and Obata, Y. (1966). *Agric. Biol. Chem.* (*Tokyo*) **30**, 420.
Wojcicka, A. R. and Massicot, J. (1965). *C. r. hebd. Séanc. Acad. Sci. Paris* **261**, 4540.

Author Index

13

Subject Index

A

Acacetin, 40, 97, 243
  glycosides, 47, 48, 146
Acaciin, 208
Acylated anthocyanins, 27
  structures of, 28
Afrormosin, 93, 169
Alkannin, 226
Alpinetin, 249
Amentoflavone, 96, 118, 222, 338
Amurensin, 180
Anemone anthocyanins, 128, 149
Anthocyanidins,
  chromatographic properties, 7
  distribution, 102
  identification, 6
  in the free state in plants, 4
  isolation, 6
  sources, 3
  spectral properties, 7, 9
Anthocyanins,
  and flower colour, 282
  biosynthesis, 268
  chromatographic properties, 19, 30
  general distribution, 101
  glycosidic variation, 15, 103
  identification, 15
  in flowers, 286
  in fruits and vegetables, 297
  in the Archichlamydeae, 127 *et seq.*
  inheritance, 256
  in the Monocotyledons, 234 *et seq.*
  in the Sympetalae, 186 *et seq.* 337
  isolation, 13
  oxidative destruction, 299
  pseudobases of, 105
  sources, 30
  spectral properties, 17
Apigenin, 39, 75, 97, 285
  7,4'-diglucoside, 47
  7-glucoside, 230
  7-glucuronide, 49
  O-glycosides, 46, 48, 145
  C-glycosides, 48
  methyl ethers, 39
Apigeninidin, 13
  glycosides, 35, 131

Apiosylglucosides
  of flavones, 46
  of isoflavones, 92
Apple anthocyanins, 155
Artimetin, 228
Asebotin, 198
Aspalathin, 51, 145
Atanisin, 227
Aucubin, 226
Aurantinidin, 12
  glycosides, 35, 131
Aureusidin, 93
  glycosides, 83, 86
Aureusin, 83, 86, 223
Aurones, 83
  and flower colour, 284
  biosynthesis, 270
  chromatography, 85
  identification, 84
  inheritance, 257
  spectra, 84
Awobanin, 28, 33
Ayanin, 54, 164
Azaleatin, 54, 112, 133
  as taxonomic marker, 309
  glycosides, 70, 136
Azalein, 196

B

Baicalein, 44, 217, 219
Balsam anthocyanins, 252, 259
Banana pigments, 247
ψ-Baptigenin, 92
Bayin, 51, 145
Betacyanins, 105
  distribution, 308
  metabolism, 293
Betanidin, 106
  glycosides, 106
Betaxanthins, 106
Biflavonyls, 96
  chromatography, 96
  distribution, 118
  spectra, 96
  taxonomic significance, 307
Bilobetin, 97
Biochanin-A, 94

Index of Orders and Families

Index of Genera and Species